Integrated Pest Management: farmer field schools generate sustainable practices

WAGENINGEN AGRICULTURAL UNIVERSITY PAPERS
93-3 (1993)

Integrated Pest Management: farmer field schools generate sustainable practices

Elske van de Fliert
Wageningen Agricultural University

Wageningen Agricultural University

CIP-GEGEVENS KONINKLIJKE BIBLIOTHEEK, DEN HAAG

Fliert van de, Elske

Integrated Pest Management: farmer field schools generate sustainable practices/Elske van de Fliert - Wageningen : Agricultural University. - (Wageningen Agricultural University papers,
ISSN 0169-345X ; 93-3 (1993)
ISBN 90-6754-321-7

Printed in the Netherlands by Veenman Drukkers B.V., Wageningen

Authors abstract

Integrated Pest Management: farmer field schools generate sustainable practices.
A case study in Central Java evaluating IPM training.
Elske van de Fliert, 1993.

An evaluation study of the National Integrated Pest Management (IPM) Programme in Indonesia was conducted in one Central-Javanese district looking into processes and effects occurring at the village level when sustainable practices in rice cultivation, which contrast in many respects with the prevailing high-external-input technology, are introduced through nonformal farmer training in conditions created by policy measures. The IPM training contents consisted of a set of principles, instead of preset recommendation, providing the farmers with a tool for decision making. Training processes were field-oriented and based on experiential learning. Main objective was that farmers become independent decision makers and managers of their farms. Trainers performed as facilitators of the learning process. As a result of training, farmers took better-informed pest management decisions, pesticide use and expenditures on pest control decreased, yields increased, and yield variability became smaller. Horizontal communication on IPM was hampered by the non-representativeness of trained farmers in the farming communities.

The nonformal training approach appeared to be consistent with the ecological approach of IPM. The experience of the Indonesian IPM Programme showed interesting perspectives for extension supporting sustainable agriculture.

Keywords: Integrated Pest Management, nonformal education, sustainable agriculture, agricultural extension, facilitation, rice cultivation, Indonesia, Central Java

Preface

The book in front of you evolved in the course of the past three years, a period that was rich in diverse experiences for me. Assigned in Indonesia as an associate expert to the 'Inter-country Programme for the Development and Application of Integrated Pest Management (IPM) in Rice in South and Southeast Asia' of the United Nation's Food and Agriculture Organisation (FAO), I was given the opportunity to do my doctoral research within this dynamic programme. The Inter-country IPM Programme had just initiated a National IPM Programme in Indonesia (in full the 'Programme for Training and Development of Integrated Pest Management in Rice-based Cropping Systems') that was going to embark upon a new course with respect to IPM farmer and staff training, and therefore provided ample interesting opportunities for study. The Wageningen Agricultural University was willing to supervise the research. The operational costs of the study were supported by the FAO Inter-country IPM Programme and the National IPM Programme, whereas the Department of Communication and Innovation Studies of the Wageningen Agriculture University provided the facilities to write the dissertation. This material support was crucial for the successful completion of the book.

The study would also not have been possible without the mental and physical support of many people. First, I am greatly indebted to Dr. Russel Dilts, Dr. Kevin Gallagher and Dr. John Pontius of the Indonesia National IPM Programme for their many valuable ideas, guidance and confidence, and to Dr. Peter E. Kenmore of the FAO Inter-country IPM Programme for his initiative and efforts allowing me to do this research. Further, I wish to express my gratitude to the Government of Indonesia for the opportunity which the National IPM Programme provided to be involved in a significant and progressive activity. I hope that this study may be a contribution to the development of the country.

Of great value for the research have been my two supervisors at the Wageningen Agricultural University, Dr. Niels G. Röling, former professor of Extension Science, and Prof. Dr. J.C. Zadoks, professor of Ecological Phytopathology, who both had the opportunity to visit me in Indonesia and taste the atmosphere of the study villages. Their often opposite perspectives on science were highly complementary to my research. I express my sincere thanks to Niels for his unlimited enthusiasm, the many discussions that strongly inspired and motivated me, and the broader insights he gave me (being a biologist) in social sciences. My sincere thanks also go to Prof. Zadoks for showing me many new aspects and perspectives of (Integrated) Pest Management, and for his detailed, accurate and adequate commentaries during field work and writing of the dissertation.

Further words of thanks are addressed to:

- Ms. Rini Asmunati for being the best assistant and friend I could ever get;
- the survey enumerators, Ms. Nastiti Tri Winasis, Mr. Sri Budi Santoso, Ms. Siti Mulyani, Ms. A. Yovita Wishnu Subando, Ms. Kusdaru Widayati, Mr. Sampurno Alfapriyandi, Mr. Muhammad Musiyam, Mr. Ahmadi, for the great job they did in the field;
- the village study assistants, Ms. Murmiati, Mr. Agus Mantono, Mr. Mustofa Lutfi, Mr. Muh Wachid Hasim, for doing the tedious field observations, and helping me to understand their villages;
- all the people in the villages and agricultural institutions in Grobogan who participated in the study activities or in any other way made the study possible and enjoyable; in particular I want to thank the families who provided boarding for the study team during the field activities, making us feel so much at home;
- staff of the Yogyakarta and Jakarta IPM Secretariats for their assistance and friendship, in particular Mr. Triyanto PA who made the illustrations for this book;
- the many people who provided inspiration and motivation through discussions, comments, or distraction, of whom I can only mention here the staff, secretariat and students of the Wageningen Department of Communication and Innovation Studies, and Mr. F. Nangsir Soenanto, Mr. Herry Soewartoyo, Dr. H.A.J. Moll, Dr. William Settle, Dr. Thom Gillespie, Mr. Karel Van Elsen, Dr. Frans Hüsken, Mr. Taco Bottema and Dr. Patricia C. Matteson.

A special word of gratitude is addressed to my parents who have always allowed, encouraged and supported me to develop in the direction I wanted. Their company and special care during the past year is a valuable memory that will stay forever. Most of all, I am grateful to Yogi for sharing his life with me the way he does, a feeling which I can only express with 'matur nuwun sanget'.

After a simultaneous 'growth period', the 'delivery' of this dissertation precedes the delivery of our first baby with one month. In a way, the two deliveries are contrary: one is an autonomous process after eight months of hard work, the other will be hard work after an autonomous process of nine months. On the other hand, the latter autonomous process was highly contributive to the former hard work, by giving motivation, energy and sense of relativity. The two deliveries are similar in that both will supposedly put my life on a new track.

Ede, March 26, 1993

Contents

List of tables

List of figures

List of boxes

List of Boxes

1 Introduction

IPM, farmer field schools, and evaluation
'Where's this bus heading? A little strategic look....' This was the title of an informal paper sent by the programme leader of the National IPM Programme in Indonesia to the programme staff, a year and a half after the 'bus had departed'. The paper listed goals and existing resources, and suggested strategies for the coming period. Its title and content stand for the extraordinary dynamics of a programme manoeuvring within the Indonesian bureaucracies. An innovative training programme is breaking through the traditions of Green Revolution agricultural development and top-down extension systems. Indonesian farmers and extension staff are learning about Integrated Pest Management (IPM) in farmer field schools, a nonformal training approach. Is this programme the onset for more sustainable practices in intensive food crop production, and for more farmer-oriented extension methods in the archipelago?

The introduction and implementation of a large-scale training programme that is innovative in its approach of both technology and extension, within a conventional context, is a unique event and, therefore, an interesting object for study. This book is about a village study conducted in the district of Grobogan, Central Java, to evaluate IPM training. The study is longitudinal: it looks at processes and effects that take place at the village level before, during and after the introduction of IPM through farmer field schools. Purpose of the study is:

1. to deliver useful management information with respect to characteristics and diversity of the intended and actual programme beneficiaries, and to actual

achievements considering behaviour of farmers and farm level effects as a result of IPM training and dissemination; and

2. to describe and analyse the processes and effects taking place at the village level under the impact of an unconventional extension programme in a conventional context, as a case of scientific interest.

Definition of terms

Throughout this book, various terms and names will be used consistently that need to be defined first. As far as possible, abbreviations and Indonesian terms are avoided in the text, and when used they are explained in the 'List of terms'.

The Republic of Indonesia is subdivided, in descending order of administrative level, in provinces ('*propinsi*'), districts ('*kabupaten*'), subdistricts ('*kecamatan*'), and villages ('*desa*'). Although not completely appropriate equivalents, the terms district and subdistrict for '*kabupaten*' and '*kecamatan*' are used in the text. All names of villages and subdistricts are fictive. The Javanese meanings of the names describe more or less typical characteristics of the villages or their residents.

The term 'farmers' comprises both women and men, and owner-operators as well as tenants. The main criterion for being considered a farmer is decision making in rice cultivation. Farmers are defined as those who take cultivation decisions, as opposed to farm labourers who work on the instructions of the farmers. Tenant farmers are those operating either on a fixed rent or a sharecropping agreement. A farmer group is the administrative unit consisting of all farmers from a certain limited rice area, designated by the government's Agricultural Service for extension purposes, as described in Chapter 2. For privacy reasons, all people referred to in this book are given fictive names.

When talking about 'pests' without specification, the whole pest complex in the crop ecosystem is meant, including pest insects, diseases, rodents, weeds and other organisms damaging crops. Major types of pests occurring in rice cultivation in the study area are insects and rodents. Diseases are of minor importance, and weeds, although abundant, seldom cause yield loss as a result of intensive weed control.

Indonesia enjoys a tropical monsoon climate. Cultivation seasons, however, vary from place to place. The definitions for seasons used in the following chapters refer to (the irrigated parts of) Central Java, in particular the study district Grobogan. Within one year, a distinction is made between a wet, a dry and an intermediate season (Figure 1.1). The **wet season** is defined as the (first) rice-growing season during the main part of the wet monsoon, normally from November to February ('*musim rendhengan*' in Javanese). The **dry season** is defined as the (second) rice-growing season during the first months of the dry monsoon, normally from March to June ('*musim ketigo*'). The **intermediate season** is defined as the secondary food crop season during the last months of the dry monsoon, normally from July to October ('*musim labu*'). Slight shifts in the seasons over a year are possible due to late rains or late irrigation. The study seasons in this book mostly relate to wet and dry seasons, since rice cultivation is primarily considered.

Extension science has undergone some major changes in the last couple of years.

Month	Nov.	Dec.	Jan.	Febr.	March	April	May	June	July	Aug.	Sept.	Oct.
Season	wet				dry				intermediate			
Rainfal	little	abundant			little	none						
Irrigation	yes								no			
Crop	rice (first crop)				rice (second crop)				secondary food crops			

Figure 1.1: General calendar for season, rainfal, irrigation and crops in irrigated areas in Central Java.

Whereas previously the 'transfer of technology' was considered as the model for extension, fitting well in the 'hard systems' thinking of agricultural development, the introduction of 'soft systems' thinking led extension science to take a broader perspective (Röling, 1992b). The terminology has, however, not yet been fully adapted to this perspective, and is still strongly imbued with linear thinking. In this book, it has not always been possible to use language which is consistent with the new perspective.

Organisation of the book

This book describes and analyses the introduction, implementation and effects of the National IPM Programme in one district in Central Java. It is a case study, and although comparisons are made with findings in other areas, where possible, the study does not allow generalisation across the National IPM Programme's field implementation. Such a generalisation is unwarranted in view of the high diversity in cultures, social structures, and crop cultivation behaviours among the provinces covered by the programme. The study aims to deliver useful information for extension programme implementation by pointing out specific linkages and effects, and by emphasising existing diversity among farming communities, even within limited areas.

The broader Green Revolution context which determined Indonesia's agricultural development to date, the history of IPM and IPM training, and the National IPM Programme's training model, culminating in the justification for this study, are described in Chapter 2. Chapter 3 gives a theoretical framework for this evaluation study, and describes the research questions and methodology. A description of the research area emphasising diversity among apparently similar village communities is the topic for Chapter 4. This diversity should be kept in mind while reading the other chapters. **It is of major importance for understanding the arguments in the book that the names and profile characteristics of the eight different study villages are remembered.** To help the reader, a flipchart is provided in the back of this book which gives an overview of the study villages. Chapter 5 continues with a description of rice cultivation in the study area. In addition to displaying the whats, hows and whys of cultivation practices by 'Green Revolution' rice farmers, it serves as a baseline for measuring the effects in farmers' practice after having followed IPM training. In Chapter 6, a portrait is given of the IPM farmer field school, mainly based on the observations in the study area. Effects of the IPM training on farmers' rice cultivation behaviour are described in Chapter 7, whereas Chapter 8 discusses the effects at the farm level as a result of changed behaviour. Finally, a discussion of the findings and conclusions are given in Chapter 9.

2 Contrasting approaches in agricultural development

2.1 The Green Revolution context

Rice in history and culture

Man first domesticated rice some 10,000 years ago in river valleys in South and South-east Asia (Chang, 1976). Rice cultivation began in Indonesia around 1600 B.C. (Ward, 1985). Long before modern scientists, Indonesian farmers utilised the natural variability of rice obtained from wild species, natural hybrids and seeds imported by travelling traders through saving seed of superior plants. Rice takes a central place in the daily life of most Indonesians. If available and affordable, it is the staple of every meal. Rice is a major element in legends and traditional beliefs, and rice is revered in many religious ceremonies.

In the Indonesian archipelago more than ten million hectares of rice land are planted annually. The majority of rural people, especially in the island of Java where more than 60% of the nation's rice is produced (Biro Pusat Statistik, 1988), are in some way involved in rice production. Although not always the most profitable enterprise, rice growing is a way of life for many Javanese farmers. For instance, farmers who repeatedly experienced severe crop loss as a result of rat damage, would still establish their rice crops rather than shift to other enterprises, even if another failure is virtually certain (Van de Fliert et al., 1993).

Population growth, rice production and the state

During the past century, the population in Java increased continuously reaching about a 100 million people living on an area four times the size of the Netherlands.

Population densities, already high at the end if the nineteenth century, have become enormous (Booth, 1988). Despite the increasing production of rice and other food crops, until the mid-sixties food production growth could not keep up with population growth. The result was a long-term decline in per capita rice output from about 0.20 ton in 1880 to 0.13 ton in 1965 (Booth, 1985). Yields remained stable at a level of around 2.1-2.2 tons per hectare. The production growth of rice during the past century can be divided into three periods:

- before 1900, when almost all production growth was achieved by expansion of the area of irrigated land;
- between 1900 and 1960, when further area expansion was accompanied by increases in double cropping;
- after 1960, when large-scale introduction of new input-intensive technologies allowed considerable yield production increase.

In the period prior to the technological innovations in rice production, rice imports had increased to a level of 10% of domestic consumption (Mears, 1981). The fifties and early sixties were a period of economic and political instability in Indonesia in which rural development stagnated (Collier et al., 1982). In the crisis year of 1965, rice imports drastically dropped causing severe food scarcities, especially in the cities. After a turbulent period with transition of political leadership in the late sixties, efforts were made by the 'New Order' government to stabilise the economy and slow down inflation. The survival of this government depended on two imperatives: (1) the need to achieve and maintain political control in rural areas, and (2) the need to ensure food availability in the cities (Hüsken and White, 1989). The sometimes heavy handed agricultural and rural development strategies of the government should be seen in the context of these conditions.

In the 'New Order' government's first five-year plan for national development (REPELITA I, 1969-74), agriculture took a central position. The need for intensification of food crops had become obvious because of severe shortages, accompanying urban riots, and forecasts of serious overpopulation. As the Green Revolution swept over a larger part of Asia, a board for the intensification of food crops, the BIMAS Secretariat ('*Bimbingan Masal*' or Mass Guidance), was established. BIMAS was defined as 'an intensive extension campaign supported by inputs and credit facilities delivered down to the village level, stimulated by market prices favourable to the farmers' (Adjid, 1983).

The mass guidance idea was derived from some pilot activities by students in the period of 1963-66 (Hüsken and White, 1989; Adjid, 1983). Through guidance programmes rice farmers were given credit and extension that had to encourage them to use improved rice varieties and chemical fertilisers. The success of this pilot programme was considerable.

These activities were modified in 1968 into the first large-scale intensification programme called '*BIMAS Gotong Royong*' ('*gotong royong*' = mutual cooperation). Through this programme, high-yielding rice varieties from IRRI, nitrogen and phosphorus fertilisers, and pesticides to be sprayed on calender-basis were introduced as a package. Repayment of the input package was, initially, in kind by giving one-sixth of the harvest to the national collecting agency. This scheme was made possible by a contract of the Indonesian government with multinational corporations, which were paid US$ 50 per hectare for the provision of the necessary inputs, extension and management. Extension agents from collaborating banks were sent to the villages to promote the package, supported by letters from higher governmental institutions. Because of failure due to mismanagement and corruption, this scheme was terminated in 1970 (Hansen, 1971).

The next BIMAS programme BND ('*BIMAS Nasional yang Disempurnakan*' = National Improved BIMAS) was directly controlled by the state. This programme used a more individual approach. The command strategy as applied in the previous years shifted towards commands supported by subsidies (Mears, 1981). Village Units, consisting of the village extension worker (PPL), the retailer of inputs, the Rural Bank of Indonesia (RBI) Village Unit, and the KUD ('*Koperasi Unit Desa*' = Village Unit Cooperative), were developed as local activity centres to serve individual farmers (Bureau of Agricultural Extension, 1986). Recommendations regarding the use of the technology package were conveyed to the farmers by the extension worker, inputs were made available through the KUD and retailers, KUD took care of marketing of farm products, and credit could be obtained from the RBI. The government supported the village programmes through (1) improvement of infrastructure and irrigation, (2) price subsidies of inputs (fertilisers and pesticides), (3) procurement of rice under the floor-price protection plan, (4) credit funds and subsidies for the RBI Village Units, and (5) funds to (partly) offset bad debts and repay loans to farmers who suffered total crop loss. The price support programmes were considered important to provide adequate incentives for rice farmers to intensify their production (Birowo and Budianto, 1983). The agricultural extension service received an extra injection in 1976 when the World Bank, introducing its Training-and-Visit system, started to fund a series of extension projects (section 2.2). The individual approach of the BND scheme was not considered sufficiently effective by the government since only one-third of the farmers responded actively to the programme's stimuli (Adjid, 1983).

As a result of the introduction of high-yielding varieties and the more intensive use of cheap, subsidised fertiliser, rice production started to increase at annual rates of 4.7% during REPELITA I, and 3.3% during REPELITA II, which were above the 2.3% population growth. The expected gigantic yield increase, however, failed to materialise. Reasons for this were the actual shortage of fertiliser, and the unequal distribution of fertiliser under the BIMAS schemes. A few, usually rich and influential, farmers appeared to be the main recipients of the limited supply, while the majority did not have sufficient access. A yield reducing factor were outbreaks of the brown planthopper, *Nilaparvata lugens* (Stal.), in the late seventies.

The expected sharp yield increase, displayed for the wet land rice areas in Central Java in Figure 2.1, occurred from 1979 onwards after several events had taken place under the third five-year plan (REPELITA III, 1979-84). The principal objec-

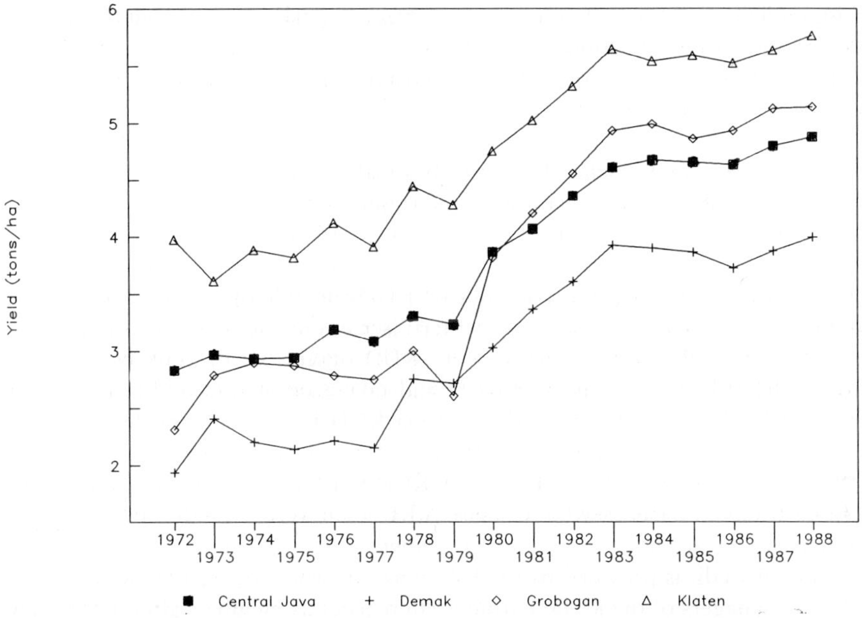

Figure 2.1: Rice yields in tons/ha from wet land in Central Java in the period 1972-88. In addition to the province average (Central Java), lines for the highest (Klaten), the lowest (Demak), and an average (Grobogan) rice producing district are given. (Source: Central Bureau of Statistics/database CGPRT Centre)

tive of REPELITA III was to attain self-sufficiency in food crop production, especially rice, through four main efforts: intensification, expansion, diversification, and rehabilitation. As a result of the construction of three new fertiliser factories, made possible by extra state revenues generated after the 'Oil Boom', a sufficient supply of fertiliser became available. The fertiliser was more efficiently distributed to the rural areas through a new model intensification programme INSUS ('*Intensifikasi Khusus*' = Special Intensification), and its successor SUPRA-INSUS, the latter applying a farmer group approach again. The more efficient distribution allowed adequate use of fertilisers by all farmers. This coincided with the availability of brown planthopper resistant rice varieties from IRRI (IR 36 and IR 38) for Indonesian farmers which helped to overcome yield reduction caused by this disastrous pest. Better fertilisation and prevention of yield loss due to brown planthopper resulted in higher average yields.

Koperasi Unit Desa (KUD)

Cooperatives have existed in Indonesia since the beginning of the twentieth century after the Dutch colonial government introduced the concept (Soedjono, 1986). From the beginning, the influence of the (first colonial and later independent Indonesian) government in the cooperatives has been very large. Cooperatives have often been used as a policy tool to support development. The most obvious example in this respect is the KUD (Village Unit Cooperative) that was brought into being under the BND programme as an essential part of the Village Units to encourage food crop intensification in rural areas.

The KUD as a village cooperative is given a threefold responsibility in the intensification programmes:

1. provision of inputs such as fertilisers, pesticides and certified seed;
2. management of credits for farming and trading;
3. purchase, storage and trade of rice via BULOG (National Logistics Agency).

Other activities frequently carried out by KUDs include operation of rice mills, and electricification of rural areas. All fertiliser in the country is distributed and traded via the KUD. In villages without KUD offices, small kiosk owners buy stocks of fertiliser at the nearest KUD and take care of further distribution at the field level. Compulsory distribution via KUD is not the case for pesticides which are also promoted directly by salesmen from chemical companies.

KUDs operate on a regional basis. A KUD working area is usually less than a subdistrict, but covers several villages. A KUD unit can be established by people in a village who are able to guarantee investment and fulfil certain other conditions. Additional credit is provided by the RBI. The investors are usually the ones who become managers of the KUD. Farmers can register as members after which they may enjoy the KUD's services, such as contracting loans. Purchase of fertilisers and pesticides is possible for non-members at the counter of the KUD kiosk on cash payment. According to the design, the KUD is owned by its members. In practice, however, the organisational structure is usually very top-down, and the influence of the members is nil. The KUD's activities and influence have declined since the mid-eighties (Collier et al., 1988). Some of the reasons might be the many cases of corruption within KUD management, and the fact that KUD's services often do not meet farmers' needs. Rural communities lost their confidence in the KUD (Suwandi, 1986).

A special role in the relation between KUD and farmers is performed by the village extension worker. Initially being the messenger of the agricultural extension recommendations and purposely not involved in input distribution, the extension worker turned out to be the intermediary between KUD and farmers. In addition to advising the farmers on what inputs they need, these officials have become an impassable link for farmers to obtain input packages on credit. By funnelling the input packages from KUD to farmers, the extension workers obtain a certain percentage as a commission which they actually need to supplement their low civil servant income. Through this double role, their position as extension worker supposedly serving the farmers can easily be affected, as described in Box 2.1.

The Green Revolution success

The Agro Economic Survey (1972) showed that by 1971 either national improved or IRRI varieties accounted for 30% of the cultivated rice area in the country, implying a wider use in the major rice growing areas such as Java. This is considered a fairly rapid adoption of the innovation, proving the effectiveness of the

Box 2.1
Extension worker between KUD and farmers

In one of the subdistricts covered in this study, an extreme case was encountered of a village extension worker (lets call him Sujono) bending all the opportunities within the input distribution system towards his own profit, regardless of any harm done to the farmers through his practices.

During the 1990 dry season when the young rice crop was infected by bacterial leaf streak, Sujono advised the farmers to spray a certain type of fungicide (not available in the local kiosk) of which he sold small packages for a reasonable price. The tricky point in this intervention was that a fungicide can in no way cure a bacterial disease, and that this particular disease is usually cured by compensation of the rice plant itself, two facts known by Sujono. Many farmers, however, in good confidence spent their money on a completely useless fungicide application.

Except for his regular practice of strongly (and often wrongly) promoting pesticides and foliar fertilisers to farmers, sold at his house, Sujono's first opportunity to take great advantage of the farmers was in the intermediate, secondary food crop season of 1991. In that season, the Agriculture Service launched a soybean intensification programme ('*OPSUS Kedelai*'), and Sujono was ready to implement it in his working area. In each village under his jurisdiction, one farmer group was appointed to obligatorily participate in the programme. All farmers in the appointed farmer group area's were told by the village government to go to the KUD and register for a credit package. Sujono provided the packages, (received his commission) and was not seen any more in the villages throughout the season. Due to low quality seed and an immature technology package, most crops failed and the farmers were left with debts at the KUD.

These debts forced them to take a new loan for the next rice crop, creating Sujono's new chance. He distributed enrolment lists for KUD credit packages to the farmer group leaders who were given the task to get the names of farmers wanting to apply for credit. Although from 1986 onwards, provision of inputs through KUD, even on credit, should go according to farmers' needs (which they can fill in themselves on the forms), Sujono insisted that everybody had to take the complete package according to the recommendations. He anticipated this by 'providing the service of' filling out the forms himself. For every farmer's name listed, he ordered the complete package for 0.5 hectare, regardless of the actual (often smaller) size of the farmers' fields. In addition to seed, urea and TSP fertilisers, this package included KCl and ZA fertilisers, pesticides and foliar fertiliser that most farmers did not want to purchase. Sujono even lied openly to the farmers who objected to take the extra chemicals, saying that it was an instruction from the national authorities to take the whole package. The result was another failure. Due to heavy stemborer attack during this rice season, which apparently could not be controlled by the additional pesticides in the package, farmers' debts only grew. Neither the local KUD managers nor Sujono himself took action to claim compensation from the government to repay the farmers' debts (be it without the attractive interest for the KUD), which is possible in case of pest outbreak. Many of the farmers began to realise that they were cheated.

Because of the increased debts, the farmers in Sujono's work area were no longer allowed to contract new loans at the KUD. As new actions this extension worker undertook aggressive input promotion campaigns in collaboration with pesticide and foliar fertiliser companies during the following rice season, in order to prepare the farmers for his own special programme with hybrid corn during the next secondary food crop season.

Although an extreme case, Sujono is not an exception. Most village extension workers are pinched between their own insufficient financial situation and the opportunities offered by the system they live in. Apparently, they all have to grasp some of these opportunities to survive within the system, but it shows personality if this is done without disadvantage to the farmers.

BIMAS programme. A resurvey in 1981 in several main rice-growing areas in Java reported that almost all farmers used the modern seeds and fertilisers, except for a small percentage planting traditional varieties for special occasions and special clientele (White and Wiradi, 1989).

Adoption of the package by rice farmers was initially (under the BIMAS Gotong Royong programme) not always on a voluntary basis. Input packages were distributed through the village governments which allowed easy social control. In some areas, the introduction of the new varieties coincided with outbreaks of brown planthopper, the most feared rice pest, in traditional rice crops, which strongly influenced farmers' attitude towards innovative practices. In other places, pressure to promote the package was exerted in various ways, ranging from moral pressure by blaming farmers not to cooperate in the nation's development, to enforcement by the army. It was not uncommon in Javanese villages that seedlings of traditional varieties were pulled out by government officials to force farmers to use the new seeds (Box 5.6). Less oppressive measures to encourage farmers to apply the new technologies included demonstration plots cultivated by village officials (Hüsken, 1988). But the coercive nature of the introduction of the new technologies should not be overstressed: farmers soon discovered the benefits of the increased yields and readily adopted (at least a part of) the package. Although farmers feel that rice production with the use of the new technologies is riskier and more of a hassle, they say they are much better off now.

The result of the food crops intensification programmes at a national level is obvious: Indonesia became self-sufficient in rice in 1983 after being the world's largest rice importer for years, although it appeared difficult to maintain that status. According to FAO statistics, Indonesia's growth in per capita cereal production (especially contributed by Javanese rice farmers) between 1975/76 and 1984/85 was the second highest in Asia after Myanmar (FAO, 1985). This success became possible through efforts to introduce the technology package, supported by investments in the supply of inputs and credit, and in infrastructure (Sayogyo, 1987), in combination with the government interventions in price and marketing policies (Ifzal, 1987). Besides self-sufficiency, two other objectives of the government's food programme were achieved, namely protection of consumer interest and increase of farmer income. The rapid agricultural growth contributed to increased employment in agriculture, even in densely populated Java (Manning, 1988). Prices are still carefully managed so that most farmers continue to make a minimal living, while rice remains cheap and allows low urban wages.

Green Revolution effects

The other side of the technological success started to be increasingly felt from the late seventies onwards in the areas of farm economics, human health and environment. Production growth seemed to have reached its limits since yields were plateauing. Figure 2.1 shows that the plateau in rice yields in Central Java was reached around 1983, though at a level twice that where farmers had started fourteen years before. A decline in productivity growth has already been recorded in many areas in Asia (Byerlee, 1987; Barker and Chapman, 1988). In general, the use of high-yielding rice varieties is well established, land use intensity is high, and input use, especially fertilisers, is high but inefficient. Scientists at IRRI discovered that there is no longer a yield gap between farmers and experiment stations, but rather between farmers themselves (Pingali et al., 1990). This gap is explained by differences in farmer ability and in access to irrigation water, of which the former is of main importance for extension programmes. A stagnating, or even declining, production growth will soon become problematic considering the continuous population growth (Rosegrant and Pingali, 1991).

The introduction of the input-intensive technology brought about the use of pesticides. Production of pesticides in Indonesia dramatically rose from 9,128 tons in 1878/79 to 39,000 tons in 1983/84, a 4.3 fold increase (IOCU, 1987). Nationally produced and imported pesticides were primarily distributed through the BIMAS intensification programmes (Oka, 1987). As was foretold in 'Silent Spring' by Rachel Carson (1962), these chemicals began to cause more problems than they solved. Vast yield losses, caused by brown planthopper, rice leaffolder, stemborer and gall midge, threatened food security in Indonesia. Nationwide brown planthopper outbreaks occurred in 1975-77, and again in 1985-86 destroying an estimated 275,000 hectares of rice. These outbreaks were the invariable result of the indiscriminate use of (mainly broad-spectrum) pesticides which induced resistance in pests and destroyed natural enemies (Van den Bosch, 1980; Gallagher, 1988; Litsinger, 1989). Even the larger natural enemies such as frogs and snakes are commonly observed to be affected by pesticides applied in rice fields (Guan-Soon and Seng-Hock, 1987). The best known example of a pesticide-induced rice pest is the brown planthopper which reproduces very fast in the absence of predators, but has little chance under balanced ecosystem conditions where natural enemies are not disrupted by pesticide use (Kenmore, 1980). Resistance of brown planthoppers to various organophosphates, carbamates and pyrethroids was detected in West and Central Java (Soejitno, 1991). Pest resurgence as a result of the elimination of natural enemies played an important role in causing the outbreaks. Hazards by pesticides to the natural environment, including loss of food sources such as fish, frogs and ducks, and poisoning of drinking water supplies, are becoming another increasing problem. In terms of human health, high exposure is observed among farmers and farm labourers who usually do not wear any protective clothing. Pesticide are often badly (re)packaged, and sometimes do not contain labels, or illegible labels (IOCU, 1987).

A study in the island of Bali showed that the introduction of the Green Revolution technology led to a radical decrease of energy efficiency of production, an increase of yield variability (thus higher risk for farmers), and an increase in material inequality (Foley, 1987). These effects caused a transition from a sustainable society into one characterised by dependence on imports of goods and services. Several studies in Java also showed that the technological innovations with concomitant rationalisation and mechanisation in rice cultivation led to growing social inequalities and rural polarisation (Hansen, 1981; Sayogyo, 1987; Booth, 1988; White and Wiradi, 1989). Slower increases in rice production will provide less agricultural employment opportunities in the near future (Manning, 1988).

Through the large-scale introduction of modern, high-yielding varieties, many traditional rice varieties appear to have been lost, and with them a store of genetic diversity which took thousands of years to develop. Large areas covered by crops of the same genetic make-up create conditions for pests and diseases to spread rapidly. Excessive fertiliser application was found to destroy the soil structure (Ibusno and Petil, 1988). In various places, organic matter levels in soils have dropped, soil acidity has increased, micronutrients have become unavailable, and useful soil micro-organisms have been destroyed.

Indigenous knowledge (e.g. Warren et al., 1991) about the major components of rice farming seems strangely lacking among the generations of young and middle-aged farmers. This knowledge, for instance on seasonal influences, probably declined under influence of the command strategy of the government's food intensification programmes. Indigenous knowledge about names and life cycles of many pest insects and their natural enemies is virtually absent, partly due to the fact that the main pest problems now are second generation problems, induced by the indiscriminate use of pesticides, but also as a result of the input-oriented technology advocated by extension.

Recent developments indicate that agricultural profits are becoming smaller as world rice prices drop, and input prices rise because of inflation and reduction of subsidies (Hüsken and White, 1992). Especially for tenant-operators, in general the relatively small farmers, who have to pay an increasingly high rent for land or share the harvest with the landlord, rice cultivation is no longer a profitable enterprise. It seems time for the next 'wave' of innovation which will allow farmers to increase efficiency, while consolidating and enhancing what has been achieved during the Green Revolution. New ways of farming are expected to use fewer external inputs than current farming, but will be more knowledge-intensive (Zadoks, 1991; Reijntjes et al., 1992). The US National Research Council (1989) contributed to the emergence of the concept of sustainable agriculture by clearly defining the goals of, what they called, alternative agriculture (Box 2.2). Now,

Box 2.2
Definition of alternative agriculture

Any system of food or fibre production that systematically pursues the following goals (National Research Council, 1989):

1. more thorough incorporation of natural processes such as nutrient cycles, nitrogen fixation and pest-predator relationships into agricultural production processes;
2. reduction in the use of off-farm inputs with the greatest potential to harm the environment or the health of farmers and consumers;
3. greater productive use of the biological and genetic potential of plant and animal species;
4. improvement of the match between cropping patterns and the productive potential and physical limitations of agricultural lands to ensure long-term sustainability of current production levels; and
5. profitable and efficient production with emphasis on improved farm management and conservation of soil, water, energy and biological resources.

the farming communities and governments need to change to more sustainable forms of agriculture.

INPRES 3/86

The first effort of the Indonesian government favouring more sustainable agricultural practices were some major policy measures relating to pest management. After the second nationwide brown planthopper outbreak in 1985-86, seriously affecting the recently achieved self-sufficiency in rice production, the negative side effects of pesticides were finally recognised as a cause of pest outbreaks. In November 1986, the Government of Indonesia announced a presidential decree, INPRES 3/86, that primarily aimed at the control of the brown planthopper by declaring:

- the ban of 57 broad-spectrum pesticides on rice crops;
- Integrated Pest Management (IPM) as national pest control strategy;
- the creation of 1,500 new pest observer positions.

The ban of the fifty-seven brands of pesticides left ten brands with only four different active ingredients available for rice treatment. The number of pest observers (PHP), who are field officers employed by the Directorate of Food Crops Protection (DITLIN) and responsible at subdistrict level for reporting pest damage, totalled 2,900 persons throughout the country after INPRES 3/86. A first activity resulting from the presidential decree was a pilot training for master trainers, pest observers, village extension workers and farmers in Integrated Pest Management (see section 2.3). In addition, brown planthopper resistant varieties (IR 36 and IR 64) were more actively promoted, and *'POSKO's'* or commando-posts at the subdistrict level were established to take immediate and adequate action with specific pesticides in case of brown planthopper occurrence.

Prior to INPRES 3/86, the Government of Indonesia had reduced state subsidies on pesticide from 85% to 75%. In those years, the subsidies consumed 130-160 million US dollars of the country's annual budget. More radical steps were taken after INPRES 3/86 by gradually reducing these subsidies until complete abolition was realised in January 1989. Accordingly, pesticide production decreased. Contrary to the popular belief fanned by the pesticide industry, total rice production increased, proving that pesticide use is not an imperative for rice cultivation (Figure 2.2).

The various policy measures of the Government of Indonesia were enough to bring to an end the threat to food security from massive brown planthopper resurgence, to save a considerable annual outlay for insecticide subsidies and vastly reduce pesticide imports, and to make farming more cost-effective, a benefit passed on to urban consumers. Moreover, a favourable climate for the implementation of a large-scale IPM training programme was created which started in mid 1989. Before dilating upon this programme (section 2.4), the agricultural extension system in Indonesia, and the history of IPM and IPM training will be discussed first in the following sections.

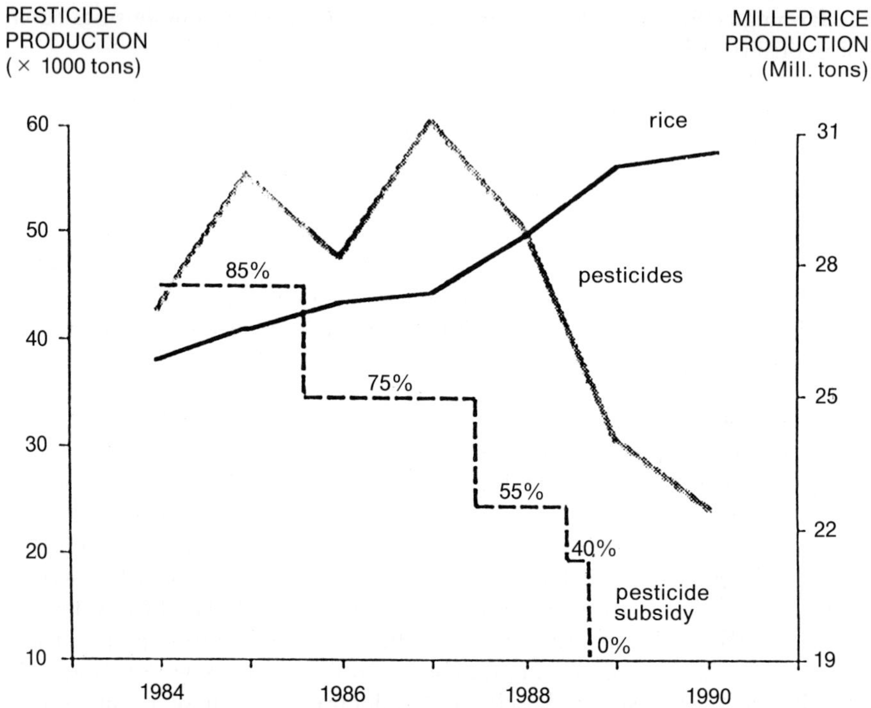

Figure 2.2: Pesticide production, rice production and pesticide subsidies in Indonesia over the period 1984-90. (Source: National IPM Programme)

2.2 Agricultural extension

In the Dutch literature on colonial Indonesia, activities of some sort of 'agricultural extension' can be found, showing clearly the attitude of superiority of the 'extensionists' who think they convey valuable, innovative recommendations, and the indifferent attitude of farmers towards these recommendations not relevant to their current objectives and habits. A translated quotation which might evoke the thought that, in many cases, not much has changed to date, is given below (Opheffer, pseudonym meaning 'uplifter', about 1914):

> 'We, the ethicists, activists, who wanted to make the Javanese rich and wealthy at all costs, we in particular are the ones who are distrusted! [...]

I had those folks plant coconut and cassava and coffee and tea and pepper; I provided them with water and roads; I had them plant rice in rows; only for them I searched for wisdom in many books. [...] The only result was that those folks beat like crazy on their 'gamelan' at the time I left, relieved that they got rid of that tiresome fellow.'[1]

The Indonesian extension service

Agricultural extension activities in Indonesia can be traced back to the establishment of the first Agricultural Senior High School in Bogor in 1903 (Bureau of Agricultural Extension, 1986). A considerable impulse to the development of the agricultural extension service was given only in 1970 by the creation of Village Units under the Improved BIMAS Programme (BND). Before, extension services were carried out by agricultural officers at the subdistrict level ('*mantri kecamatan*') and village officials in charge of agricultural matters. As part of the Village Units, village extension workers (PPL) were assigned to convey the recommendations for the technology packages to the farmers. These village extension workers were backstopped by subject matter specialists (PPS) at district and provincial levels.

All extension officers were directly employed by the BIMAS Secretariat which formed an independent unit of the Ministry of Agriculture. The other five units of the Ministry were commodity-line departments (food crops, estate crops, forestry, animal husbandry and fisheries), each headed by a Director General. In 1974, two new agencies were attached to the Ministry of Agriculture, the Agency for Agricultural Education, Training and Extension (AAETE), and the Agency for Agricultural Research and Development (AARD) (Ward, 1985). The extension divisions of the five departments were closely attached to AAETE. The extension officers, however, remained administratively under the BIMAS Secretariat. Appendix I shows the structure of the Ministry of Agriculture, as far as it is relevant in the context of this study.

In 1976, the World Bank started to fund the Indonesian agricultural extension system through the National Agricultural Extension Programme (NAEP). Under the first and second phases, NAEP I (1976-82) and NAEP II (1980-88), 1,300 Rural Extension Centres (REC) were established throughout the country. An REC work area covers, on average, three subdistricts, and houses 10-15 village extension workers, 1-2 pest observers, and their supervisors. Routine training of the extension workers took place at the REC. Agricultural Information Centres (BIP) were built at the provincial level in 1979 to backstop the RECs. Agricultural In-Service Training Centres (BLPP), also at the provincial level, were established for upgrading all levels of extension officers. Technical Committees at subdistrict, district, province and national levels were set up. In the late seventies, the World Bank

[1]Original Dutch text: 'Wij, ethici, opheffers, die den Javaan met alle geweld rijk en kapitaalkrachtig willen maken, wij vooral worden gewantrouwd! [...] Ik heb de lui klappers laten planten en cassave en koffie en thee en peper; ik heb ze water bezorgd en wegen; ik heb ze de padi op rijen laten planten; om hunnentwille wijsheid uit vele boeken opgediept. [...] Het eenige resultaat was, dat de lui als razenden op de gamelan sloegen, wanneer ik wegging, blij dat ze dien vermoeienden vent kwijt waren.

introduced its Training-and-Visit extension approach (further called the T&V system) in Indonesia through NAEP.

The T&V system

The Training-and-Visit (T&V) system, introduced in at least forty countries by the World Bank, is a form of extension management that relies heavily on diffusion processes (Benor and Harrison, 1977; Benor and Baxter, 1984; Benor et al., 1984). It emphasises regular training of village extension workers and designated 'contact farmers'. In addition, it seeks to enhance linkages between extension and research. The system aims at upgrading the technical content of field extension activities, while making the field workers' contacts more predictable and thus more accessible to farmers, and more enforceable to the supervising ranks of the extension service.

The T&V system for agricultural extension applies a 'Transfer of Technology' model. In this model, information is supposed to be generated by research activities, extended via extension channels, and finally utilised by the farmers. The extension channels comprise several levels, in that village extension workers who operate at the field level are supervised by extension officers at subdivision level, who in turn are supervised by officers of the next administrative level, and so on until direct supervision by the extension headquarters. Subject Matter Specialists (SMS), in close contact with research institutions, are designated to support the various levels with the latest information on agricultural technology developments.

The village extension workers are trained on a biweekly schedule by their direct supervisors supported by SMSs. They receive chunks of information to pass on to their farmers in the next period of two weeks. Farmers in the jurisdiction of a village extension worker are divided into farmer groups. About 10% of the farmer group members are selected to become 'contact farmers'. These farmers have to be visited once in two weeks to receive the recommendations for that period. The contact farmers, in turn, are supposed to convey the extension messages to 'follower farmers'. With this system, information is expected to flow from research via subject matter specialists and (village) extension workers, to contact farmers and, finally, to follower farmers.

T&V in Indonesia

The T&V version in Indonesia is applied in accordance with the technology transfer approach of the World Bank's model. The model fitted well in the hierarchal, top-down structure of the Ministry of Agriculture, and the Extension Service in particular. In addition, the strategy of providing technology packages for food crops intensification required a linear extension approach.

At the national level, AAETE together with the extension divisions of the various commodity-line departments are responsible for extension activities at lower levels, whereas the BIMAS Secretariat is in charge of the administration of extension officers (recruitment, promotion, salaries, etc.). At province and district level, the Agriculture Service (DIPERTA Tk I and DIPERTA Tk II) has extension sections where Subject Matter Specialists (PPS) and district extension officers are housed (Appendix I). Below the district level are the Rural Extension Centres (REC) as the central operational units for all agricultural extension activities. An REC is staffed with an REC head, an administrative officer, (at most) five commodity-line mid-level extension workers (PPM), at least ten village extension workers (PPL), and 1-2 pest observers (PHP). In total, Indonesia employs about 26,000 village extension workers. Biweekly trainings at the REC are given by the PPMs, and occasionally PPSs, to the village extension workers. Farmer trainings are sometimes given at the REC, otherwise in the villages.

Each village extension worker has a working area of 1,000-3,000 farm families, divided into sixteen farmer groups. A farmer group ('*kelompok tani*') consists of all the farmers operating in a certain tract of rice fields ('*hamparan*'), usually marked-out by natural boundaries. The number of farmer groups per village varies from one to fifteen, as does the number of farmers per farmer group from approximately 50 to 150 farmer-operators. Farmer group members are predominantly men, even though women are actively involved in farming. Extension activities specially organised for women only relate to family and household affairs.

A farmer group is headed by a farmer group leader ('*ketua kelompok tani*'), also called contact farmer ('*kontak tani*') who is selected by the group members. At this point, the system as it works in practice differs a little bit from the T&V model. As observed in most places, it is not the 10% of farmer group members who are designated as contact farmers, but usually only the farmer group leader. Most farmer groups also appointed a secretary and a treasurer. The terms 'farmer group' and 'farmer group leader' will be consistently used in this book with the above defined meanings.

According to the T&V model, village extension workers are supposed to meet representatives of each of their farmer groups once in two weeks in order to convey the recommendations received in their own training. In practice, this is seldom realised, not only in Indonesia but also in many other countries applying the T&V system (e.g. Wijeratne, 1989; Moris, 1983). Often heard complaints of village extension workers are that the working area is unmanageable, and that incentives are not equivalent to the tedious work of paying field visits to farmers in often remote locations, day after day.

In Indonesia, village extension workers' salaries are, indeed, very low and require other income sources to provide a minimum living standard to an officer with a family. Extra incentives are obtained from a credit point system determining the career advancement of extension staff. One must have a certain number of points to be promoted from one career level to the next. It becomes increasingly difficult: as one gets higher, one needs more points to make the next step. Points are awarded by superiors for specific activities. This credit point system does not favour services to farmers. A visit to a farmer group gives only 0.007 points, whereas making a seasonal work plan gives two points, and participating in a 60-hour training course at central in-service training centres, one point. With such a system, it is not hard to imagine that extension workers can sooner be found at the REC writing reports than in the field. The necessary, additional income-generating activities, such as trading and farming, often take so much time of the extension worker that there is little time (and motivation) left for visits to farmers' fields.

Other commonly observed weaknesses of the T&V approach are that messages conveyed to farmer group leaders are not implemented, and/or do not trickle down to the follower farmers as expected according to the model. Several reasons have been mentioned in the literature to explain these phenomena. The biweekly chunks of information passed on to village extension workers are often not appropriate to the conditions in the field and the practices of farmers by the time the recommendations reach them (Wijeratne, 1989). Farmers having contact with extension workers are usually the better-informed, more innovative and wealthier farmers anyway (Havelock, 1973; Röling, 1988), who can afford to try out new technologies. Dissemination through communication between these and other farmers is not guaranteed, intentional or not. The 'trickle-down' approach is often criticised for increasing the gap between rich and poor (Adams, 1982).

The division into farmer groups of the Indonesian agricultural extension system appears to be an artificial categorisation, in most cases existing on paper only and not functioning in practice. Some groups organise regular meetings for the 'progressive farmers' once in thirty-five days ('*selapan*'), which is more exception than rule. Quite common in irrigated areas are pre-season extension meetings to discuss and plan synchronised, and sometimes uniform, planting. These meetings are usually not more than a get-together of the extension worker with some selected farmers, who are usually the more prominent farmers in a village attending village meetings anyway. The decision process ('*musyawarah*') always reaches a consensus, usually in line with the proposals from the officials, but this does not guarantee that everybody agrees. Final decisions are not uncommonly made on an individual basis. According to a high-ranked official in the Extension Service, the only operational group in Indonesian society is the family. All decisions taken in the farmer group still have to be approved by other family members, especially when it comes to financial matters, since usually the household money is managed by the women. One farmer group leader once expressed that he had very low knowledge about the functioning of the farmer group and of the village extension worker. He had only seen on television that farmer groups in other areas functioned well, but he had no idea how this could be organised in his village. He did not dare to initiate something afraid of not doing it in accordance with the government's plans.

A carefully calibrated scale, based on ten criteria, is used to 'grade' farmer group members, allowing them to advance to 'progressive farmers'. Similarly, farmer groups are classified according to a set of criteria considering the individual members and the group's administration. By 1990, 51% of the farmer groups in Central Java still belonged to the lowest class, and less than 1% had advanced to the fourth or top class (Source: KANWIL DEPTAN JATENG). Apparently, the farmer group as an institution designated by the current extension system does not fit the needs of the farmers. Especially the complicated administration required by the system – a group has to keep accounts in more than ten files in order to obtain qualification points – has nothing to do with farmers' daily practices and interests. In addition, many farmers do not value the recommendations of the extension worker so much since they consider the officer an inexperienced, urban nitwit when it comes to farming. Their prejudice is confirmed by the fact that the extension workers can hardly ever be seen in the field. If they visit the village at all, they are mostly seen in the village office. In some instances, however, farmers are dependent on the extension worker, that is when they need credit from the KUD (Box 2.1).

Major changes of the extension system's organisational structure at the subdistrict and village levels occurred in 1992, not significantly changing the top-down structure, though. All village extension workers, now with commodity-line responsibilities (e.g. only food crops, animal husbandry, etc.), and pest observers were moved from the RECs to the subdistrict offices, and REC head and PPM positions were abolished. Direct supervision of village extension workers became the responsibility of the subdistrict agricultural officer ('*mantri tani*'). Since the new structure was not yet in effect during the study period, it will not be discussed here.

Alternative extension approaches

In extension science, alternative approaches different to the transfer of technology model have drawn a great deal of attention in recent years (e.g. Röling, 1988). Instead of promoting the transfer of ready-made technologies, the emphasis is on building of farmers' capacity to access external information when they need it, on developing farmers' ability to experiment and draw conclusions, on enhancing farmers' individual and collective ability to take sound decisions, and on 'empowerment' to improve their socioeconomic position vis-a-vis other groups. A few of these approaches are discussed below (Röling and Van de Fliert, 1991).

1. Chambers and Jiggins (1986) reversed conventional thinking by emphasising that farmers are experimenters and technology developers themselves, instead of only users of the findings of scientists. In fact, the entire development of agricultural systems before 1850, including irrigated rice systems in Java, was developed by farmer-researchers. An important approach to developing agriculture is, therefore, to assist farmer experimentation by (1) creating stable networks of experimenting farmers as platforms for reaching consensus about problems for research, and for discussion of experimental design and its results, and (2) by training farmers to set-up and draw conclusions from experiments. This approach has been called **participatory technology development** (ILEIA, 1989). Systematic attempts to assist farmers in becoming better researchers are emerging across the globe (e.g. Chambers et al., 1989; Goebbels, 1988; World Education, 1992).

2. A related perspective is the concern with **indigenous knowledge** (e.g. Warren et al, 1991). Farmers are seen as researchers who have, for centuries, developed effective survival strategies suited to their environment. Otherwise they would not be around. It is extremely important, from a development point of view, to study such indigenous knowledge systems and to design methodologies for analysing and using indigenous knowledge. Modern agricultural development

tends to destroy indigenous technologies and knowledge, however important they are as sources of future development, especially for low-external-input technology. Hence the interest in establishing national centres for indigenous knowledge which can record what still exists and help farmers maintain what is left over.

3. A third important development in recent years is the recognition that innovations usually do not only require external supply of new technology, but also the development of internal capacity to innovate. In other words, it is not only the supply side that counts, but also the demand side. External input of technology and capacity development are complementary essentials for innovation. Most public sector extension agencies focus only on transfer of technology. As was described above, this often leads to innovation among those farmers who have the capacity to innovate already. Especially small farmers are bypassed and eventually squeezed out of farming. In response to this world-wide phenomenon, another approach has been developed, especially by NGOs. This approach, coined the **five element model** (Röling and De Zeeuw, 1983), builds on work of Colin (1978) and Jiggins (1983) and recognises that five elements form an essential 'mix' (Box 2.3).

Box 2.3
The five element model

The 'five element model' comprises a mix of the following five elements essential for rural development (Röling and De Zeeuw, 1983):

1. **mobilisation** to create awareness and commonality of purpose;
2. **organisation** to create platforms for consensus building and for taking communal action;
3. **training** to empower, capacitate and provide skills (e.g., leadership, bookkeeping);
4. **tangible opportunities**, e.g. credit, markets, technologies, which allow improvement of livelihoods;
5. **mix management**, usually by an NGO, or an apex organisation of farmers, to coordinate the elements and help protect the initial stages of the process from vested interests.

Most effective agricultural knowledge systems are characterised by a high degree of farmer control over research, extension and other institutions (Röling, 1988). In the Netherlands, for instance, farmers routinely form part of programming committees for research and extension. Through their organisations, they wield considerable power at the political level. They run the (cooperative) agricultural banks and auctions. Through their organisations, they determine how the budgets of experiment stations, experimental farms and the recently privatised extension service, to which government contributes only 50% of the costs, are to be spent.

4. A alternative perspective on extension to discuss last is **Nonformal Education** (NFE). NFE is defined as 'the fostering of quality-of-life enhancing learning outside the formal school system' (Dilts, 1983). It explicitly recognises human values as a pre-requisite for learning and is based on Paulo Freire's (1972) perspective on education as a problem-solving, consciousness raising strategy for empowerment. Education as an empowering process places importance on how educational processes and relationships affect the learners, not only on the contents (Kindervatter, 1979). Therefore, NFE emphasises experience-based learning linked to living problems. It seeks to empower people to actively solve those problems by fostering participation, self-confidence, dialogue, joint decision making and self-determination. Group dynamics exercises are an important part of learning because they improve the capacity to learn and take decisions. For NFE to support social change, LaBelle (1976) stated that not only ideology should be emphasised, but that also technological and structural interventions are required. Acquisition of skills and provision of structural supports are needed to really liberate people and realise change.

The last approach is particularly relevant in the context of this evaluation study since it was applied by the National IPM Programme in Indonesia.

2.3 IPM and IPM training

The IPM concept

Integrated Pest Management (IPM) was practised long before the term existed. Reports are available on efforts using the same principles as IPM for the control of boll weevil on cotton in the United Stated in 1923 (Flint and Van den Bosch, 1981), and on 'harmonious control' in Canada in the 1950s (Pickett and Patterson, 1953; Pickett et al., 1958). Inspired by Pickett's ideas of harmonious control, the 'Working Party on Harmonious Control' was founded in 1958 in The Netherlands, changing its name into 'Working Party on Integrated Control RNLO/TNO' in 1967 (Anonymous, 1969; Minks and Gruys, 1980). IPM in rice was pioneered in Japan in the 1960s (Kiritani, 1972). The first publication clearly defining 'integrated control' and using economic threshold as an operational concept was that of Stern et al. (1959), which is considered the starting-point of IPM.

The discovery of the insecticide DDT in the 1940s and of its later successors temporarily displaced previous integrated pest control approaches, since the effects of the new-found pesticides seemed miraculous. Emergence of the nowadays well-known side effects of indiscriminate pesticide use, however, did not take long. Pesticide resistance developed, pest resurgence and outbreaks of formerly secondary pests occurred, not to mention environmental contamination and human health problems (e.g. Ito et al., 1962; Luck et al, 1977; World Commission on Environment and Development, 1987; Litsinger, 1989). A solution to the problems caused by indiscriminate insecticide use began to be developed by entomologists through integrated control practices (Zadoks and Schein, 1979). Only by 1975, when systemic fungicides were more commonly used causing similar side effects as insecticides, IPM, and particularly threshold, concepts emerged in the field of plant pathology (Zadoks, 1989a).

Several definitions for IPM have been given, of which two are presented here:

1. IPM means a pest management system that, in the context of the associated environment and the population dynamics of the pest species, utilises all suitable techniques and methods in as compatible a manner as possible, and maintains the pest populations at levels below those causing economically unacceptable damage or loss (FAO, 1968);

2. IPM is an ecologically based pest control strategy that relies heavily on natural mortality factors such as natural enemies and weather, and seeks out control tactics that disrupt these factors as little as possible (Flint and Van den Bosch, 1981).

Important in both definitions given above are the ecological approach of pest management and the integrated manner of applying all control techniques available. A consequence of these two aspects is that pesticide use is allowed as a control measure, but only as a last resort and using application methods least disruptive to the natural environment. IPM seeks to consolidate the achievements of the Green Revolution, but to remove its negative consequences by reducing cost of production and helping farmers become better managers. It seeks to incorporate natural processes into farming, and reduce off-farm inputs, leading to a more profitable and efficient production, and to better human and environmental health. IPM relies on farmers' increased knowledge, active monitoring and analytic decision making with respect to pest management, active use of improved genetic and biological potential of cultivars, and, in some areas, to better rotation. Pest control decision making is based on frequent and systematic field monitoring considering pest populations, natural control factors, crop status and climatic conditions. Measures preventing the development of pest populations, such as the use of resistant varieties and cultural practices, are an important premise in IPM.

An important concept used in IPM is the 'economic threshold' or 'action threshold' (Zadoks, 1985). Initially, thresholds were developed that considered only pest population scores per rice hill or square meter, which was useful for crop loss assessment purposes. Later applications of the concept made it more sophisticated by considering economic factors relating yield loss to control costs. The 'economic injury level' or 'damage threshold' was defined as the point at which expected yield loss (in terms of damage (kg/ha) multiplied by price per kg) equals estimated pest control costs. Application of economic thresholds in chemical pest control proved to be profitable in comparison with prophylactic or no treatments (Smith et al., 1989; Herdt et al., 1984).

Applying the economic threshold concept in farmer conditions, however, appeared difficult, which will be the reason that it is not practised by most rice farmers (Waibel, 1986). An economic threshold considers one pest only, whereas a pest complex occurs in crops. Expected yield loss depends on many factors, including the development stage of the crop, the ability of the cultivar to compensate injury, the trends in population dynamics of both pests and predators, and the expected weather and market conditions. Pest control costs might even fluctuate as a result of changing pesticide prices and labour wages, not to consider possible opportunity costs. In addition, diversity in socioeconomic status of farmers brings about a variety in objectives and makes the use of just one threshold level ineffective, whereas a variety of thresholds will create confusion (Zadoks, 1985). Only recently it is recognised that the economic threshold is not a workable tool in IPM programmes for farmers. Farmers themselves usually possess the necessary farm-specific information to make the right decision whether a control measure is economically justifiable. In addition, they themselves have many years of experience with the consequences of ever-changing market conditions for their own farms.

Field studies in rice and other crops have shown that through IPM implementation pesticide use can be reduced and farmers' profit can be increased (e.g. Litsinger, 1984; Bottrel, 1987; Matteson, 1986; Kenmore, 1987). Constraints to farm implementation of IPM, however, were also identified, including continuing subsidies on pesticides and free pesticide schemes (Waibel, 1989), farmers' deficiencies in knowledge about pests and their ecology, in monitoring and forecasting (Zelazny et al., 1985), and in technology transfer (Goodell, 1984). As IPM evolved, it became more and more obvious that emphasis should be placed on the farmer himself or herself as the manager of the crop who takes decisions based on field evaluations, previous experiences, and knowledge about ecological processes and effects of their actions. This makes IPM not only knowledge-intensive, but also farm-specific, in which case the term 'technology' seems no longer appropriate, as will be described below. Extension of IPM to farmers, especially to those who have been exposed for years to package technologies such as the Indonesian BIMAS farmers, therefore requires adequate and extensive training.

The Inter-country IPM Programme

Major efforts in IPM training in irrigated rice have been carried out in Asia by the United Nations' Food and Agriculture Organisation through the 'Inter-country Programme for the Development and Application of Integrated Pest Control in Rice in South and South-east Asia'. In the pilot phase of the programme in 1978-80, basic training principles were developed and a pilot training programme was conducted in the Philippines (Matteson et al., 1992). Farmers were trained in groups throughout a growing season. Weekly two-hour sessions were organised that were dominated by practical field activities and group discussions. Real plants and insects were used as training materials instead of handouts. Lectures were avoided and technology was greatly simplified. The result was notable: farmers were able to identify pest problems better and trusted more in their own decision making ability. Pesticide use, and thus pest control costs, by trained farmers decreased (Adalla and Rola, 1986; Kenmore et al., 1987). In several cases, it was found that rice yields increased, probably as a result of increased fertiliser use that could be purchased from the money saved on pesticides.

In the first and second implementation phases of the Inter-country Programme (1981-90), similar IPM trainings were delivered in several Asian countries through T&V extension services (Matteson et al., 1992; Whitten et al., 1990). Training intensity and quality, however, deteriorated as the programme scaled up. In the Philippines, it was obvious that the same enthusiasm that had inspired the pilot group of IPM trainers could not be achieved on a larger scale. Trainers fell back on their conventional ways of lecturing in the classroom. In Sri Lanka, T&V village extension workers did not deliver more than half of the scheduled training sessions. Farmer group trainings were supplemented by multimedia strategic extension campaigns after farmers' training needs and information channels had been identified (Adhikarya, 1989; Van de Fliert and Matteson, 1989, 1990). Evaluation findings about the impact of the campaigns were not very encouraging (Matteson et al., 1992). The influence of the persistent pesticide promotion through the same media is considered a major factor having caused this low impact.

Training efforts through local NGOs complemented national IPM extension activities. Satisfactory results were achieved, since activities were implemented on a small scale with highly motivated staff. Although contributive, especially in generating innovative ideas, NGOs are not likely to become the main channel for extending IPM to the millions of Asian rice farmers, among others because of the small-scale operation of most NGOs.

IPM training through Indonesia's T&V system

Efforts to introduce IPM in Indonesia had started as early as 1979 when the Directorate of Food Crops Protection (DITLIN) adopted the concept in its policies (Oka, 1990a). The attempts followed the technology transfer approach which had been so successful in the rice intensification programmes. IPM training activities focused on packages and prescriptions, and were incorporated in the routine extension meetings. Abundant spraying of pesticides was applied over whole areas at once with government pesticide aid, based on pest status assessments of agricultural officers. No impact of these activities on farmers' behaviour has ever been reported.

In 1984, a field evaluation of IPM demonstration areas was conducted by the FAO Inter-country Programme and DITLIN, and yielded the astonishing result that pest populations in some of these areas appeared to have increased. A workshop was organised in which university entomologists presented data of more field studies concluding that occurrence of high pest populations was strongly related to an intensive use of pesticides with respect to both number of applications and number of types used. As described before, nationwide severe brown planthopper outbreaks occurred in 1985-86, after which INPRES 3/86 was announced.

As a reaction to INPRES 3/86, declaring IPM the national pest control strategy, a crash IPM training programme was conducted through the T&V extension system. The Indonesian Government requested the World Bank's permission to use US$ 4.19 million, remaining for the second phase of the National Extension Project (NAEP II), for IPM training. FAO's Inter-country Programme provided specialist training for the new recruits of DITLIN. A tremendous effort was made to train trainers and to develop trainers' guides, flipcharts, slide-audio modules, leaflets and pamphlets of which 150,000 copies were distributed by NAEP II. Coordination meetings at several official levels were organised, and two-day farmer trainings were planned (DirJen Pertanian Tanaman Pangan, 1987). The total budget was spent in seven months on travel money, honoraria, vehicles, subsistence and pocket money for farmers, and other monies, an amount which would have totalled US$ 7 million if calculated on an annual basis.

Though the activities had Presidential priority and special facilitation by the Finance, Planning and Economics Ministries, training funds did not reach training centers earlier than four days before the courses began. Training materials arrived mid-way through courses, often too late to be used. Years later, hundreds of these materials can still be found neatly wrapped in their original plastic containers at provincial training centers. In spite of repeated meetings on the subject, only 25% of the training groups actually entered a rice field. The goal of the crash training programme was to train 125,000 farmers, but only 10,300 actually received training. Only 8.5% of the allocated resources could be delivered to the field to train less than 10% of the farmers targeted. Where farmers were reached, trainers used top-down approaches and did not use fields or farmers' own experiences.

In a preliminary study, the remnants of the crash IPM training programme were observed in a few farmer groups of two Rural Extension Centre (REC) work areas in the provinces of Central Java and Yogyakarta in early 1989. Ex-trainees reported to be taught about pest identification, balanced fertilisation, water management, spraying pesticides, weeding, and soil preparation during one day of theory teaching and one day of field work. Most of them, however, had not learned many new things. Farmers mentioned that as a result of the training their crop cultivation practices changed a little relating to balanced fertilisation, crop monitoring, and pesticide use. Pests identified in the field were reported to the pest observer who, in turn, arranged pesticide aid for the farmer group. These pesticides were sprayed by the IPM trained farmers who had formed a special crew to do this. The trained farmers reported that they did no longer spray indiscriminately as before. However, they did not feel confident yet to take their own decisions, and relied fully upon assistance from the pest observer and the village extension worker. They actually only learned to look at and recognise pests in the field, to apply the preset economic threshold levels for brown planthopper, and to report high populations to the REC. Natural enemy populations were not considered at all.

According to the immediate evaluation of the Indonesian crash IPM programme, trained farmers' pesticide use decreased and their yields increased (FAO, 1988). However, desired changes in their behaviour did not occur, which makes the sustainability of the results doubtful. The FAO Mid-term Review Mission summed up the experience (Whitten et al., 1990):

> 'A rigid system equipped to move simplistic messages to a large number of passive farmers could not absorb the energy of IPM's field skills training. A transformation from within was needed to meet the new challenges from outside.'

The top-down extension system had proven again not to be suitable for changing farmers' behaviour necessary for sustainable IPM implementation, a finding conform to theories stating that internal rather than external forces are needed to cause change in people (Axinn, 1988). A new training approach was designed building on the achievements of the Philippine pilot programme but including the lessons learned from scaling-up through T&V systems. Thanks to the policy measures of the Government of Indonesia in 1986-87 favouring IPM, implementation of the National IPM Programme could be realised as of 1989. The first phase of this programme (1989-92) was financially possible by reallocation of the money from abolished pesticide subsidies which had previously been granted by USAID. The Indonesian training approach became the model for other Asian countries participating in the third and current phase of the FAO Inter-country IPM Programme.

2.4 Indonesia National IPM Programme

A different approach to training contents and process

With its broad experience in IPM training in several Asian countries during ten years, the Inter-country IPM Programme was the appropriate partner to assist in the design and implementation of a large-scale National IPM Programme in Indonesia. Heaving learned from these experiences, the Indonesian model embarked upon a new course, with regard to its approach to both training contents and process.

As described above, IPM is knowledge-intensive and farm-specific for which preset recommendations in a package had proven to be inadequate. Instead of mechanical instructions for field sampling and spraying based on centrally determined economic threshold levels, the National IPM Programme in Indonesia began to apply a more ecological approach of IPM. IPM was provided with a set of principles instead of instructions. This is the reason why we no longer want to use the term 'technology' for IPM, but will speak of principles, approach, or practice(s) in this book. The IPM principles include:

1. Grow a healthy crop;
2. Observe the field weekly;
3. Conserve natural enemies, and
4. Farmers become IPM experts.

With these principles, a set of agronomic and ecological concepts are provided to farmers as tools for their decision making. The farmer remains the central manager and independent decision maker. One obvious example of the different approach is the changed concept of economic threshold. Instead of presenting a precise but unworkable coefficient, the programme developed the concept of 'experience threshold' which develops as farmers learn and experience, and is applicable under specific farm conditions (Gallagher, 1990). A threshold that builds on farmers' expertise and experience fits better into the consideration of farmers as independent decision makers. The behaviour which IPM promotes subsequently include regular observation and informed decision making about pest control.

This different approach to IPM requires a different approach to extension. Farmers are not considered as passive receivers and acceptors of external recommendations, but as active learners and expert masters in their own fields. Therefore, the principles of nonformal education (section 2.2) seemed appropriate as an approach to training farmers in IPM. The principles applicable in this respect are learning through experience, focusing on field problems, and empowering farmers. This approach to farmer training, in turn, has important implications for extension staff training and for the institutional design of the extension service.

Farmer behaviour supported by IPM

The IPM Programme does not focus on transferring specific technologies or bits of information. Rather it seeks to capacitate farmers to take sound decisions by providing some basic principles. Specifically, as derived from the four IPM principles, the programme emphasises the following behaviours:

- focus on a **healthy crop**, resistant to local diseases and able to compensate for pest damage;
- a **good knowledge of pests and their natural enemies**, not in terms of their Latin names, but in terms of function, what they do to plants and to each other at what stage of the crop. Such knowledge also includes the developmental stages of a pest and their recognition. This knowledge is expected to be updated and improved by farmers' own observation and experimentation, and by farmer-to-farmer exchange of experience;
- **regular and systematic observation** in the field, using systematic procedures (random selection of sample rice hills) to asses the occurrence of pests and natural enemies in relation to the crop's development stage;
- **sound decision making** and discussion with other farmers about such decisions. The process of decision making is more important than the decision itself;
- **experimentation** with planting times, varieties, soil cultivation practices, fertilisation, rotations and biological control for their effect on pest populations;
- **use of relevant, science-based knowledge**, such as the regenerative capacity of rice varieties after pest damage, or on parasites in the egg masses of stemborers.

With such priorities, farmers' own expertise and mastery is fostered rather than only their adoption of external information. More specific competency objectives will be given in Chapter 7.

Actors in the programme

The **National IPM Programme** is a temporary structure that will be continued for a limited number of years. The first phase (1989-92) was financed by donations from USAID to BAPPENAS, the national planning agency, that were originally meant for pesticide subsidies. The second phase (1993-98) is sponsored by the

World Bank. The programme is implemented with technical expertise from FAO. It is run by both expatriate and local experts. There are no senior counterparts in the traditional sense of the word. A Steering Committee, an Advisory Board, and a Working Group with members from various government institutions and universities were called into being to assist the programme management. For management and curriculum development, special **secretariats** were established in Jakarta and Yogyakarta. The programme works intensively within the country's existing framework, putting strong emphasis on creating linkages and contracting for specific jobs, such as curriculum development and training. In addition to training of extension staff and farmers, the programme supports research activities such as a field laboratory in West Java focusing on white stemborer problems, a health impact study in a pesticide-intensive area in Central Java, insect habitat studies, studies on IPM in secondary food crops, and several training evaluation studies.

At the regional level, the programme operates from twelve **Field Training Facilities** (FTF) in the eight main rice-growing provinces where, in total, around three-quarters of the nation's rice is produced. Existing Agricultural In-Service Training Centres were partly transformed into IPM FTFs. Primary trainers at the FTFs are the **Field Leaders I** (PL I, 21 in total) who are assisted by **Field Leaders II** (PL II, 129 in total), and by some training experts of the training centres where the FTFs are hosted. Most of the Field Leaders belong to the group of pest observers upgraded in the crash IPM training in 1986 to become IPM master trainers. The development of a strong and qualified cadre of trainers is given high priority in the IPM programme, since it is the key factor for the training of the millions of Indonesian rice farmers. Field Leaders I assisted in designing the final curriculum and field guides for staff and farmer training.

Pest observers, whose numbers were doubled by INPRES 3/86, were assigned to become the trainers at the field level. They were all previously trained in the crash IPM programme in 1986-87. To date, some 1,120 pest observers received intensive IPM training through the National IPM Programme and function as the IPM trainers at the Regional Extension Centres to train both farmers and extension staff. This means that all officers intensively involved in the IPM programme (PL Is, PL IIs and pest observers) belong to the Directorate of Food Crops Protection (DITLIN), and **not** to the departments involved in extension activities through NAEP (BIMAS Secretariat and Agricultural Service). The Rural Extension Centres (RECs), however, which fall under the responsibility of the Agricultural Service, are the base from where all IPM training activities executed by the pest observers at the field level are organised. It was obvious from the beginning that the village extension workers are not very suitable candidates for introducing IPM. They have many tasks, among which pest control extension is a relatively minor one. They are heavily involved in the T&V routine and in input distribution activities which conflict with the nature of IPM.

Rice farmers are, obviously, the ultimate beneficiaries of IPM training activities. At first sight, they seem similar in their cultivation practices. However, as will be described in Chapter 4, farmers differ a great deal in terms of their use of inputs, farm size, tenure status, the type of off-farm jobs they engage in, the activity in farmer groups and so on. The villages in which farmers live also show great diversity due to geographical and infra-structural isolation, leadership, history, and other factors. Farmers are formally organised by the Agriculture Service in farmer groups, but seldom are these groups active.

Vested interests in pesticide use are not immediately apparent beyond the agrochemical companies. But pesticide production is an enormous industry with a turnover valued at some US$ 1.5 billion per year. Involvement in this industry can be found in various sectors and levels, including salesmen, organisations such as the KUD, village officials and extension workers. Influence of these interests on the result of IPM training and implementation should not be underestimated.

Policymakers are critically concerned with food security. Varying individual interests, however, sometimes result in a mixed support for the IPM Programme. But increasingly, senior policymakers recognise that the programme has energised farmers, given them new confidence, and captured their imagination. Many consider this a welcome change from the existing farmer groups and extension approaches which fail to engage farmers beyond token and formal participation.

Research institutes and Universities have only been marginally involved sofar. The universities have trained the pest observers for a few months to provide them with a diploma which allows them to advance in salary scale. Much greater involvement of research institutes and universities is expected in the future.

Farmer training in IPM farmer field schools

Farmer training is organised in so-called **IPM farmer field schools** (Photo 2.1). The philosophy behind this name is that farmers go back to school, a place reputed

Photo 2.1: The farmer field school is located in the rice field.

for **learning**, where they can obtain a diploma (a certificate) allowing them to be IPM experts and trainers. Key ingredients of the IPM farmer field school are the following:

- a field school group consists of **twenty-five farmers**, selected either from one farmer group, or across the farmer groups from the same village;
- in the field school, farmers work in **subgroups of five**, the optimal size according to NFE principles;
- training starts with a **pre-test** and ends with a **post-test of knowledge**. Twenty five 'ballot boxes' with three slots representing multiple-choice answers are made of carton and placed in the field. A question is posted above the box concerning a field problem, and the three possible answers are indicated with ropes leading from the box to a damaged leaf, an egg mass, or a similar item. Farmers enter a chit with their number on it in the slot with the appropriate answer for each box. The scores of the tests, in themselves fairly meaningless, are a great motivational device for the participants, and give an important diagnosis of trainees' ability;
- the field school lasts the main part of an **entire season**, so that farmers can work with each stage of rice plant development. This is very important, because pest problems change with the stage of the crop. The groups meet once a week for some ten weeks;
- each field school group has a **demonstration field**, consisting of an IPM plot where IPM principles are used to take pest control decisions, and a plot where the INSUS recommended package is applied;
- there is hardly any lecturing during the training. The pest observers have been carefully trained not to allow themselves to be forced into the position of an expert, but to be a **facilitator of the learning process**. They are not to answer questions directly, but instead, make the farmers think themselves. 'What do you think?', 'What did you find?', 'What did it do?'. This is called the '*Apa ini?*' principle, Indonesian for 'What is this?'. A question answered with a clear answer is a lost opportunity for learning.
- farmers meet somewhere **in or close to the field** under a tree or in a small shack which provides some shade;
- the main activity, and first in the morning, is to step into the demonstration fields in groups of five and **observe** sample rice hills, usually chosen at random along a diagonal across the field. Notes are made of insects, spiders, damage symptoms, weeds and diseases, observed at each hill. The growth stage of the plant is carefully observed, as is the weather condition. Interesting insects and other creatures are caught and placed in small plastic bags.
- in subgroups, the observations are collated in drawings, the **agroecosystem analysis**. On large sheets of cheap paper fixed to a sheet of plywood, using different coloured crayons, farmers draw the rice plant at its present growth stage, with pests and natural enemies of the moment. A leaflet with pictures of pest insects and natural enemies, distributed to each subgroup, is used as a reference. A conclusion about the status of the crop and possible control measures is drawn

by the five members together and noted down on the paper;
- the subgroups' agroecosystem analyses are **presented** to the whole field school group. The conclusions drawn from the field observation with respect to pest control are discussed in the entire group. The field has become the main training material and farmers' own observations the source of knowledge for the group;
- during each session, **special subjects** are introduced. The trainers' training provided the pest observers with a substantial repertoire of carefully developed modules. Special topics relate to field problems, such as rat population growth, effects of pesticides on natural enemies, and life cycles of rice field inhabitants;
- **group-dynamic exercises** enliven the field school and create a strong sense of belonging to the school;
- farmers often keep an 'insect zoo', plastic netting around four bamboo poles set around a rice plant. Inside this insect zoo, various pests and predators are introduced, and watched by farmers. Through own **experiments and observations,** farmers gain ecological knowledge;
- active group members are encouraged to train other groups. This **farmer-to-farmer training** is an important strategy for mass replication;
- a **field day** is organised at the end of the season in which the result of the farmer field school is presented to the surrounding community, including village and subdistrict heads in order to obtain (financial) support for follow-up activities.

Farmers participating in the IPM field school receive a compensation of Rp 1,000 (approximately US$ 0.50) per day from the programme to remunerate possible loss of income when spending time in the training. Many groups use these monies for buying uniform caps and T-shirts, decorated with the emblem of the programme and their farmer group name, visibly increasing group spirits. Some groups also use (a part of) the compensation to go on excursions, for instance to experiment stations or training centres.

Horizontal communication on IPM was encouraged in various areas through folk theatre. Some IPM field school groups were given additional training on the principles of theatre by a local NGO active in the field of communication ('S.A.V. PUSKAT'). The groups were guided in designing and performing their own IPM play for the whole village community.

An impact study executed in 1991 by the National IPM Programme among more than 2,000 field school graduates showed that trained farmers evidently changed their pest control behaviour (Pincus, 1991). Reduced pesticide applications, lower use of banned insecticides, and decreased expenditures were measured. The study also indicated that these results were directly linked to the IPM training. Pest targets were rationalised, and a radically different decision making structure of trained farmers was observed.

Training of trainers

The type of farmer training discussed above requires different staff training. The training of trainers (pest observers) and of extension worker is based on the same principles as the farmer training. The NFE approach to train the first pilot batch of fifty-two field trainers began in earnest in July 1989 at the FTF in Yogyakarta, followed by training cycles in, first, six, and later eight provinces throughout the country. As for the farmers, the two main principles of staff training are (1) agroecosystem analysis based on careful field observation; and (2) dialogue instead of lecturing. Key elements of the training of pest observers are the following:

- training takes fifteen months and consists of the following components:
  - rice IPM training (3.5 months);
  - extension training in IPM farmer field schools (3.5 months);
  - dry secondary food crops IPM training (3.5 months);
  - a diploma course at the university (4.5 months).
- one FTF training group consists of fifty people, divided into subgroups of five;
- rice and secondary food crop IPM training takes place at the FTF where the pest observers grow their own crops. They have to become farmers first before they can face farmers in a position as trainers;
- training curriculum is completely field-oriented. The '*Apa ini*?' principle is the basis for learning. Problems discovered in the field become topics for discussion. Carefully designed field experiments, such as comparison of varieties, fertiliser treatments, and variations in pesticide treatments, as well as a range of special topics are executed and discussed;
- extension training takes place in the home areas of the pest observers where they conduct four IPM farmer field schools each during one season. In this training, two village extension workers per pest observer are trained in IPM on-the-job;
- Field Leaders I and II facilitate the FTF trainings, and supervise the extension training in the field;
- a field day is organised at the end of the rice IPM season on which the result of the training is presented to policymakers at regional and national levels in order to obtain (financial) support for follow-up activities.

The goal of training the pest observers is to make them confident IPM experts, instill an attitude of self-learning through experimentation, and to develop a cadre of effective trainers of farmers and village extension workers. Since 'the methods we learn from are the methods we fall back on when we teach others' (Pontius, 1990), the methods used during pest observer training are those they are expected to use with farmers. During their training, pest observers work in their fields every morning, a rare event for civil servants. The special topics are presented in a set of modules which pest observers feel confident to handle with farmers or extension workers. Pest observer training is supported by elaborate manuals.

During the extension training, one pest observer has to choose two village extension workers from her or his REC to form a team with for farmer training. The extension workers are given a one-week introductory training at the FTF in which they are acquainted with the principles of IPM and with the farmer field school training approach. The trainer teams formulate work plans for farmer training. One team conducts two farmer field schools, which implies four field schools per pest observer. During the implementation of the farmer field school, the pest observer is the main facilitator, whereas the extension worker assists where necessary and, at the same time, learns on-the-job to become IPM facilitator.

With this set-up, one training cycle at one FTF, which takes about a year, delivers 50 trained pest observers, 100 extension workers, and 5,000 field school farmers. In addition to this main training model of the programme, various other models were developed and tried out in order to expand IPM to more extension staff and farmers (Van de Fliert et al., in prep.). Since the model described above was the subject for the present study, other models will not be discussed here.

2.5 Two contrasting approaches

In the above sections, the Green Revolution context of food crops intensification and agricultural extension (T&V) in Indonesia, and the introduction and approaches of the National IPM Programme were described in detail. Looking at the two 'models' of agricultural development, a structural contrast can be discovered with respect to (technology) contents and extension approach. The philosophy and practice underlying the introduction of Green Revolution technology was necessary considering political and food security problems in Indonesia in the sixties. Adoption of high yielding varieties and inputs was compulsory. Consequently, extension had to follow a transfer of technology model that relied on instruction and blanket recommendations. Decisions about insecticide use, varieties, and water management were usually taken by officials, and farmers often had little room for manoeuvre under strong social pressure from local officials.

In contrast, the National IPM programme follows a completely different approach, emphasising empowerment of farmers to make their own decisions based on knowledge, field monitoring and experimentation, and decentralised deployment of a highly trained and autonomous cadre operating from local centres. Farmers are not told what they have to do but through facilitation they learn to learn from and build on their experiences. Through this facilitation model, the IPM programme aims at making informed decision makers and independent farm managers. Indonesia is ahead of most other countries in attempting such an approach on a massive scale.

The introduction of IPM through policy measures and nonformal training as a public sector endeavour within the conventional Green Revolution context creates a unique situation. This 'marriage' should lead to some interesting effects, supposedly including conflicts, especially at the village level where the ultimate beneficiaries of IPM, the rice farmers, have to liberate themselves under the counteracting conventional forces of package technology and input distribution. Therefore, evaluation of the National IPM Programme at the village level was considered an interesting object for study.

3 Evaluation concepts and methodology

3.1 Programme evaluation

Definition, purpose and strategies

Programme evaluation is a form of applied social science which intends to assist in improving the quality of human services (Posavac and Carey, 1989). A more specific definition is given by the same authors (1989: 3) stating that 'programme evaluation is a collection of methods, skills and sensitivities necessary to determine whether a human service is needed and likely to be used, whether it is sufficiently intense to meet the need identified, whether the service is offered as planned, and whether the human service actually does help people in need without undesirable side effects'.

Programme evaluation as a scientific activity emerged in the 1960s and expanded during the 1970s (Shadish et al., 1991). The need for evaluation studies followed on the emergence of many public programmes, often with large investments but doubtful effectiveness. Financial, political, managerial and intellectual concerns increasingly pushed towards evaluation research. Various strategies and methods evolved in the course of time (e.g. Campbell, 1971; Weiss, 1972; Cook and Campbell, 1979; Scriven, 1980; Cronbach, 1982). As evaluation strategies and methods matured, evaluation theory developed towards one focusing on methodology, but in a broad context (Shadish et al., 1991). Evaluation theory comprises the many possible decisions about the shape, conduct and effects of evaluation, and is, therefore, far from focusing on methods only. Inevitable ramifications when discussing methodology within the context of evaluation include the philosophy of science, public policy, validity, and utilisation.

The broad, overall purpose of programme evaluation is to provide feedback to human actors. Specific sub-purposes always need to be defined for specific evaluation studies. Commonly, two main types of evaluations are distinguished, called formative and summative evaluation (Scriven, 1980). Formative evaluation aims at providing feedback to improve a programme, whereas summative evaluation provides information for decision makers to base decisions on whether to fund, continue or terminate a programme.

In the early days of evaluation research, policymakers and managers appeared to be the primary beneficiaries of mainly summative evaluation activities. They were the source of questions and the audience for the evaluation results. Objectives stated in advance and expected outcomes served as the standard for measuring programmes' successes and effectiveness. With the growing consideration of multiple stakeholder groups (Weiss, 1983), the context for evaluation practice and evaluation utilisation became of greater concern. The formative types of evaluation gained importance. A higher degree of participation of beneficiaries in programme evaluation can take place by incorporating evaluation within programme activities (e.g. Feuerstein, 1986; Groot and Boon, 1992). This approach is likely to lead to (1) more reliable and utilisable evaluation outcomes, and (2) a programme enhancing self-evaluation by the beneficiaries. Such evaluation activities need prior and very careful planning by all parties involved in programme design.

Another division between types of evaluation can be made according to the kind of questions asked about a programme, distinguishing evaluation of (1) need, (2) process, (3) outcome, and (4) efficiency (Posavac and Carey, 1989). In addition to focusing on different types of questions, this division also differentiates between moments of evaluation in programme design and implementation. An evaluation of need (or feasibility study) takes place before programme implementation in order to assist programme design and preparation. Process evaluation is done during programme implementation to assess whether the programme is implemented as planned, in other words, whether inputs result in the expected outputs through the expected processes. This activity is also called monitoring, and is not always considered a type of evaluation but a separate research strategy (Groot and Boon, 1992). Evaluation of outcomes and efficiency can take place both during and after programme implementation, through mid-term or ongoing, respectively terminal evaluations. Evaluation of outcomes measures whether expected outcomes have been achieved, and assesses to what effects on farmer practice, consequences for farm management, and impacts these outcomes have led. Evaluation of efficiency considers the relation between programme costs and outcomes.

Summarising, evaluation is an effort to make visible the social consequences of purposeful action (such as training programmes) to different stakeholders, at different times (before, during and after), for different purposes (technical, managerial or political), and using different criteria (significance, effectiveness, efficiency, sustainability, stability).

Utilisation of evaluation

In the first decade, programme evaluation research often did not lead to major changes in social programmes or policies. As a reaction to this so-called 'utilisation crisis', a new focus emerged in evaluation research: utilisation (e.g. Alkin, 1985; Patton, 1986). Although utilisation is hard to define since perceived usefulness depends highly on the person who perceives it with her or his individual values and objectives, an implicit definition is given by Patton (1986: 30) as follows:

> 'Utilisation occurs when there is an immediate, concrete and observable effect on specific decisions and programme activities resulting directly from evaluation findings.'

Within the context of utilisation-focused evaluation, the purpose of programme evaluation needs to be complemented by deliberate statements on the use of information gathered for programme decision making. With such purposes, the evaluation outcomes need to answer questions of the programme's stakeholders, not of the evaluators. Evaluation is then aimed more at action rather than at delivering information only. This does not necessarily mean that evaluations should cause revolutionary changes in programme management of implementation. On the contrary, more realistic, and often more useful, impacts of evaluation are more likely to deal with the slow processes of evolutionary programme development, such as reducing uncertainty, speeding things up or getting things started.

For evaluation to be utilisable, the main emphasis is put on **people** and not on programme goals (Patton, 1986). A first step in evaluation design should be the (careful) identification of the primary intended users of the evaluation outcomes. These intended users are not (impersonal) organisations, but individuals or groups of people who are likely to have an interest in the information generated. These can be programme managers who seek for information to base their decisions on, but also end beneficiaries (e.g. farmers in a training programme) who want to know whether the programme is worth their valuable time and effort to participate. Particular questions of particular stakeholders need to be the guideline for the evaluation design. Summarising, Patton (1986) lists five premises for utilisation-focused evaluation:

1. The concern for utilisation should be the driving force in evaluation;
2. From the beginning to the end of the evaluation, this concern is ongoing and continuous;
3. Evaluations should be user-oriented, aimed at the interests and information needs of specific, identifiable people;
4. The identified evaluation users should be personally and actively involved in making decisions about the evaluation;
5. It should be recognised that there are multiple and varied interests around any evaluation; intended usefulness for users is the first priority when attention to interests must be limited.

In addition to the personal factor highly emphasised by Patton, Alkin (1985) considers two other factors of importance for evaluation use, those of context and of evaluation itself. Context factors include organisational and programmatic arrangements, as well as the social and political climate of the programme evaluated. Evaluation factors refer to the conduct of the actual evaluation, comprising its procedures, the information collected, and the reporting of the information. It should be noted that the conduct of the evaluation study and the position of the evaluator can strongly influence (and bias) its outcomes, and thus its utilisation.

The case study method in evaluation research

Since World War II, the case study was neglected as a potential research method in social sciences, in favour of sample surveys and continuous data series (Felstehausen, 1982). This underestimation was based on the belief that case studies were merely exploratory, using a single test design only, and not suitable for making generalisations. Revaluation of the case method emerged in the late seventies and eighties, and cases were increasingly used for programme evaluations (Kennedy, 1979). Perceptions that case studies can only use certain, qualitative methodologies, are merely exploratory and unable to make causal inferences, were disproved by examples of case studies serving explanatory and descriptive functions, and permitting testing of and elaborating upon existing theories (Yin, 1981a). Some authors even claim that case studies may be in better harmony with people's experience, and thus provide the natural basis for generalisation (Stake, 1978).

Previous experience in conducting a large-scale sample survey had left me with the unsatisfactory feeling that it was often impossible to explain information obtained from single-visit, quantitative data collection only when the context of the information and reasons for practice are not known. Detailed observations, rather than generalisations of large data sets, might considerably contribute to the understanding of situations, practice and processes. With good understanding, phenomena are likely to be also recognised in new contexts, which is a form of generalisation that develops within a person as a product of experience. When research aims at achieving understanding, extension of experience, and increase in conviction in that which is known, the case study seems to be a suitable investigation method.

Smith (1974) defined case studies as being 'credible, holistic, particularistic, individualisable, process-oriented, ego-involving, and blending of behavioural and phenomenological methodologies'. In case studies, a variety of sources of evidence are relevant, and, therefore, both quantitative and qualitative data collection methods can be applied which are likely to be complementary. It is important to ensure whether evidence from different sources converges on a similar set of perceptions (Jick, 1979). Two types of case study design are distinguished (Yin, 1984): (1) the single-case design to test theories, and (2) the multiple-case design that draws conclusions from a group of cases, appropriate when a phenomenon is thought to exist in a variety of situations, such as training programmes.

Considering case methods a separate research strategy, Yin (1981b) stated that the need to use cases arises whenever (1) an empirical inquiry must examine a contemporary phenomenon in its real-life context, especially when (2) the boundaries between phenomenon and context are not clearly evident. The strength of a case study to cover both a phenomenon and its context has the likely consequence that a great number of variables are of interest. This, in turn, means that few (advanced) statistics will be relevant for analysing the data, since static computations are mostly subordinate to the critical role of context variables in case studies. The separate, quantitative methods applied within the study, however, can justifiably make use of simple statistics. Case study analysis should be carefully planned to make the utmost use of the possibly large amount of valuable data generated through a variety of methodologies.

3.2 Perspective for evaluating IPM training

A framework for training programmes

Evaluating a training programme requires insight in the relations between aspects of the programme's starting-point conditions, design and implementation, and effects. To analyse these relations, a framework illustrating processes in training programme design and implementation is suggested (Figure 3.1).

DESIGN/CURRICULUM DEVELOPMENT	IMPLEMENTATION
1. Appreciation of an undesirable situation	10. Impact
2. Nature of training contents	9. Consequences
3. Intended farmer practice	8. Changed practice
4. Training strategy/ curriculum for farmers	7. Training of farmers
5. Training strategy/ curriculum for trainers	6. Training of trainers

Figure 3.1: Processes in training programme design and implementation.

Stage 1 is the actual situation in which problems have occurred giving the impetus for planning an intervention to change. Stages 2 to 5 on the left are the elements and processes in the planning, design and curriculum development of a training programme to meet these problems. Nature of the training contents and the intended practice of the beneficiaries are defined, from which training strategy and curriculum for, first, farmers and, then, trainers are deduced. The stages on the right describe what happens when the programme is implemented: training activities in stages 6 and 7, and outcome of training in the stages 8, 9 and 10. Outcome distinguishes three different levels: effects of the training at the level of farmer practice (including perceptions and decision making) in stage 8, consequences of changed practice at the farm level in stage 9, and the impact of the implementation at a higher level of communities and their environment in stage 10. Elements in the implementation phase relate to elements in the design phase on the same horizontal line.

The framework described above should be seen as a bare frame of stages in training programme design and implementation. In the analysis of the (relations between) factors and processes, especially those on the implementation side of the framework, more dimensions of the individual stages need to be recognised. Examples of important dimensions within this study were the (cultural, political, socioeconomic) context, and the position and role of the various stakeholders. Discerning these dimensions is considered crucial for programme design, and for evaluations, especially those focusing on process and outcomes.

Contrasting extension models, contrasting evaluations

Extension models are likely to vary with respect to the individual stages mentioned in the analytic framework above, as a result of different objectives and conditions underlying the models. Therefore, evaluation of different extension models is assumed to be different, too. The evaluation of the National IPM Programme within the contrasting (T&V) extension context in Indonesia forms an interesting case for comparison in this respect.

The World Bank's efforts to introduce the Training-and-Visit (T&V) system in many developing countries, the biggest extension programme sofar, consumed immense investments in the course of time. Therefore, considerable attention was paid to the evaluation of these programmes. Previous activities of the World Bank in evaluating its T&V system focused on farm level effects such as yields, production and farm income (stage 8 in the analytic framework above), and tried to show causal relationships between these effects and the extension services (stage 6) (Murphy and Marchant, 1988). An important variable for process evaluation was the number of visits by the extension worker to the farmers, whilst no attention was paid to more crucial aspects such as farmer practice and decision making (stage 7) leading to the farm level effects measured.

A more recent strategy used by the World Bank considers four main concepts for evaluation: inputs, outputs, effects and impact (Röling, 1992a). The inputs are the resources made available, whereas outputs are the extension activities actually implemented. These lead to the direct effects of the intervention, which cause a certain (long-term) impact. In evaluation, **efficiency** of resource use is determined by comparing inputs and outputs. The **effectiveness** of the activities is assessed by comparing output and effect, and **significance** of the intervention (e.g. for policy objectives) by comparing effects and impacts (Patton, 1986). Intended practice in T&V interventions is the adoption of the technology offered. Therefore, the key factor in measuring extension effects is the adoption rate of the innovation (Murphy and Marchant, 1988). Röling (1992a), however, argues that adoption is not a good measure of effect, since a farmer must have access to factors (such as credit, markets, etc.) other than those provided by the intervention only (technology and knowledge) to be able to adopt. The World Bank's evaluation approach does not consider this. Evaluation is primarily seen as an activity useful for programme management and indispensable for funding agencies.

In contrast to the objectives of the T&V extension model, the National IPM Programme's principal aim is that trained farmers become more able to take considered decision. 'As long as they make an informed decision, it does not matter whatever the decision is.' As described in Chapter 2, such an aim requires a different approach to all aspects of programme design and implementation. Applying the stages of the analytic framework for training programme design and implementation to the National IPM Programme, shows more clearly why the model for IPM training is different.

1. **Appreciation of an undesirable situation:** research creates awareness that pest outbreaks occur because of pest resistance to pesticides and destruction of natural enemies, as well as negative effects of pesticide use on human health and environment; the current extension approach overlooked many of the farmers' problems and interests, therefore did not work effectively;
2. **Nature of training contents:** defined by the four IPM principles: grow a healthy crop, conserve natural enemies, observe the crop weekly, and farmers become IPM experts;
3. **Intended practice:** informed pest management decision making takes a central position; desired field activities and perceptions of farmers, defined in the competency objectives, are deduced from the four IPM principles;
4. **Training strategy/curriculum for farmers:** farmer training focuses on experiential, field-based learning;
5. **Training strategy/curriculum for trainers:** consistent with farmer training strategy: technical, communication and management skills needed to facilitate farmer training;
6. **Training of trainers:** pest observers are trained at regional IPM Field Training Facilities, using methods to replicate with the farmers;
7. **Training of farmers:** farmers are trained season-long in IPM Farmer Field Schools by pest observer/extension worker teams;
8. **Changed practice:** farmers' practices, knowledge and perceptions relating to pest management, in particular pest management decision making;
9. **Consequences:** effects at farm level economy and ecology as a result of a changed practice of farmers;
10. **Impact:** possible economic, ecological and health impacts at community, ecosystem and national level.

Evaluation of the IPM Programme at the village level is most concerned with the stages 7, 8 and 9, and their contexts. With the stated programme objectives and training characteristics, important aspects of effects to be assessed in evaluation are adaptability of skills learned, and resilience to react to changing conditions.

The focus for process evaluation, then, becomes how people have changed. With people are meant both farmers who become independent IPM experts in their field, and pest observers who become trainers prepared well enough to manage the farmer field schools. Detailed measurements of aspects of training process, changed practice and consequences should finally lead to the assessment of the extent to which things changed in the desired direction and improve the initial problem situation.

The study described in this book evaluated the first, large-scale training implementation cycle of the IPM Programme, as conducted in Grobogan. This first training cycle was preceded by a pilot cycle, and followed by more implementation cycles. This means that (most) stages of the analytic framework were passed through during each cycle, adapting the implementation according to the experiences in previous cycles.

Evaluation of IPM in perspective

Monitoring and evaluation of the Indonesian National IPM Programme comprises many activities ranging from nationwide economic surveys to anthropological farmer group studies. Since the study reported in this book specifically focused on the examination of processes and effects at the village level upon the introduction and implementation of the National IPM Programme, it can be categorised as an evaluation of process and outcomes, not of need and efficiency. The study is formative, not summative.

Evaluation at the stages of training process, changed farmer practice and consequences contains both qualitative and quantitative elements. In an interpretative evaluation as required for the IPM Programme, quantitative measurements of practices and effects should always lead to explanations why things occur in a certain way. Phenomena and processes observed have to be analysed and explained within their contexts. In this respect, it seems important to consider the configuration of the following aspects of training programmes throughout all stages of evaluation (Röling, 1992c):

- policy context;
- perception of the farmer by extensionists;
- characteristics of training contents;
- type of farmer behaviour considered desirable;
- nature of intervention;
- institutional framework.

These aspects seem to form a consistent pattern. The pattern and its consistency depend on the extension model used. In the case of the Indonesian rice intensification programmes, the consistency is evident. Within the political context of need for food crops intensification, a high-external-input technology is applied and transferred through a top-down extension system by preset recommendations. It expects farmers to apply the package. In contrast, the National IPM Programme was induced by some major policy measures supporting IPM. Farmers are considered active learners and experts. Farmer training applies farm practices based on ecological principles, it uses facilitation of experiential learning as its training approach, it involves farmers actively in problem identification, and it defines better decision making by rice farmers as its main purpose.

The consistency of the aspects mentioned above has been a leading supposition in conducting the evaluation study of the National IPM Programme, partly determining what information needed to be collected. In a village level study looking at the introduction of IPM, aspects of training contents, farmer practice and intervention are important. But the total phenomenon can not be explained when focusing on these aspects only. Explanations of processes observed can not be divorced from the policy context, the way farmers are perceived in the extension model, and the institutional framework. A multidisciplinary approach was applied to cover the many aspects in farming, farm communities, and contexts relating to the configuration of a training programme mentioned above. Four disciplinary perspectives were used to observe and analyse the processes occurring in the study villages: (1) extension/communication studies, (2) agronomy, (3) socioeconomics, and (4) ecology. To look at processes before, during and after IPM training, the study needed to have a longitudinal character.

The study departed with fairly broad objectives and few clearly defined hypotheses, due to initial lack of knowledge about the programme's objectives and approaches. As an internal evaluator assigned to the programme, but not having been involved in programme design, I started immediately in the field and learned about the programme by observing its field implementation. With this experience, more ground existed to discuss objectives and approaches with the programme management and field implementors, after which evaluation objectives were redefined. Although not initially, little by little, premises for utilisation of evaluation findings, as described in section 3.1, were fulfilled. Guided by the various stakeholders, the study determined its own needs in the course of time, made possible by few limitations to study methods, and by the longitudinal design allowing to adjust to experiences obtained throughout the seasons. The evaluation study evolved towards something consistent with its subject of study: an experiential learning process, hence providing many opportunities for self-evaluation.

3.3 Problem statement and research questions

As pointed out before, the introduction of IPM practice required a different extension approach. With its training based on nonformal education principles, implemented as a government programme through the existing extension system, the IPM programme creates a situation that is unique, but might induce conflict. Since both extension workers and farmers have for long been exposed to the existing system, it is interesting to investigate what happens if a programme with a completely different approach is introduced, and whether its consistency can be maintained during implementation at the field level. Therefore, the research problem of this study is formulated as follows:

What processes and effects result at the village level when sustainable practices, which contrast in many respects with the prevailing high-external-input technology, are introduced through nonformal farmer training in conditions created by major policy measures?

Research questions that served as guidelines throughout the implementation of the study, were formulated taking into consideration the objectives, research problem and theoretical perspective of the study. The questions are divided into six main issues. Contextual factors exclusively hold for the situation in Indonesia, and the research area in particular.

1. Conflict between technology of Green Revolution versus sustainable agriculture
a. Villages have for long been exposed to the Green Revolution technology and the concomitant extension approach. In order to identify possible constraints to the introduction of more sustainable practices, it is important to know how the system currently functions at the village level, which actors are involved, and to what extent new practices and perspectives are incorporated into current ones.
b. A clear conflict is visible between goals of the Government pushing for maximum production through Green Revolution technology necessary to support national food security, and goals of the farmers who can be expected to have more interest in profits and continuity of their farm enterprises. Sustainable agricultural practices are assumed to be more consistent with the interests of farmers. How does this conflict manifest itself at official and farmer levels, when a more sustainable technology is introduced?
c. The extension workers are the promoters of the Green Revolution technology at the farm level, being actively involved in, and also benefiting from input distribution. They have been assigned as co-trainers in the IPM farmers field schools. How do they perform, and what conflicts do they encounter?
d. It is assumed that the nature of training process is consistent with the nature of the training contents (Röling and Van de Fliert, 1993). This is the case with the Green Revolution technology and the T&V extension system. It is of interest to explore to what extent this consistency can be observed in the actual implementation of the IPM farmer field schools at the village level.

2. Empowerment of farmers
Green Revolution technology has been imposed on farmers. Recommendations were often more like instructions. In this respect, there seems to be a historical continuity through feudal, colonial and independent governments. Have farmers become docile and submissive, losing their dignity with their independence, or have they accommodated to centuries-long coercion, behaving submissively and feeling independent at the same time? As IPM implementors, farmers have to observe, analyse and be independent decision makers. Based on nonformal education principles which owe much to Paulo Freire's (1972) approach in the Latin American context of oppression of peasants by the ruling elite, the IPM training

approach seeks to empower farmers to become independent farm managers. It is interesting to investigate whether this approach works in Indonesia, where the reaction to oppression and exploitation on the part of farmers is different from Latin America.

a. Considering the cultural and historical context in Central Java, are farmers really submissive and docile? And what strategies do and did farmers employ to cope with Government coercion?
b. Experimentation by farmers, e.g. on fertiliser application, rice varieties and pesticides, in the coercive Green Revolution context can be regarded as an indicator of non-submissive behaviour. Do experimenting farmers exist, and what are their characteristics and practices?
c. Are the empowerment approach of the IPM programme and its assumptions suitable for the situation of Javanese rice farmers?

3. Diversity and selection of beneficiaries
During the early implementation cycles, the National IPM Programme was implemented in irrigated areas in the six main rice growing provinces in Indonesia. Villages were selected on the basis of accessibility and activity of T&V farmer groups. Twenty-five farmers deemed suitable for participation in the farmer field schools were selected locally. The selection of villages and farmers raises questions with respect to (1) the selectivity of the selection procedure, (2) the characteristics of the actual beneficiaries, and (3) the consequences for training, implementation and horizontal communication. Factors to be investigated are:

a. diversity among villages;
b. differences between IPM and non-IPM villages;
c. composition and characteristics of field school groups;
d. actors in the selection procedure; and
e. communication channels, networks and occasions that are used locally.

Additional topics relating to the issue of diversity and selection include:

f. It is assumed that felt needs or problems differ across categories of farmers. Introduction of IPM assumes certain felt needs. Does IPM have differential relevance for different categories of farmers?
g. What is the estimated suitability of the IPM farmer field school model for categories of farmers not selected sofar?

4. Changed practice of trained farmers
What do farmers do, and what do they know after being trained in the IPM farmer field school? The extent to which farmers' knowledge and actual practices reflect the intended practice is an indicator for the effectiveness of the training. Various degrees of mastering IPM can be measured:

a. internalisation of ecosystem analysis: knowledge and skills in rice field monitoring and analysis;
b. decision making processes;
c. pest management practices;
d. practice in the case of outbreaks (stemborer, rats);
e. application of IPM principles to secondary food crops.

Two additional questions relating to the possible changed practice of trained farmers are investigated:

f. What is the perceived relative advantage of IPM for farmers?
g. Did farmer field school participants experience an increase in expertise, self-esteem and job satisfaction?

5. Horizontal communication

Horizontal communication from trained farmers to the environment is required to popularise IPM because not all Indonesian farmers can be trained directly. Since IPM focuses on principles and field practices, and seeks to enhance such farmer practice as field observation and decision making, it can be expected that dissemination of IPM is very different from that of conventional technologies. The following questions relate to this issue:

a. What is being communicated?
b. From whom do others hear about IPM? Do (certain) members of the field school undertake special initiative with respect to telling others about IPM?
c. To which types of farmers does communication take place?
d. What is being perceived and practiced as a result of horizontal communication?
e. What is the perceived relative advantage of adopting IPM practices?

6. Consequences

The IPM farmer field school model focuses on enhancing farmers' skills in field observation, analysis and decision making. Changes in these behaviours constitute the direct effects of the training programme. These effects are to be distinguished from the consequences, both intended and unintended, that occur at the farm and village level as a result of changed farmer practice. Consequences of importance to investigate in an evaluation study comprise:

a. yields and yield variability (as a measure for risk);
b. expenditures on pest management;
c. return from rice production;
d. pesticide load and ecological balance in the rice ecosystem.

Possibilities of generating reliable information on issues relating to environment and human health are restricted when conducting a village study. Therefore, these issues and related topics are not emphasised in the study.

3.4 Methodology

A longitudinal case study evaluating IPM training

Focusing the study at the village level was considered particularly interesting because the programme applied innovative, unconventional extension and technology approaches within a conventional, in many respects conflicting, context. Considering this strong influence of the context on evaluation outcomes, the case study seemed to be the method par excellence to apply. The various aspects of extension programme configuration (policy context, perception of the farmer in extension, characteristics of technology, type of desirable farmer practice, nature of intervention, and institutional framework), served as a leitmotif throughout the conduct of the study. Coverage of all aspects requires a variety of research methods and a satisfactory level of in-depth exploration, for which, again, the case study is the most suitable method of investigation.

In correspondence with its purposes, this study evaluating IPM training at the village level had to apply a multi-case design to leave ample room for comparing processes and effects over time and space. The study design included several villages within a limited area, but with varying conditions and varying study intensity. Each village was considered a separate case. The study was longitudinal in order to cover periods before, during and after IPM training, and was conducted over five consecutive rice-growing seasons in 1990-92.

Area selection

Characteristic for a case study is that the study area is limited. Considering the choice for a multi-case design with the emphasis on the individual village, eight study villages within the boundaries of two subdistricts in the district of Grobogan, Central Java, were selected. Maps of the Indonesian archipelago, the island of Java, and the province of Central Java are given in Appendix II. An overview of the study villages displaying the locations in the subdistricts and describing the village profile characteristics is given on the flipchart in the back of this book. The selection procedure of study villages was as follows:

- step 1: **selection of district:**
  Grobogan was purposely selected as a district that is representative for the province of Central Java. It is a main rice-growing area with a cropping pattern of two rice crops and one fallow season or secondary food crop per year. The average land area per farmer (0.5 ha) and the average yield (over 5 tons/ha in 1988) equal the averages for Central Java. Although Grobogan is known as an area where rice stemborer is endemic, the average pest profile for rice cultivation does not show continuously occurring levels of extreme infestation by any pest;
- step 2: **selection of pest observers/Rural Extension Centres:**
  Two pest observers from Grobogan who had followed the rice IPM training during the wet season of 1989/90, and were to conduct their farmer field schools during the following 1990 dry season, were selected according to the location

of their Rural Extension Centres. The two pest observers had to originate from two different Rural Extension Centres which were not too far from each other.

– step 3: **selection of villages:**
Each pest observer had to conduct field schools in four different villages within the work areas of two different village extension workers who became IPM apprentice trainers. One of the two extension workers per pest observer was chosen randomly, at the same time determining the two villages to be observed per trainer team. Since the work area of a village extension worker consists of about one-third of a subdistrict, the two IPM villages per pest observer/extension worker team were located in the same subdistrict. The villages Senengsari and Sugihsari in the subdistrict Sumbersari, and the villages Kuduagung and Sumberagung in the subdistrict Sumberagung were selected.
Comparison (non-IPM) villages within the same subdistricts were randomly selected using the following criteria: (1) located in the work areas of non-IPM trained extension workers, and (2) having water supply from irrigation schemes. The irrigation criterion was applied for comparability of rice cultivation conditions with the selected IPM villages. The villages Jayasari and Plosoksari in the subdistrict Sumbersari, and the villages Mulyoagung and Lempungagung in the subdistrict Sumberagung were selected.

Period
In order to explore processes and effects in farmers' practice as a result of IPM training and farmer-to-farmer dissemination, the study required a longitudinal character, expanding over growing seasons before, during and after the intervention. Five consecutive rice seasons were covered:

- 1989/90 wet season: baseline season;
- 1990 dry season: IPM training, first monitoring season;
- 1990/91 wet season: second monitoring season;
- 1991 dry season: third monitoring season;
- 1991/92 wet season: fourth monitoring season.

An overview of the seasons with study phases, main activities, and season profile characteristics is given on the flipchart in the back of this book. During the intermediate seasons of 1990 and 1991, when secondary food crops were grown or the fields were left fallow, no activities were executed in the study area. The study started at the end of the baseline season which allowed only recall methods for baseline data collection

Intervention and study intensity
Two types of 'treatments' of the study villages can be distinguished, one concerned with the intervention by the IPM training, and the other with the study intensity. The IPM treatment involves the IPM farmer field school versus no field school, further referred to as IPM villages and non-IPM villages. The study intensity treatment differentiates between intensively studied and control villages.

The intensively studied villages were frequently visited by the study team, and subject to intensive monitoring over five rice-growing seasons, whereas in control villages only a baseline survey in the first season and an evaluation survey after one year were conducted. It was assumed that frequent presence of the study team in the villages, including people from those villages, might influence farmers' rice cultivation and, particularly, pest management practices. In addition, IPM training effects might be influenced by participatory observation of the investigator. To estimate these influences, control villages, both IPM and non-IPM, were included in the study design to apply rapid, quantitative data collection method before and a year after IPM training without any long lasting contact with the farmers.

Combining the two treatments, four types of study villages can be defined of which each type is represented in each subdistrict. The study villages are further referred to as:

- intensively studied IPM villages (Senengsari and Kuduagung);
- control IPM villages (Sugihsari and Sumberagung);
- intensively studied non-IPM villages (Jayasari and Mulyoagung);
- control non-IPM villages (Plosoksari and Lempungagung).

It should be repeated here that it is important to remember the names, treatments, and profiles of the eight villages in order to understand the arguments throughout this book. The flipchart in the back has been provided to facilitate this understanding.

Methods of study

When choosing for a case study investigating processes, in-depth, qualitative methods are the most obvious study methods to apply. But when looking at the assessment of effects in programme evaluation, some quantitative data are required by policymakers. The solution of solving the dilemma between qualitative and quantitative methods lies in combining elements from both approaches, likely to provide complementary information (Casley and Lury, 1982; Bulmer and Warwick, 1983). Therefore, both qualitative and quantitative data collection methods were applied in the Grobogan study. Qualitative group methods often served to raise topics to be probed in individual interviewing. An overview of all research activities over five seasons and over eight villages is given in Table 3.1. Description of the activities is divided into three categories: (1) quantitative surveying and record keeping, (2) qualitative observations and interviews, and (3) field observations. Preliminary, semi-structured interviews with farmer group leaders, village heads and officials, pest observers, village extension workers, and Rural Extension Centre staff were conducted in all eight study villages before planning and designing further research activities.

1. Quantitative surveying and record keeping:

- **Household census** in order to compose a sampling frame of rice farmers in the eight study villages, since no completely reliable sampling frame was available.

Tabel 3.1: Research activities in Grobogan over five seasons. (1989/90 = baseline; 1990 = IPM training; 1990/91, 1991 and 1991/92 = post training; non = non-IPM)

Season	Activity	Intensively studied villages						Control villages						Total	
		Senengsari		Kuduagung		Jaya-sari	Mulyo-agung	Sugihsari		Sumberagung		Plosok-sari	Lempung-agung		
		IPM	non	IPM	non	non	non	IPM	non	IPM	non	non	non	IPM	non
1989/90	Preliminary study	*		*		*	*	*		*		*	*	*	
wet season	Census (households)	504		860		499	322	889		593		939	726	5332	
	Baseline survey (respondents)	25	65	25	64	85	85	25	63	25	62	85	85	100	594
1990	IPM field school (participants)	25		25				25		25				100	
dry season	Observation field school (sessions)	8		7										15	
	Group interviews (sessions)	3	2	3	2	3	3							6	10
	Crop monitoring (fields)	5	5		10	16	16							5	47
	Obs. pest control behaviour	*		*		*	*							*	
	Follow-up survey (respondents)	25	65	25	64	85	85							50	299
1990/91	Follow-up of field school	*		*										*	
wet season	Crop monitoring (fields)	10	10	10	10	20	20							20	60
	Obs. pest control behaviour	*		*		*	*							*	
	Record keeping (farmers)	2		2		2	3							4	5
	Follow-up survey (respondents)	23	59	23	61	80	82	24	55	23	58	74	80	93	549
1991	Follow-up of field school	*		*										*	
dry season	Crop monitoring (fields)	10	10	10	10	20	20							20	60
	Obs. pest control behaviour	*		*		*	*							*	
	Group interviews (sessions)	1	1	1	1	1	1	1		1				4	4
	Qualitative data collection	*	*	*	*	*	*	*		*			*	*	*
	Follow-up survey (respondents)	21	55	21	60	72	76							42	263
1991/92	Group interviews (sessions)	2		2		2	2							4	4
wet season	Crop monitoring (fields)	10	10	10	10	20	20							20	60

* = non-quantifiable activities conducted

A reliable sampling frame is a premise for random sampling, which in turn is necessary to ensure representativeness of the sample (Bulmer, 1983);

- **Seasonal surveys** through individual interviews with farm families (Birgegård, 1980; Fink and Kosecoff, 1985; Casley and Kumar, 1988). The surveys explored demographic, socioeconomic, and agronomic aspects, relating to rice farmers' practices, knowledge, and perspectives of rice cultivation in general, and crop protection in particular. One baseline and three evaluation surveys were conducted in the four intensively studied villages over four consecutive rice seasons, including two wet and two dry seasons, whereas the control villages were only subject to a baseline and one evaluation survey, both during wet seasons. Since inconsistent, local area units were used by farmers, all fields of respondents were measured for per hectare conversion of production and economic data.
  Respondents were (non-IPM) rice farmers randomly selected from the sampling frame, plus all IPM farmers selected to participate in the farmer field school. In the baseline survey, sample size per village consisted of eighty-five respondents (including IPM farmers in the IPM villages), or between 14% and 38% of the farm families. In the evaluation surveys, respondents who had stopped farming were not replaced, consequently reducing the sample size. The survey samples in the non-IPM villages were found representative for the farming population when survey data on farm size distribution, tenure status and farm employment were compared with census data. In the IPM villages, sampling frames for selection of non-IPM respondents were not completely representative because the IPM farmers, operating on relatively large farms, who had to be interviewed anyway, were removed from the sampling frame.
  The survey team consisted of the author and a fixed group of eight to four enumerators, mainly anthropologists and all Javanese-speaking, who were also in charge of correcting and coding the questionnaires, and data entry. A questionnaire was used, with for the larger part highly structured, and for the smaller part open-ended questions. The questionnaires were pretested by the enumerators in non-participating villages in the research districts. Quantitative data were processed using database, spreadsheet and statistics software packages, whereas the answers to some of the open-ended questions were processed qualitatively. Only simple statistical procedures were used, and these even with certain reservations, since the use of statistics is not completely justifiable considering the not completely randomised, and thus biased sampling of villages and of farmers in the IPM villages;
- **Farm record keeping** with nine farmers in the four intensively studied villages was done during the second wet season. Main purpose of the record keeping was to obtain detailed, continuously updated and thus highly reliable data on (economic) farm management as a reference for the retrospective data obtained through survey interviews. The farmers participating in this exercise were purposely selected in order to cover various farm styles with regard to size, tenure status, and IPM management. In order to facilitate the daily note-taking, they were provided with a booklet consisting of forms with each page covering seven days in a week. All activities related to rice cultivation were noted down with

corresponding labour and input involvement. The records were discussed with the farmers once in two or three weeks. Data of the entire season were processed, analysed and discussed with the farmers.

2. Qualitative observations and interviews:
- **Participatory observations** (e.g. Jorgensen, 1989; Spradley, 1980) of IPM farmer field schools in the two intensively studied IPM villages were conducted during the training season. Eight of the ten field school sessions were attended and recorded by the author, and two by an assistant. Follow-up IPM meetings in IPM villages, and regular farmer group meetings in non-IPM villages were attended during the post-training seasons;
- **Group interviews** (e.g. Kumar, 1987; Krueger, 1988; Morgan, 1988) with IPM farmers, and with non-IPM farmers in both IPM and non-IPM villages were conducted during the IPM training season and two post-training seasons one year after training. The interview team consisted of a facilitator (hired communication experts and/or the author) and minutes secretaries (the village study assistants and some of the survey enumerators), mostly of varying composition. Non-IPM respondents were selected randomly from the sampling frame, whereas all IPM farmers were invited to participate, although separately from non-IPM farmers. Interview groups consisted of 10-20 people. The following themes were topics for the discussions:

  - Perception of the rice ecosystem, including a healthy crop, pests, natural enemies, and pest management (both IPM and non-IPM);
  - Decision making process in pest management (both IPM and non-IPM);
  - Introduction of Green Revolution and adoption of innovations (only non-IPM);
  - Lessons learned from IPM training (only IPM, immediately after training);
  - Implementation of and horizontal communication on IPM (only IPM, two seasons after training);
  - Risk and reward of IPM (only IPM, three seasons after training);
  - Qualifications of (good) IPM farmers (only IPM, three seasons after training).

  Various methods were used in the group interviews, to mention a few: picture ranking and story telling, theatre, drawing, and small group brainstorming;
- **Informal interviews** with farmers in the field and in their houses, wives of IPM farmers, spray labourers, village heads and officials, subdistrict heads ('*camat*'), pesticide dealers, KUD staff and managers, IPM field school trainers (pest observers and village extension workers), subdistrict agricultural officers ('*mantri tani*'), staff of Rural Extension Centres and district Agriculture Service, Field Leaders I and II, and IPM Programme management staff.

3. Field observations:
- **Village study assistants** were assigned in the four intensively studied villages (one per village). They were young men and one woman actively involved in

farming and residing in the villages. Their task was to assist in the field observations, both crop monitoring and observations on farmers' pest control behaviour. In the later seasons, they executed these observations mostly on their own. These assistants proved to be very valuable in facilitating contact with various people in the villages, and served an important role as informal information source;

- **Weekly crop monitoring** was done in randomly selected fields in the four intensively studied villages over four consecutive seasons. Purpose of crop monitoring was to obtain a reference for farmers' perceptions of crop condition and pest occurrence, and to measure possible effects of IPM management in the rice ecosystem. Twenty fields per village dispersed over the village's whole rice area were monitored for crop condition and pest and natural enemy populations. The sample per field included five randomly selected rice hills at a mutual distance of about ten steps in a diagonal line through the field. Monitoring was done on two fixed days per week before 10:00 a.m. Scores were recorded on specially designed forms. The size of the sample fields were measured, records were kept of all treatments applied to the field, and the harvest was weighed immediately after threshing (moisture contents averages 15% for dry season and 20% for wet season crops);
- **Observations on farmers' pest control practices** were done during weekly crop monitoring and two additional late afternoon visits to the field. Purpose of these observations was to obtain qualitative data on farmers' pest control behaviour, in particular application methods, and to obtain a reference for the retrospective survey interviews. Farmers met along the crop monitoring route through the fields, who were applying a pest control measure, were observed and briefly interviewed on their practices. Notes were taken on specially designed forms. Data were processed in a qualitative manner.

The longitudinal character of the study enabled to repeat most of the methods for several seasons. Based on experiences, evaluated with the team, implementation of the methods could be improved in the course of the seasons. Another advantage of the longitudinal study design was that gaps in data collection, for instance when respondents had temporarily left the village or data obtained were not clear at the moment of data entry, could easily be filled during follow-up visits.

To inspire confidence among the people in the study villages, the study team stayed as much as possible in the villages. In all eight villages, lodging facilities were provided in the house of the village head, village secretary or a village school teacher. In addition to the survey interview periods at the end of each survey season, the intensively studied villages were visited weekly during the training season, once in two or three weeks during the first and second post-training season, and once a month during the last post-training season. At every visit, at least one day and one night were spent in each village. A good relation with all the village communities was established in the course of the study period.

Focus of study
Four units of investigation were distinguished in the study: (1) the individual farmer, (2) the farm, (3) the farmer group, and (4) the village. The individual farmer, either woman or man, forms the principal focus for study. In the text, generalisations are sometimes made to address farmers by male pronouns, because the large majority of respondents in the study were men. In addition, and logical in the cultural context, where husband and wife managed the farm together the man was usually called the one responsible for the farm. The role of women in cultivation practices and decision making, however, should not be underestimated and will be underlined in several sections in this book.

The village forms the second important focus for study, which is considered essential based on the assumption that farmers' practice is influenced by place-dependent, external factors in a village, such as local history, tradition, and role of local authorities. The farm as unit of investigation is mainly of importance for the economic analysis of farm management, and for measurement of farm level effects as a result of IPM training. The farmer group as the fourth focus of study is relevant in the context of extension process evaluation.

Comparative studies
As a preparation to the activities in Grobogan, a preliminary study in Central Java and Yogyakarta among two farmer groups having received IPM training in the 1986-87 crash IPM programme, and among two control groups was conducted. Purpose of this study was to understand the operation of the extension system at the village level, and estimate possible effects of previous IPM training efforts. Methods used in this study included semi-structured group and individual interviews with farmers (men and women separately), farmer group leaders, village officials, pest observers, extension workers, and KUD staff and management.

In order to compare and estimate representativeness of phenomena observed in Grobogan, many other farmer field school groups and FTFs in various parts of Java, North and West Sumatra, and South Sulawesi were visited. In Sumatra in particular, involvement of women in IPM training was examined since relatively many women farmers were participating in field schools in these provinces, in contrast to Java. Exclusively qualitative methods were used in investigating these field school and FTFs, including informal individual and group interviews, participatory observations, and analysis of existing data.

Two additional farmer groups in the province of Yogyakarta, that had organised IPM farmer field schools on their own initiative and expense, were observed throughout all training sessions. One school was a normal rice IPM training that followed the model of the IPM farmer field school as described in Chapter 2. The other school comprised a training of IPM in soybeans designed by the participating farmers themselves following previous experience in the rice IPM field school. Participatory observations, group and individual interviews were applied as methods to investigate these two training groups.

4 A different village, a different profile

4.1 This chapter

One of the surprising experiences in conducting the study was to encounter an immense diversity among villages. Although the study area is considered to be homogeneous regarding climatical, environmental, and socioeconomic conditions, and some of the study villages are even adjacent, each village has its typical atmosphere, its own history, its direction of rural development, its unique social structures. This diversity makes it impossible to generalise but, as stated in section 3.2, it is questionable whether generalisations really serve an important purpose of evaluation studies. A large diversity encountered in a case study is an indication of the complexity of the real situation. Large-scale training programmes, such as the National IPM Programme, have to deal with this complexity. In this regard, it might be interesting to programme management not only to be informed through evaluation studies **what** is happening before, during and after intervention, but also **what factors and processes** are determinant at the different stages. More attention to diversity occurring within and among places may identify more of these factors and processes.

This chapter describes the study area by outlining profiles of the eight villages, and highlighting aspects that determine the diversity among these villages. The purpose is twofold: first, to present the context for interpretation of the chapters below, and second, to generate hypotheses on the consequences of large diversity for trainingprogramme and evaluation activities.

4.2 The district Grobogan

Grobogan is one of the twenty-nine districts of Central Java, located east of the provincial capital Semarang (Appendix II). Its area is 1,976 km^2 divided into eighteen subdistricts (Source: Diperta Dati II, 1989). The more than one million inhabitants of Grobogan are living in 284 villages where agriculture is the main source of income. Rice is cultivated on an area of 61,426 hectares with mainly brown clay soils (vertisol), which is two-thirds of the district's arable land. More than one-quarter of all rice fields, especially in the flat western and northern parts of the district, are supplied with irrigation water. The more hilly landscape in the east and south with rainfed water supply contains an important soybean production area. Maize is an important crop grown in the intermediate season on both wetland and dryland fields. In addition to several other secondary food crops such as cassava, groundnuts and mungbeans, mainly grown on drylands, the district contains some rubber estates and grows some minor other estate crops.

Since the introduction of the short-duration, high-yielding varieties in the late 1960s, three rice crops a year were grown in areas where water supply allowed so. Because of this intensive cropping pattern and the increasing use of pesticides, rice production was seriously affected by brown planthopper outbreaks in 1979 and 1985-86. On the authority of a declaration from the governor (the province head) in 1986, the district government announced an instruction prohibiting farmers to plant three rice crops a year, which became effective in April, 1988. Since then, rice is cultivated two times a year only, usually from November to February and from March to June, which is under strict control by the district irrigation authorities through irrigation water supply (Figure 1.1). During the months of July to October, no irrigation water is provided so that farmers are forced to grow other crops or leave the land fallow. Initially, most farmers did not want to take the trouble to try out other crops. When rice was no longer grown continuously, higher yields per crop were attained because soils could recover their fertility and labour could be saved allowing farmers to go for off-farm employment during the fallow season. Since 1989, when crop diversification was emphasised in the new five-year plan (REPELITA V, 1989-94), the District Agriculture Service began to encourage farmers more strongly than before to plant secondary food crops ('*palawija*') during the intermediate season, for example by conducting a special soybean intensification programme ('*OPSUS Kedelai*'). Seed companies were allowed to promote secondary food crop packages in villages, e.g. hybrid maize. The area of rice land cultivated with other crops during the intermediate season considerably increased since 1989.

Over the eight villages studied in Grobogan, the average area of rice land cultivated by one household was 0.56 hectare during the baseline season, a figure in line with the average farm size for Central Java. One household contains five members, on average. Rice yields obtained in 1990, averaged for the whole district and over one wet and one dry season, equalled 5.4 tons/hectare, which is comparable to average yields for the whole province of Central Java (Source: Diperta Dati II). Figure 2.1 also shows that over the fourteen-year period of production growth beginning with the introduction of high-yielding rice varieties, the rice production curve for Grobogan closely followed the Central Java curve.

Regularly occurring pests in the district of Grobogan are rice stemborer and gall midge. Although some damage by these pests is reported to the District Agriculture Service every season, severe infestation generally takes place at intervals of six to ten years. In addition to stemborer and gall midge, rats are an ever-occurring pest in rice crops, causing varying levels of damage depending on the activity of the villagers to take timely action. According to the head of the Agriculture Service, rats are the main pest problem of rice in the district.

The census conducted among all households in the eight study villages recorded that 56% of households earn (part of) their living by rice cultivation, with higher percentages in the remote villages and lower ones in the villages close to the main road. On average, these farmers obtain around 70% of their total net income from rice production. Another 19% of households are involved in farm labour. The remainder consists mainly of tradesmen and off-farm labourers. These figures reflect a picture of the daily life in the villages dominated by agricultural activities.

The two study subdistricts Sumbersari and Sumberagung are not principally different. Climatical and environmental conditions are the same. The landscape is flat, and the view is dominated by rice fields. Irrigation water in Sumbersari comes from the river Tuntang. Sumberagung is, since 1989, supplied by the large irrigation scheme of Kedungombo. Both areas have a fairly reliable water supply. The distance between the two subdistrict towns is almost twenty-five kilometres.

4.3 The villages

A Javanese village

As said before, large diversity exists among villages, even within a limited area. A few general characteristics of a Javanese village can be identified, though (Tjondronegoro, 1984; Fox, 1983). The description below especially holds for the district of Grobogan.

Dwellings in villages are clustered. Looking at the countryside from an aeroplane gives an impression of a patchwork quilt with patches of built-up areas, recognisable by the high coconut trees in every yard, surrounded by vast areas of arable land. The patchwork is traversed by irrigation channels and small roads. Hamlets belonging to one village are sometimes scattered.

The village is the lowest administrative unit in the Indonesian government hierarchy. The village government consists of a village head, a village secretary, hamlet heads, and a number of village officials, varying between one and fifteen, each keeping a certain portfolio. The village head is elected by the villagers, which puts him or her in an ambiguous position: under direct command of the central government but also representing the constituency. The village secretary and other officials are assigned by the village head but subject to the approval of his or her seniors at the subdistrict level.

Typical for Java, unlike many other areas in Indonesia, is that the members of the village government get their salary in the form of land ('*bengkok*') at their disposal during the term of office. The size of these salary lands depends on the position held and the amount of land available in the village. The land is either cultivated by the officials themselves or (partly) rented out. It has become a common practice that village officials have to pay considerable amounts of slush money to enter the job. Village heads in particular need a lot of money to become a serious candidate. This implies that their net income during the first years of assignment, when these loans have to be amortised, is negligible, especially for officials with lower positions and accordingly smaller salary lands. This situation encourages fraudulent practices, which sometimes seriously affect the community. One example is the embezzlement of a part of the money meant for the construction of a new mosque in one of the study villages, which money was for 25% contributed by the villagers themselves. Remarkably, villagers seldom revolt as if such practices are inherent to power structures.

A village is divided into hamlets. A hamlet ('*dusun*' or '*dukuh*'), which is usually determined by natural boundaries, represents a social unit of village life. The hamlet head, accordingly, fulfils a more social function than the village head who is an administrator and responsible to higher governmental levels. The hamlet head is assigned for life, while the village head has to be elected every eight years. Another division of villages consists of neighbourhoods, again with an administrative function. Large neighbourhoods ('*rukun warga*' or RW) are split into small neighbourhoods ('*rukun tetangga*' or RT) of around 30-50 households. The neighbourhood head (either RW head or RT head), who takes this position voluntarily, is the eminent medium to pass on village government messages to the community.

In each village, a village meeting hall can be found, which is usually a spacious, open building where people gather at certain occasions. These include meetings of neighbourhood heads, farmer groups, women welfare associations, and so on. The village office where all village officials have their desk is attached to this hall, with a separate room for the village head. People have to come to this office for all administrative arrangements. Other buildings to be found in every village are one or more primary schools, a mosque and several muslim praying houses. Secondary schools are usually only in subdistrict towns, and sometimes in the somewhat larger and better accessible villages.

Eight profiles: a description

An overview of the eight villages selected for this study, with brief profile descriptions and maps indicating the location of the villages in the two subdistricts, is given on the flipchart in the back of this book. Seven of the eight villages are, at first sight, in no way conspicuous compared to average Central-Javanese villages. Obvious differences, like an urban- or village-like appearance, are strongly related to the location of the village, particularly the distance to the main road, as anywhere. Remote villages, like Plosoksari, Mulyoagung and Lempungagung, look traditional with many bamboo houses and unpaved roads. Few, small kiosks can be found there. Villages on the main road, like Sugihsari and Sumberagung, have more wooden and stone houses, more and bigger shops, and, therefore, give a more urban impression. An interesting observation in this respect can be made in Sumberagung: one hamlet which is not directly connected to the main village and cannot be reached by motorised transport, exudes a completely different atmosphere, one of a traditional village, in contrast to the hamlets located along or close to the main road. The one outstanding village is Jayasari which is located far from the main road, in between very traditional villages, but is very modern itself. Box 4.1 describes what is outstanding about this village and how it could have become outstanding.

Some descriptive, demographic and agricultural variables of the eight villages, as far as relevant for this study, are listed in Table 4.1 in an attempt to outline the villages' profiles. Accepting some subjectivity, general qualifiers are ascribed to each village relating to its accessibility and appearance. Using these qualifiers, the eight villages can be divided into three categories: (1) the more urban villages close to the main road: Sugihsari and Sumberagung, (2) the villages somewhat off the main road, but well accessible, still having a more village-like appearance but under direct influence of the facilities in town: Senengsari, Jayasari and Kuduagung, and (3) the remote and traditional villages: Plosoksari, Mulyoagung and Lempungagung. It should be noted here that the villages selected for IPM training are exactly the four villages located along or close to the main road (Senengsari, Sugihsari, Kuduagung and Sumberagung), not necessarily being the villages that need training most since usually many other off-farm enterprises are undertaken where transport allows it. This is obviously the case for Sugihsari where 52% of the inhabitants sampled in the census are not involved in agricultural activities.

Distance to the main road (which practically equals the distance to the subdistrict town given) does not always explain the accessibility of a village. Two contrasting places are Jayasari which is far but easily accessible because of the good road leading to the village, and Plosoksari which is actually close to the main road but almost inaccessible for motorised transport, especially during the rainy season. The accessibility has strong consequences for the development of a village, among others through an either facilitated or hampered supply of agricultural inputs. Farmers in Plosoksari are in no way able to obtain fertilisers that are not commonly sold in the local kiosks, other than loading a fifty-kilo sack on their bicycle and pushing the sacks, one-by-one, over the three-kilometre muddy road. The same holds for Mulyoagung and Lempungagung, although these villages can still be reached by trucks provided that there have not been heavy rains. Jayasari farmers, on the contrary, can easily charter a horse-buggy from the subdistrict town to transport whatever they need in whatever quantities along the eight-kilometre, well-built road.

Related to the quality of the roads in and leading to a village is the availability of public transport. Motorcycles are the only means of transportation that can be used everywhere, although after rain it is still difficult to charter someone to bring you to Plosoksari, Mulyoagung or Lempungagung. All study villages except Plosoksari and Lempungagung have fair, although mostly unpaved, roads passable for tricycles ('*becak*') serving public transport within the village and to neighbouring villages. Tricycles are often used for transportation of goods, for instance the

Box 4.1
The village of Jayasari

After a ride of 7 km on a relatively good asphalt road through three rather traditional villages, passing the border of Jayasari one gets the impression of entering a model village. Small white-painted bamboo fences and ornamental trees have been placed along the roads, and three huge, urban-style buildings can be found: a mosque, a modern village hall, and a junior high school. The Indonesian People's Bank and the Primary Health Care Service have branches in Jayasari. Once in five days there is market on the junction at the far end of the village, which is also the place where the good road stops and turns into an ordinary, unpaved village road. The built-up area of the village is 1 km long and 300 meters wide, pressed-in between the road and the river bank. Crossing the road and the irrigation channel that runs along the road, a vast area of rice fields stretches out as far as the eye can see.

In contrast to most other villages, the house of the village head, Pak Lurah, is not very big, but it is a modern brick construction. This village head is the main reason why Jayasari is so forward and relatively prosperous. He was elected for the first time in 1975, only 30-odd years old. After a few years, he founded a social organisation based on religious principles, the '*Lembaga Mo'at*' or Share Foundation. He had experienced in the previous years that people always run to him with their problems, mostly financial difficulties. He realised that he could not function well as the leader if he had to spend most of his time solving individual problems. After reading an article in a magazine about a social village organisation in an East-Javanese village, Pak Lurah organised a study tour to this place with a group of religious leaders, around twenty people in total. Realisation of Jayasari's own '*Lembaga Mo'at*' did not take long. '*Mo'at*' is the 10% share that moslem people have to give to the poor, equivalent to the tithe in christianity. The system in Jayasari works by collecting shares of the rice harvest for social purposes. After each harvest, a form is distributed to the better-off people (which is nowadays around 75% of all villagers) to fill out the yield that was obtained and the amount to be contributed to the village. The 10% share is a guideline but everybody is free to contribute less or more. The whole system is based on moral responsibility. In a meeting of the foundation's committee, in which the heads of all neighbourhoods are represented, it is decided to whom and for what the contributions are used. Two-thirds are given directly to poor people. Each neighbourhood can propose a few families who are, then, helped to improve their houses or pay the school fees for their children. Through decentralisation at the neighbourhood level there is a very strong social control and hardly an opportunity for misuse or fraudulent practices. One-third of the contributions is kept for funds to be requested in case of calamities, e.g. the death of a family member in which case Rp 20,000 is contributed. People who are actually well-to-do but at a certain moment need money that they do not possess, can get an interest-free loan from the association that has to be reimbursed after the next harvest. This system has been applied for more than ten years now without any major problems. The result is obvious: no real poverty exists, there are no crumbling houses, and the number of well-to-do people contributing a share is increasing.

Another concern of Pak Lurah that favours the village development considerably is his special attention to agriculture. In several ways, he tries to create advantageous conditions for farmers to cultivate as optimally as possible. One example is the supply of high-quality seeds which he arranges directly from a seed farm in another part of Central Java. Because of this seed, which farmers can purchase on an interest-free credit to be paid back after the harvest, higher yields are achieved. What is more, higher prices are attained because the harvested seeds still have a high quality. In the same way, pesticides purchased with money from the village's funds are provided to farmers to be reimbursed without interest after the harvest. This policy is unique, especially since it is competitive with another position of Pak Lurah: as manager of the KUD which hardly any Jayasari farmer is a member of.

Table 4.1: Profiles of the eight study villages.

	Senengsari	Sugihsari	Jayasari
A. Description			
– Qualifiers	within reach, village-like	close, urban	far but well accessible, modern
– Distance to subdistric town (km)	2	1	8
– Accessibility for motorised transport	easy	very easy	very easy
– Availability of public transport	tricycles, motorcycles, horse-buggies, local buses at 1 km	tricycles, motorcycles, horse-buggies, local buses	tricycles, horse-buggies, motorcycles
– Availability of electricity	yes	yes	yes
– Possession of radio (as % households)	47%	57%	58%
– Possession of TV (as % households)	27%	27%	28%
– Distance to nearest secondary school (km)	2	1	0
– Important enterprises (other than rice cultivation)	sawmill	10 rice mills, KUD brickworks, blacksmith	2 rice mills, rice trading, market
– Village government policies	indifferent, fraudulent	progressive	progressive
B. Demographics			
– Population (households)	514	1,492	553
– Neighbourhoods (RW/RT)	3/16	9/32	3/12
– Profession (as % households)			
• farmers	55%	34%	58%
• farm labourers	25%	14%	21%
• off-farm labourers	7%	13%	6%
• tradespeople	7%	23%	8%
• unemployed/retired	2%	5%	1%
• other[1])	4%	10%	5%
– Education (as % inhabitants)			
• no education	36%	32%	4%
• not (yet) finished primary school	33%	23%	76%
• primary school	23%	38%	12%
• secondary and higher education	8%	7%	8%
– Education (as % farmers reporting)			
• no education	17%	13%	16%
• 1-3 years primary school	26%	27%	15%
• 4-6 years primary school	47%	52%	36%
• secondary and higher education	11%	8%	32%
– Average education of farmers (years)	4.2 ± 2.7	4.3 ± 2.6	5.9 ± 3.9
– Age of farmers (as % farmers reporting)			
• younger than 30 years	8%	16%	16%
• 30-49 years	44%	54%	45%
• 50 years and older	48%	30%	39%
– Average age of farmers (years)	40 ± 11	42 ± 13	43 ± 13
C. Agriculture			
– Area of rice land (ha)	253	240	190
– Irrigated water supply (as % farmland)	100%	100%	100%
– Farm size (as % farmers reporting)			
• smaller than 0.2 ha	21%	24%	26%
• 0.2 – 0.49 ha	38%	54%	49%
• 0.5 – 0.99 ha	22%	16%	14%
• 1 ha or more	19%	6%	11%
– Average farm size (ha/household)	0.66 ± 0.83	0.41 ± 0.38	0.49 ± 0.54
– Tenurial status (as % farmers reporting)			
• owner cultivators	58%	49%	62%
• tenants + sharecroppers	42%	51%	38%
– Farmer groups	occasionally active	partly active	not active

[1]) 'Other' includes civil servants, teachers, private sector employees.
Sources: Subdistrict and village data, census, baseline survey.

Plosoksari	Kuduagung	Sumberagung	Mulyoagung	Lempungagung
isolated, traditional	within reach, village-like	partly urban, partly village-like	remote, traditional	remote, isolated
3	3	0	7	9
very difficult	easy	very easy	difficult	difficult
motorcycles	tricycles, horse-buggies, motorcycles	tricycles, motorcycles, horse-buggies, local + long-distance buses	tricycles, motorcycles	motorcycles
no	yes	yes	partly	no
48%	56%	48%	53%	47%
8%	18%	24%	20%	5%
3	2	0	7	9
rice mill	3 rice mills, 1 village rice barn	rice mills, market, KUD	1 rice mills, 1 village rice barn	rice mills
fraudulent	orderly, fairly progressive	fairly progressive, strict	traditional, confidential	indifferent
951	840	615	322	735
5/29	3/25	4/20	3/11	3/13
65%	56%	52%	70%	69%
18%	22%	15%	15%	20%
6%	8%	4%	2%	4%
6%	5%	16%	5%	3%
3%	5%	6%	5%	3%
2%	5%	8%	3%	2%
20%	12%	14%	3%	19%
55%	36%	32%	30%	27%
24%	40%	20%	58%	49%
1%	12%	34%	9%	5%
19%	15%	13%	11%	19%
19%	12%	31%	22%	18%
49%	59%	42%	55%	56%
13%	15%	15%	12%	7%
4.5 ± 3.0	5.0 ± 2.9	4.5 ± 3.0	5.2 ± 2.9	4.4 ± 2.8
27%	7%	13%	28%	19%
45%	44%	47%	36%	53%
28%	49%	40%	35%	28%
40 ± 14	48 ± 14	44 ± 13	40 ± 15	42 ± 13
275	337	258	107	259
98%	100%	45%	100%	86%
35%	7%	6%	20%	13%
42%	29%	37%	39%	41%
14%	49%	44%	28%	27%
8%	15%	13%	13%	19%
0.42 ± 0.47	0.68 ± 0.51	0.69 ± 0.62	0.58 ± 0.64	0.61 ± 0.49
65%	69%	61%	65%	74%
35%	31%	39%	35%	26%
irrigation groups active	not active	not active	not active	not active

weekly shopping from the market. Only the people from the villages close to the main road (Senengsari, Sugihsari and Sumberagung) can easily make use of long-distance bus transport.

Although widespread, electricity is not yet a common commodity in Central Java. Five of the study villages, and one partly, enjoy the benefits of electricity. In Plosoksari, Lempungagung, and the smaller part of Mulyoagung houses are still lighted with kerosene lamps. Facilities as television are limited to a few rich people who can afford to buy and maintain a battery. It is not surprising that the non-electrified villages are the more remote ones since electricity networks expand via the main roads. Lack of electrification shows painfully how remote villages become increasingly isolated and retarded in development. When electricity has entered a village, immediately followed by illumination, amplifiers, TV sets, and even videos, as happened only recently in Mulyoagung (1988) and Jayasari (1985), social life changes rapidly with all kinds of activities in the evenings. People watch TV or video programmes together, prayers in the mosque are amplified and can be heard throughout the village, meetings in the village hall are organised, or small groups of youngsters hang around in the streets. This contrasts sharply with places without electricity where everything is dark and quiet after 8:00 p.m.

All villages possess one or more primary schools. (State) junior high schools can only be found in Sumberagung, which is to be expected since it is the subdistrict town, and in Jayasari. The latter was built in 1985 after the persevering negotiations by the village head to obtain permission from the national authorities. In order to get a state school built, the village should provide an adequate piece of land, and guarantee enough clientele, i.e. enough children of schooling age from reasonably prosperous families in the area, and no other school nearby. The presence of the school is another proof of and an extra driving force for the progressive development policies of Jayasari.

The level and direction of development of a village seems to be strongly dependent upon the policies and commitment of the village government, in particular the village head. After a long history of feudal and hierarchical power structures, paternalistic relationships between leaders and followers exist. Collective activities at the village level, e.g. rat control, stand or fall with the active involvement of the village head (Van de Fliert et al., 1993). This dependency on commitment is obvious in the eight study villages. A clear example is Jayasari (Box 4.1) that became relatively prosperous because of the determined policies for social welfare and agricultural development. Contrasting situations can be found in Senengsari and Plosoksari where policies are often directed to serve the personal interests of the village head first and foremost through fraudulent practices. Although villagers are usually aware of these practices, and even complain about them to outsiders such as the survey enumerators, no action is taken against the leaders, out of respect for the existing power structure. Another example of indifferent village development policies can be observed in Lempungagung, although there is no proof of fraudulent practices. The isolated location of the village gives fewer opportunities for progressive development which might influence the commitment of the village officials. But comparably isolated conditions can be found in the neighbouring village Mulyoagung where, in contrast, a collaborative relationship exists between farmers and village government with positive effects for the agricultural development of the village. Even though the direction of the development policies here is not very forward looking, but rather traditional, it is solid and puts farmers first, which is one of the suggested reasons why Mulyoagung always achieves the highest rice yields among the eight study villages (Box 5.3).

Two other villages with an agriculture-oriented policy are Sumberagung and Sugihsari. The village head in Sumberagung is a motivated farmer himself, which makes him popular in the farming community. He has a difficult task in this subdistrict town to serve the needs of both the farming majority and the influential trading minority. In addition, he was left with an inheritance from the former village head who had applied an indifferent policy, which he wants to reverse by ruling with a strict hand. In Sugihsari, the proportion of tradespeople is even higher (23%), and this branch seems to be encouraged by the village government. The village head here, who is more a manager than a farmer, holds an important position in the local KUD, and tries to encourage agricultural development by promoting the KUD input package through the SUPRA-INSUS programme. This effort is not free from personal interest. A partly comparable situation is observed in Kuduagung where the village head is also a KUD manager. Several intensification programmes have been implemented in the past in which the application of pressure from the village government on farmers to participate was not uncommon. The programmes have not been very successful and never resulted in a sustained change of farmers' cultivation practices. In spite of this, the village head has a good reputation in the village which is shown by the fact that he was reelected after his first term of service. This reputation is due to his solid administrative management and strong social commitment. He is as a father for his people. It is remarkable that most of his assistants, the village officials, are not much in favour among the people.

The great majority of the population in all eight villages is muslim (from 96% in Kuduagung to 100% in Mulyoagung). Religion plays an important role in Jayasari and Kuduagung where strict religious obedience is practised by some groups.

Demographic characteristics

A few demographic variables are selected to describe the study villages. Population and the number of neighbourhoods are an indication for the village size. The number of households, with one household consisting of five members on average, varies a lot among the villages, from somewhat more than 300 in Mulyoagung, divided over eleven neighbourhoods, to almost 1,500 in Sugihsari over thirty-two neighbourhoods. Walking around, all villages give an equally densely populated impression. Senengsari, Sugihsari, Sumberagung and Lempungagung have hamlets somewhat apart from the main part of the village, which are, in all four cases, more remote and look more traditional.

If the village communities are considered by professional sector, it is obvious that the majority of people in all villages, except one, earn their living in agriculture. The census recorded that over 50% of the households, and even up to 70% in the remote places Mulyoagung and Lempungagung, have one or more members who are rice farmers. Another 15-25% is involved in farming as farm labourers. Sugihsari is an exception since a small majority is involved in off-farm enterprises, which has to do with the location of the village very close to the subdistrict town and along the main road. Only 34% of household heads are farmers here. Sumberagung has two faces: being the subdistrict town, it attracts a lot of trading activities – 16% of households work in the private sector – which are located along the main road and the market, whilst the inner parts of the village are dominated by agriculture providing an income to two-thirds of the households.

The level of formal education achieved gives some indications related to opportunities for development. As said before, only the subdistrict town Sumberagung and, recently, the village of Jayasari possess a junior high school. For higher education, people have to travel long distances or find boarding houses, which involves costs relatively high for the tight budget of most farm families. It is, therefore, not surprising that in the remote villages only a small proportion of inhabitants received secondary or higher education. Illiteracy, as measured by the percentage of people without formal education (according to the village statistics which might be out of date), ranges from 3% (in Mulyoagung) to 36% (in Senengsari). The difference between the two remote, neighbouring villages Mulyoagung and Lempungagung is striking in this respect. These two villages are completely comparable when distance to the main road and difficulty in transportation are considered. The percentage of people without formal education is more than six times higher in Lempungagung than in Mulyoagung. On the whole, the percentage of illiterates in the latter village is relatively low. The reason for this is not understood, although it might say something about a certain attitude among the people in this small village, which is also visible in the usually apt reaction of the villagers to innovations.

Rice farmers, defined as independent operators of rice land, include owner operators, tenants, and sharecroppers, both men and women. Although the proportion of independently farming women in the study area is fairly small (13% during the 1989/90 wet season, usually widows, unmarried women, and wives of men working in the towns), wives of farmers actively involved in rice cultivation are also considered farmers here. Often, both husband and wife participated in interviews. The age of rice farmers surveyed in the study villages averages forty-three years, with 18% of the farmers younger than thirty years. Relatively many young farmers were found in the remote villages Plosoksari and Mulyoagung. In contrast to many other places in Java, a part of the young generation is apparently still interested in farming. Although many young people leave for the towns to look for employment, several have been observed to return to the village and start farming. The average number of years of formal education obtained by farmers is 4.8 years. Fifteen percent has no formal education, whereas 14% received secondary or higher education of whom relatively many in Jayasari (32%).

Agriculture

The area of agricultural land, excluding home gardens, ranges from 107 hectares in Mulyoagung to 337 hectares in Kuduagung. Practically all the farmland of the four villages in the subdistrict Sumbersari, and the villages Kuduagung and Mulyoagung have water supplied by major irrigation schemes. Timeliness of water supply depends primarily on the distribution schedule of the subdistricts' irrigation authorities, but varies according to location along the primary irrigation channels. Mulyoagung and Lempungagung receive their water almost two weeks earlier than Kuduagung and Sumberagung, which can be crucial in case of late operation of the irrigation dams. Fourteen percent of the fields in Lempungagung and not less than 55% in Sumberagung are dependent upon rainfed water supply, where a second rice crop is not always possible and a secondary food crop is planted instead.

In 1989, the majority of farmers operating on irrigated lands in the eight villages did not plant any secondary food crop during the intermediate season but just left the land fallow for a couple of months. Only 12% of farmers of which relatively more in the villages Kuduagung, Mulyoagung and Lempungagung, tried to get an extra income from an extra crop. Two years later, in 1991, the number of farmers planting a secondary food crop, mainly maize, had increased to 69%, of which relatively more in the subdistrict Sumbersari. Efforts of local governments to motivate farmers to plant during the intermediate season seem to have caused this increase.

The majority of rice farmers surveyed in the baseline season (63%), varying from half in Sugihsari to three-fourths in Lempungagung, own the holding they operate (Table 4.1 and 4.2). Almost one-third of these owner operators of whom relatively many in Sumberagung and few in Kuduagung, have additional access to fields that they lease, either on a rent or a sharecropping agreement. The 37% of non-owning farmers only leasing land can be roughly divided into farmers paying a fixed rent for the land, further referred to as tenants, and farmers operating land on a sharecropping agreement, called sharecroppers. In the study area, a sharecropping agreement mostly implies equal shares of the gross yield for farmer and landlord. With only a few exceptions, all inputs are paid by the sharecroppers, despite the 1959 Share Tenancy Regulations stipulating the sharing of all production cost in the same proportion as the sharing of the product (White and Wiradi, 1989). Only land tax, and sometimes irrigation fees, are paid of by the landlord. Plosoksari, Kuduagung, Mulyoagung and Lempungagung have higher proportions of tenants than sharecroppers.

Table 4.2: Tenure status of rice farmers over four seasons, as % farmers reporting.

	1989/90	1990	1990/91	1991
owner operator	44%	44%	45%	47%
tenant	19%	19%	19%	19%
sharecropper	15%	15%	8%	11%
mixed owner + tenant/sharecropper	19%	21%	27%	22%
mixed owner + sharecropper	2%	2%	0%	2%
N	*599*	*299*	*546*	*263*

Table 4.3: Average farm size (in ha) as classified by tenure status.

	1989/90 wet season						1990/91 wet season					
	own-er	tenant	share-cropper	mixed o + t/s	mixed t + sh	total	own-er	tenant	share-cropper	mixed o + t/s	mixed t + sh	total
average	0.53	0.42	0.35	0.94	0.72	0.56	0.50	0.45	0.38	0.89	-	0.59
st. dev.	0.53	0.42	0.21	0.81	0.44	0.58	0.45	0.39	0.28	0.75	-	0.56
N	*261*	*114*	*91*	*117*	*13*	*596*	*246*	*106*	*46*	*148*	*0*	*546*

Rents for land are usually paid in advance, which makes this form of tenancy only possible for people who have cash in hand or access to credit. In many occasions, the tenants even have to pay the rent in advance while the land can only be operated in several years time. The salary land of village officials is often leased on a fixed-rent basis, and not uncommonly auctioned, resulting in high prices. Large variation in rents exists, partly dependent upon the fertility of the soil, but also a result of unpredictable owner-tenant interactions. During the study period, rents between Rp 100,000 and 1,000,000 (US$ 50-500) per hectare per year were paid. An increase of the average rent was visible in the course of the four study seasons.

Farmers operating on both owned and rented or sharecropped fields (19% during the 1989/90 wet season) appear to be the larger farmers with an average farm size almost twice the total average (Table 4.3). The sharecroppers have, on average, the smallest farms. When the frequency distributions of land tenure status over the four study seasons are compared (Table 4.2), it can be noted that the number of sharecroppers in 1990/91 (including all eight villages) was almost reduced by half in a year's time, in exchange of an increasing number of farmers with mixed ownership-tenancy status. Since the same respondents were interviewed in all surveys, not replacing the ones who stopped farming, the trend above can probably be explained by the phenomenon also observed in other areas of Java (White and Wiradi, 1989), that small sharecroppers have to disappear because their farms are not profitable any more, in favour of large farmers (usually owners) enlarging their farms by contracting the sharecropping agreement. This statement is supported by Table 4.3 which shows that the average area of the sharecropped farms increased in the 1990/91 season compared to 1989/90 because the small sharecroppers stopped farming. A decrease of sharecroppers by 7% within one year time is alarming and illustrates the marginality of this type of farm management. The shift in tenure status is less obvious when the two dry season surveys including the four intensively studied villages only are compared, indicating that this shift was stronger in the four control villages (Sugihsari, Plosoksari, Sumberagung and Lempungagung).

The average area cultivated per household over the eight villages is 0.56 hectare in the baseline season 1989/90, ranging from 0.41 hectare in Sugihsari to 0.69 hectare in Sumberagung. Average areas per household in Sugihsari, Jayasari and Plosoksari are significantly smaller than those in Senengsari, Kuduagung, Sumberagung and Lempungagung (t-test, $\alpha = 0.05$). Land distribution patterns show noticeable variation among the eight villages. Land seems to be most equally distributed in Kuduagung and Sumberagung where the large majority of farmers cultivate pieces of land about the size of the village average. In Senengsari, land is most unequally distributed with relatively many large farms. The historical background of this unequal land distribution is described in Box 4.2. Plosoksari is outstanding in that it has a high proportion of farmers who have to live from pocket-size fields. It is, therefore, not surprising that a lot of farmers in this village, after establishing the rice crop, temporarily migrate to the cities to look for other employment. They return before the harvest, while the wives take care of the crop during the season.

Box 4.2
Unequal land distribution and the people in Senengsari

According to the village secretary, a 35-year old, lively sports teacher, and some of his much older colleagues in the village council, the unequal land distribution in Senengsari is due to a tradition of extravagance that reigns in this village. People like to spend money without calculating beforehand what the consequences are. A favourite source of considerable expenses are ceremonial feasts. Whilst in many other places in Central Java more frugal versions of traditional ceremonies commonly emerge, the people in Senengsari do not want to alter the dignity of their feasts. Since in Javanese society there is often a reason for a traditional ceremony, as soon as they have money, people in Senengsari will take the opportunity to organise a feast. The budget for these feasts, however, often exceeds the amount of money available. Therefore, especially after a good season when people attained quite some profit, an intensive trade of land can be observed. This looks contradictory but it is logical in the context of the story above, saying that people start spending money as soon as they have it in hand, but, then, spend more than they actually have. The only way to have quick access to money is to sell a piece of land. Apparently, ceremonial feasts are more highly valued than the possession of land which stands for security in the future. The village secretary admits this. People are generally quite indifferent about the future.

Nowadays, with more opportunities for trade, the same attitude of extravagance is visible in the trading behaviour of the Senengsari people. As soon as one has money available, this is invested in some enterprise without realistic estimates of profit. When the inevitable losses are suffered, one is forced to sell or mortgage another piece of land. A lot of the land in Senengsari is sold or mortgaged to Chinese traders in the nearby subdistrict town. Often, the former owner still cultivates the plot, but on a sharecropping agreement. Another group of people who profit from the extravagant attitude are the richer people in the village, who buy bits and pieces of land to 'rescue' their fellow villagers from debt, and get larger and larger lands themselves, over time causing unequal land distribution. This often results in feelings of envy among the small farmers, which influences people's attitude toward collectivity and, thus, hampers agricultural development.

The opposite attitude of the neighbouring village Sugihsari, were people are generally rational and businesslike, is striking. By comparing these two villages, the only factor of importance observed that might have influenced the different attitudes is the possible example given by the village head. This is likely to be a reinforcing process in the course of time, where change in attitude has been induced by a village head, as in the case of Sugihsari. People ban certain traditional habits, are rewarded by a faster development, and, next, elect a new village head who supports this development. Where no change occurred, as in the case of Senengsari, and people still feel more rewarded by the preservation of traditions, a leader will be elected maintaining this situation.

Another way of depicting land distribution is by relating the percentage of farmers cultivating an area smaller than the village average, to the percentage of the farm land that is cultivated by these small farmers. Both census and baseline data show that more than two thirds of the farmers in the eight villages have to be classified as small farmers (70% in the census and 68% in the baseline survey). These small farmers cultivate only 37% of the total farmland. Figure 4.1 depicts land distribution for the eight study village in the baseline season by giving the percentage of small farmers having less than the average area on the left leg of each bar, and the percentage of land cultivated by these small farmers on the right leg. The slope of the line between these two values is an indication for the equality of land distribution: the steeper the slope, the more unequal the land is distributed. The Gini coefficient included in the figure is a value for inequality. The higher the value, the more unequal the (land) distribution. Senengsari turns out to be the village with most unequal distribution of land with 74% of the farmers cultivating 36% of the land and a Gini coefficient of 0.51. Kuduagung shows the least unequal land distribution, with a Gini coefficient of 0.32.

Unlike many other places in Central Java, mechanisation in rice cultivation has widely made its entry in Grobogan. Favoured by the flat landscape, almost all land in the study villages is ploughed by small hand-tractors. The first tractor entered the village of Mulyoagung, for instance, in 1985, and four years later one could seldom discover a pair of buffaloes with a traditional wooden plough in the rice fields. The other villages show a similar process, with only one or two years delay. Since the introduction of the short high-yielding rice varieties, harvesting is solely done with sickles. Traditionally, a small rice knife ('*ani-ani*') was used in the belief that the rice goddess will be offended if other means are used. But the use of sickles is accepted for the new rice varieties since 'these came from overseas and, therefore, do not disturb the rice goddess' (Collier et al., 1973).

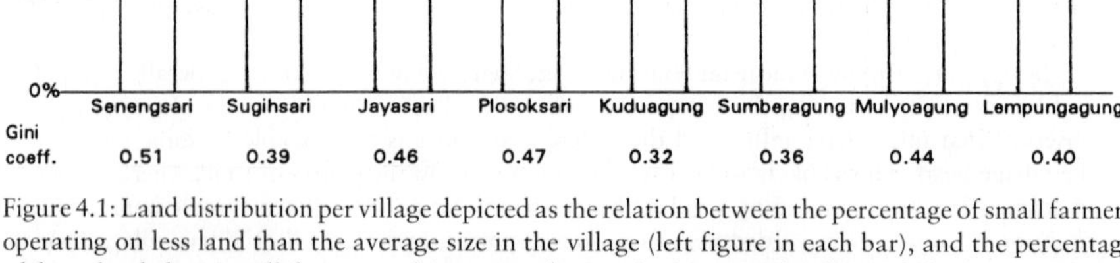

Figure 4.1: Land distribution per village depicted as the relation between the percentage of small farmers operating on less land than the average size in the village (left figure in each bar), and the percentage of farm land these small farmers cultivate together (right figure). The slope of the line connecting the two values is an indication for equality of land distribution: the steeper the slope, the more unequal the land is distributed. The Gini coefficient of inequality shows a higher value for a more unequal land distribution.

Losses of the harvest in the 'ani-ani' system were very high due to the concomitant processing techniques, the transport (including stealing), and the difficult distribution of the shares to the harvesters (NODAI, 1985), which encouraged farmers also to easily adopt the sickle system. The transition in harvest technology, however, had great social implications in many areas, especially for the poor women labourers many of whom lost an important source of income (Hüsken, 1988). The '*ani-ani*' is now only used in Grobogan when the crop has lodged.

If we draw a picture of a rice field area during the harvesting season in one of the eight study villages now, we can see groups of people, both men and women, on small pieces of land together cutting the mature rice with sickles, collecting the bundles of long-cut straw on a spot in the field where the pedal thresher is installed under improvised sunshade, and putting the freshly threshed grains in big sacks. Pedal threshers in no time replaced the more traditional threshing methods around 1985. With the introduction of pedal threshers, the whole harvest labour system changed considerably. Whilst previously individual people were allowed to participate in the harvest – not to forget that the number of people with this privilege has substantially decreased after the sickle was first used – with the new method a limited, more or less fixed, group of harvesters ('*borongan*') is formed by the owner of the pedal thresher. The group is contracted by the farmer to cut and thresh the rice. The labourers get their pay either in kind or in money. A clear distinction by subdistrict is visible in this regard. In Sumbersari, a harvest labourer is usually paid a wage per kilogram of gross yield, the amount depending on the distance of the field to the main road. In Sumberagung, harvest labour is paid by a share which varies between 8% and 12% of the gross yield.

Some of the women working in a harvest group are usually not a regular member of the group but belong to the poor women of the village who used to collect the leftovers from the field in times of the old harvest system. With a harvesting system using sickles and pedal thresher, and the changed labour arrangements, there is not much left for the gleaners to pick from the field. In return for free labour, these women are now allowed to take purposely made 'leftovers'. This practice seems to harm labour relations and work opportunities to some extent. Although the poor women are allowed to pick up a small piece of the pie, they are not paid a wage equivalent to their efforts. The group leader of the harvesters in charge of the financial arrangements with both farmer and labourers, is probably the one who profits most from this system because counting on free participation of the poor women he can reduce the number of labourers and, thus, the people who have to be paid a share.

Traders who buy the standing crop ('*penebas*') have their own harvesting groups, and pay their labourers a fixed wage. Selling the standing crop on the field ('*tebasan*') is not a very common practice in the study villages, but most frequently encountered in Senengsari and Jayasari. Farmers themselves consider this practice a stopgap solution when they urgently need cash or do not have sufficient drying or storage facilities. Farmers with many other enterprises, as is the case in Jayasari, do not want to bother themselves about arranging labour and controlling the harvest, and sell the standing crop for that reason.

The group or contract labour system (*'borongan'*) is commonly practised in the study area for planting and weeding. The system is easy for the farmer who only has to negotiate with the group leader. The group leader is responsible for arranging enough labourers to complete the operation in time. The price is determined per area, independent of the number of people working. Through this system many Javanese land labourers who previously earned a living from occasional agricultural work (usually the landless in a village and predominantly women) lost work opportunities (White and Wiradi, 1989). This disadvantageous situation has not been strongly felt yet in the villages in Grobogan, according to the people there, because there is rather a shortage of agricultural labour which requires farmers to look for additional labour in neighbouring villages.

Rice cultivation technology in the eight villages has largely adapted the Green Revolution package. Modern rice varieties are used by 100% of the farmers. Nitrogen and phosphorus fertilisers are commonly used in relatively high dosages. Pesticides are widely applied, often as a prophylactic measure. Yields equal the average for irrigated areas in Central Java: between five and seven tons per hectare. It is remarkable that over the four study seasons, the farmers in Mulyoagung always attained the highest average yields among the eight villages, a case elaborated in Box 5.3.

All eight villages enjoy a sufficient supply of commonly used inputs for rice cultivation, such as seed, nitrogen and phosphorus fertilisers, and pesticides. Uncommonly used pesticides and all kinds of foliar fertilisers, which are distributed through special salesmen, are easily obtainable in the most remote places, in contrast to a more important input such as potassium fertiliser which is distributed through the KUD. In general, rice marketing facilities are adequate in all eight villages.

It is not surprising that a more than occasional number of KUD members among rice farmers can only be found in Sugihsari and Kuduagung, where the village heads are managers of the KUD. In Kuduagung, promotion of the KUD package is also strongly supported by the village extension worker, who earns an additional income from the commissions on agricultural input distribution. Moreover, this person sells pesticides other than available in the KUD package at his house. His chemical-oriented view on agriculture, mainly based on personal interest, strongly influences farmers in his work area to which Kuduagung and Sumberagung belong. The frequency of pesticide use is much higher in these two than in the other study villages. The other village extension workers in the study area have different job commitments. The one in Mulyoagung and Lempungagung, and the one in Jayasari seem to have strong motivation to serve the farmers in the best way they can. The extension worker in Senengsari and Sugihsari often shows an ambiguous behaviour in facing either farmers or KUD staff. He has a strong leg in the KUD, and is not reluctant to profit from this, but he is also well disposed towards the farmers, and will not deliberately disadvantage them like the one in Sumberagung. In Plosoksari, the extension worker is indifferent and hardly ever visits the farmers, not unlikely because of the remote location and difficult accessibility of this village.

According to the Agriculture Service administration, all farmers belong to a designated farmer group, officially consisting of all cultivators in a certain rice tract and represented by a contact farmer. As described in Chapter 2, most of these groups only exist on paper. The passive role of the farmer group holds for most of the eight study villages. Minor exceptions are found in Sugihsari where the SUPRA-INSUS programme was successfully implemented through one of the farmer groups, and in Senengsari and Plosoksari where, previously, prizes were won with farmer group organisation in irrigation management but currently no activities are organised. After having received IPM training, the farmer group structure in Senengsari was revised and collective rat activities were organised successfully through this structure. Collective rat control and, in some villages, repair of irrigation channels, are mentioned as the only farmer group activities in addition to pre-season meetings.

4.4 Diversity and its consequences

From the profiles described above several aspects can be determined in which villages vary considerably. Most obvious is the appearance of a village, ranging from traditional, village-like to more urban with several gradations in between. A general relationship exists between the appearance and the distance to the main road.

In the context of this study, it is useful to look at diversity of aspects determining the level of development, particularly in agriculture. Although there is seemingly some logic in correlating appearance and remoteness to the level of development, this correlation does not hold exclusively in case of the eight study villages. An obvious example is the village of Jayasari which has a village-like but well maintained appearance, is located far from the main road but well accessible via a good road, and is relatively well developed. A second example is the village of Mulyoagung which is traditional, remote and difficult to reach, but where extraordinary high agricultural outputs are attained. In contrast, the village of Senengsari is close to the main road, has easy access to all kinds of facilities, but is poorly developed and not likely to develop fast in the near future.

In the three examples above, three different aspects can be observed that mainly determine these levels and directions of development. In Jayasari, these seem to be the **policies and activities of the village head**, in Mulyoagung the **coherence of the farmers**, and in Senengsari the extravagant **attitude of the people with adherence to traditional ceremonies**. Comparable specific examples can be given for the other villages. A number of aspects determining agricultural development are deducible from the examples. In addition and often related to the aspects mentioned above, the following points are thought to be important:

- availability of inputs;
- timeliness and reliability of irrigation water supply;
- frequency and direction of activities of the village extension worker and input distributors;
- farmer group structure, leadership and activities;

- land and income distribution, and
- position of the village head and other village officials, especially regarding respect received from the villagers.

It should be stressed again that all these aspects are subject to high variation for the eight study villages.

Dealing with diversity is a difficult task in both research and development programmes. In research, particularly in evaluation studies, we like to generalise, make general conclusions, and give general recommendations. Depending on the purpose of a study, general conclusions, however, might not be valid for important groups within the study population. The existence of so many factors with high variation among only eight villages in a seemingly homogeneous area, gives some insight in the complex situation to be dealt with. This diversity has consequences for the design and implementation of a study investigating effects at the village level. In particular, it questions the use of comparison (often called control) villages which are usually meant as a reference to measure effects of a certain intervention by comparing situations with and without the intervention. In the case of the implementation of the National IPM Programme in Grobogan, villages selected for training were those located on or close to the main road. Any comparison village is, then, in no way equivalent, which is an interesting observation in itself. Even if seemingly comparable villages can be found to serve a neutral control function, factors not directly observable but of great importance, such as the village head's position, are likely to cause development processes differing between villages. This is not to say that investigating comparison villages is useless, on the contrary. Interpretation of observations, however, needs to be done in the context of the diversity observed. Attention to specific situations might help to analyse reasons for success or failure.

Considering development programmes, it is easier to apply a general strategy in reaching an audience rather than to stratify and adapt to local differences. The larger the audience, the harder it is to deal with the increasing diversity. This conclusion is highly applicable to the Indonesian National IPM Programme which is implemented in eight provinces, with varying cultures, social structures, and environmental and socioeconomic conditions. Successes achieved partly depend on specific situations and conditions, e.g. the high commitment of a village extension worker, or the active involvement of a farmer group leader. This dependence holds even stronger for less successful achievements. By recognising diversity among a programme's audience beforehand, a larger audience can possibly be reached, for which a lot of flexibility and possibilities for fine-tuning is required within the necessary set of general strategies.

By paying explicit attention to diversity it is hoped that this study acquires additional value. It is hypothesised that due consideration of specific factors causing diversity helps to analyse processes and mechanisms explaining not only **what** happens but also **why** things happen in a certain way, in a certain place, at a certain time. The insight that is hoped for might be useful in anticipating generalising project strategies that will take into account both opportunities and restrictions in field implementation. Considering the methodological question on the validity of comparison villages, a second hypothesis is formulated. In a village study identifying the effects of an intervention, comparison villages have no immediate comparative value, but investigations in comparison villages can help to explain factors and processes that determine success or failure.

5 The art of growing rice

5.1 This chapter

Growing a good rice crop is an art in itself. Javanese farmers have practised this art for centuries under a variety of conditions. Since the Green Revolution, traditional rice growing changed radically. Ever since, technologies keep on changing, in an attempt to increase production where rice needs to keep up with population growth. In the end, farmers determine the speed and direction of change by implementing, or not, the new technologies, balancing constraints and opportunities.

This chapter describes how farmers in the study area grow rice anno 1990-91. Besides being a case discussing to what extent and for what reasons the modern technologies are implemented by rice farmers in Grobogan, this chapter serves as a point of reference to measure the effects of IPM training on farmers' practice in rice cultivation, in particular pest management, and on farm economics. Data are mainly taken from the seasonal surveys, complemented by information from group interviews, field observations and discussions with people from all layers in the farming community. Respondents in the surveys and group interviews were farmers who did not participate in the IPM training conducted in four of the eight study villages during the 1991 dry season. These farmers are referred to as 'non-IPM farmers'. The seasonal surveys were conducted in all eight study villages during the two wet seasons, and in the four intensively studied villages only during the two dry seasons. Therefore, the tables show blanks for the control villages in the dry seasons. The most important tables are presented directly in the text (indicated as Table 5.*), whereas some more, mainly descriptive tables can be found in Appendix III (indicated as Table III.*).

Most sections below refer to the 1989/90 wet season which serves as a baseline in the study. Data from the following three seasons are sometimes presented to show seasonal variation. Seasonal variation between wet and dry rice seasons does not so much result from the amount of rainfall, since most rice lands in the study area are supplied with irrigation water, but rather from the amount of sunshine that is higher during the dry season, causing a faster growth of the plants. Another aspect of variation can be found in pest occurrence in that the dry season, second rice crop is more likely to suffer from pest attack than the first rice crop in the wet season. Since the second rice crop is planted immediately following the wet season crop (Figure 1.1), the period in between is short enough for pest organisms to survive. An overview of the study seasons with their profiles is given on the flipchart in the back of this book.

The 1989/90 wet season and the 1990 dry season in Grobogan were average seasons with regard to weather conditions. Considerable rat attack occurred in all the study villages during these two seasons (Table III.1), a recurrent problem. Farmers reported some attack of stemborer, rice seed bug, caterpillars, and for the dry season bacterial leaf streak, but damage was negligible. The 1990/91 wet season started exceptionally well with a lot of sunshine and few pests. When the rice crop was about to flower, heavy rain and wind occurred for several days after which the farmers noticed a sudden outbreak of stemborers. Many of the panicles emerging were white, mostly due to stemborer, but some because pollination failed due to the strong wind. Despite farmers' misgivings about the following 1991 dry season, expecting high stemborer populations, this season was very good, not to say exceptional, with dry, sunny weather and hardly any pest occurrence. The 1991 season was the last survey season, but more field observation on the rice ecosystem and on farmers were made during the 1991/92 wet season which appeared to be extremely good, the best within living memory. This season was preceded by a long, dry period which is believed to be favourable for the fertility of the soil, and no pest attack of importance occurred.

Farmers' practices described in the sections below are sometimes compared with the recommendations from the Agriculture Service, i.e. the technology that is brought to the farmers by the village extension workers. This is not to indicate that these recommendations are the best for farmers to apply, on the contrary. They serve mainly as a reference, against which farmers' considerations for different practices may become clearer.

5.2 Rice cultivation practices

5.2.1 Cropping pattern and water supply

Following the District Agriculture Service's instructions, rice is grown two times a year in the study villages. The planting schedule depends on the irrigation schedule which is arranged by the District irrigation authorities (see Figure 1.1). According to local belief, the first rice crop should be established before mid December

Box 5.1
The Javanese seasons

Many Javanese farmers believe that planting beyond the 25th of December will cause damage to the rice crop by insect pests, particularly stemborers. The reasoning behind this belief is that during the seventh and eighth '*mongso*' or season, which corresponds with the months of January and February, there is usually a lot of rain. Rain, especially at full moon, brings about more caterpillars in the crop, and some people even say that caterpillars, like people, are more hungry if it is cold and rainy, resulting in more damage. With many (hungry) caterpillars in the seventh and eighth '*mongso*', timely planting should avoid that the rice crop is at a vulnerable stage during these months.

A second characteristic of the seventh '*mongso*' (January) is that there is usually much wind. Flowering of the rice crop during this month should be avoided in order to prevent that pollination is disturbed which causes empty seeds. Farmers often call this phenomenon '*walang angin*' or 'windhoppers', as if it were a pest. Some people believe that pests such as stemborer moths are transported by wind. By experience, farmers know that the best planting time for the wet season crop is November, and before end December at the latest.

The influence of the weather on insect behaviour has been investigated for years and described by Dutch colonial researchers early in this century (Koningsberger, 1915). Insects like ants and termites were observed to emerge sooner and more massively from their cocoons after being soaked by rainwater. The rice seed bug (*Leptocorisa sp.*) was observed to start swarming when heavy rainfall occurs at night. Strong wind was thought by Koningsberger to have an inhibiting influence on the activity of insects, rather than transporting them, as is believed by some Javanese farmers. Insects seem to avoid windy places. Only butterflies with a larger wing surface were seen to have problems navigating against the wind.

and preferably in November, otherwise the crop will be severely attacked by pests (Box 5.1). This situation actually occurred in the wet season of 1990/91 when a delay in the water supply until January, because of major repairs of irrigation channels, was followed by an extremely severe outbreak of yellow stemborer, *Scirpophaga incertulas* (Walker), causing great yield loss. The loss was considerably less in villages that had received water first (Mulyoagung and Lempungagung) and, thus, had been able to plant somewhat earlier.

The majority of farmers in the study villages (89%) generally have nothing to complain about water supply to their rice fields (Table III.2). The remaining farmers, mainly from the rainfed lands in Sumberagung and Lempungagung, and some from irrigated lands especially in the subdistrict Sumbersari, regularly have problems with water management, either because of droughts or of floods.

5.2.2 *Land preparation*

Tractors for land preparation were introduced into the area around 1985, and three years later replaced water buffaloes completely. Most fields are tilled twice. The first time, two weeks before planting, the soil is ploughed, and the second time, just before planting, the field is levelled. The whole treatment is executed by hired labour, not uncommonly by the owner of the tractor, at a rate of about Rp 40,000 per hectare (US$ 20.-). Hand labour (hoeing) is often needed to level larger soil accumulations in the field and to repair the bunds.

5.2.3 *Rice variety*

Although one is free in principle to select a rice variety according to his or her preference, farmers in one village, that is to say a selected group of farmers appointed by the extension system as contact farmers, usually decide together in a pre-seasonal meeting what variety could best be planted during a particular season. Reasons for collective decision and synchronisation are the dependence on the irrigation schedule which determines the duration of the variety, and sometimes forecasts of pest outbreaks which affect the choice of resistant varieties.

Although farmers often have other reasons for selecting a rice variety, little variation was observed in the study villages regarding the varieties planted (Table III.3). The favourite rice variety, and increasingly favourite over the four study seasons, is the short duration IRRI variety IR 64. In the 1989/90 wet season, 64% of the farmers had planted IR 64 on all or a part of their land, while by the 1991 dry season the proportion had gradually increased to 92%. Reasons for farmers to prefer this variety are its good yield, its pest resistance – IR 64 is moderately resistant to brown planthopper and can support some rat damage at an early stage because it soon makes new tillers –, and its good taste (Table III.4). The second variety in the 1989/90 season, the locally improved Semeru, was preferred by 43% of farmers, mainly for its good yield and good marketing through the National Logistics Agency (BULOG), but it had almost vanished in 1991 because of its bad taste and hard texture. It is noticeable that the farmers in Grobogan are particular about the taste of rice: 44% consider taste in their choice for a variety. An explanation might be that most farmers operate on relatively small fields and, therefore, grow rice for subsistence. Some farmers even grow the much tastier traditional varieties on a small plot for their own consumption, whereas the larger part of their field is planted with a high-yielding variety for the market. Others grow two different (modern) varieties to spread the risk, or to experiment with a new variety.

A seasonal influence on variety selection is not obvious in the study area. Short duration varieties are preferred for both wet and dry rice seasons, probably because of the quick succession of the two crops due to the strict irrigation schedule. A small proportion of farmers (7%) declares to select a variety for its suitability for season or location. High yield, pest tolerance and the quality of the rice are the main considerations for variety selection regardless of seasons.

5.2.4 *Seedbed*

Since only transplanting is practised in the study area, farmers usually nurse their own seedlings in carefully laid-out seedbeds. Seedbed preparation starts three to four weeks before the planned planting date, if the water supply permits. Seedbeds of, on average, 323 m^2 per hectare rice field are prepared, which are sown with an average amount of 0.44 kg/m^2 seed (Table 5.1). Although the average size of the seedbed is within the recommended range of 300-500 m^2/ha, the seed rate exceeds the recommended rate of 0.07-0.15 kg/m^2 many times, causing high seedling densities. This is especially true for Jayasari where extremely small seedbeds are made (191 m^2/ha) and extremely high seed quantities (0.55 kg/m^2) are used.

Table 5.1: Seedbed preparation by non-IPM farmers in the four intensively studied villages, 1991 dry season.

	Senengsari	Jayasari	Kuduagung	Mulyoagung	Total
Avg. size of seedbed (m^2/ha field)	430	191	300	389	323
Avg. quantity of seed/bed (kg/m^2)	0.37	0.55	0.48	0.34	0.44
Avg. quantity of seed/field (kg/ha)	77	47	73	72	66
N	*55*	*72*	*60*	*76*	*263*

Not less than 74% of the farmers studied sows a higher seed rate than recommended (Table III.5).

The amount of seed used to plant one hectare, on average 66 kg, is also higher than the recommended amount of 35-45 kg/ha, especially in Senengsari (77 kg/ha) where, in addition to high seed rates, relatively large seedbeds are used. Although it seems uneconomical, farmers definitely have their reasons for these practices. One reason might be that not all farmers sow high quality seed with maximum germination capacity. Forty percent of the farmers surveyed in the 1990/91 wet season, and even 50% in the following dry season, used seed of the third or later progeny, some on whole fields, others on parts (Table III.6). When seed of later progenies is used, a larger seed quantity helps to reduce the risk of low germination. This might explain the high seed rates in Senengsari, since many farmers there use seed of later progenies. A high seed density in the nurseries may, however, negatively influence germination and seedling growth due to shortage of space and light. If this is true, farmers can economise on seed.

A second reason for using high seed rates might be the application of higher than recommended planting density which requires more seedlings. This practice was reported by farmers in Mulyoagung where a planting distance of 16-17 cm is applied in order to prevent weed growth, instead of the recommended 20 cm. Seed density in this village is still twice too high, which is supposedly wasteful, especially since many farmers there use high quality seed.

Farmers who use progeny seed obtain this from their own or their fellow farmers' harvest. The biggests, undamaged seeds are selected immediately after the harvest and stored to be sown in the next season. New seed is obtained from either local kiosks or from the KUD. Over the four study seasons, a shift in preference for location of seed purchase towards the local kiosks is observed in all villages except Sugihsari.

5.2.5 *Transplanting*

All over the study area, as in most parts of Indonesia, rice crops are transplanted, which is a crucial practice relating to weed control. With the high-yielding varieties, the technique of rowplanting was introduced which is currently practised by the majority of farmers. The advantage of rowplanting is that it enables weeding by means of a mechanical weeder, and facilitates observations, fertiliser application and, if necessary, spraying. Mechanical weeders, however, are seldom used and not widely available. In some villages, like Senengsari, it is not always possible to get enough planting labour willing and able to plant in rows. A special case is found in Mulyoagung where already in 1953 the then village head invited somebody from another area in Central Java to the village in order to teach the farmers the technique of rowplanting. Within a short period the technique was practised by all farmers and labourers, although the planting distance was adapted from 20 cm to 16-17 cm in the course of time in order to discourage weed growth.

5.2.6 *Water management and monitoring*

An important but often underestimated occupation in tending a rice crop is water management. Water management starts with land preparation and ends at harvest. Flooding the rice field is an important weed control practice and, thus, favours the growth of the rice plants (Thurston, 1992). Especially at the vegetative stage, maintenance of the right water level is crucial for the growth of the young plants. In this period, many farmers visit and monitor their fields every other day, some even every day, especially to look after the water and repair broken bunds. If there is shortage of irrigation water, farmers might spend the whole night looking for water by opening and closing channels to provide water for their own fields, and, then, guarding that water stream. Maintenance of irrigation channels requires collaboration and collectivity among farmers, which is often difficult to organise.

The amount of time spent on regular visits to the field to monitor the crop for water and pest status is hard to quantify reliably. In the baseline interviews, farmers tried to estimate the frequency of field visits outside the labour peaks, and the average time spent per visit. From these estimates, an average of 7.6 hours per week per farmer for regular monitoring and maintenance work is calculated (Table 5.2). The majority of farmers (67%) usually does not spend more than a total of 7 hours per week in the field (Table III.7), which corresponds with a two to three hour visit every two or three days. A comparison of these figures with the more detailed and accurate data obtained from daily records kept by nine farmers during the 1990/91 wet season shows that this estimated average equals the amount of time noted in most of the records for regular field visits at the vegetative stage of the crop. At this stage, usually more time is spent in the field than at the reproductive stage.

Classification of monitoring time spent per farmer by farm size shows that large farmers do not spend that much more time than small farmers. A habitual pattern of visiting the rice field was observed: people usually go to the field at 7:00 a.m., go home by the time it gets too warm at 10:00 a.m. or a bit later, and occasionally

Table 5.2: Average time spent on monitoring and water management by non-IPM farmers as classified by farm size.

	< 0.10 ha	0.10-0.24 ha	0.25-0.49 ha	0.50-0.99 ha	≥ 1.00 ha	Total
in hours/week/farmer	5.2	6.0	6.7	8.9	10.7	7.6
in hours/week/ha	73.8	35.7	19.3	13.5	6.9	21.9
N	*21*	*140*	*202*	*156*	*77*	*596*

return to the field in the late afternoon. When the monitoring time is converted to time spent per hectare, the difference between small and large farms becomes very obvious. Much more time is spent per area unit on small than on large farms, reflecting a more intensive cultivation of the small farms, as also found in other studies in Java (Hüsken, 1989). An obvious example in this regard is reported by one farmer who spends 16 hours over a season on monitoring and water management in his one field with an area of 0.19 ha, and just a bit more (18.5 hours) on his other field of 0.28 ha, a result of his alternating visiting schedule to the two fields.

Monitoring and water management are typically jobs on which one can either save or waste labour. Some farmers, usually the ones with other occupations, will not care about the water as long as they know that there is water in the field. Others do not rest before they have assured themselves that the amount of water in the field is exactly what is needed for the coming days (Box 5.2). Farmers themselves characterise a good farmer as one who frequently visits his or her field throughout the season, which implies frequent control of the water level, water supply and pests.

Even though the influence of water management on the growth of the crop and on the final yield is hard to estimate, and impossible to quantify, farmers' experiences, made visible in their characterisation of a good farmer, suggest that good yields are positively related to good, intensive water management.

Box 5.2
Monitoring and water management practices of two farmers

Two of the daily records kept by nine farmers during the 1990/91 wet season showed contrasting attitudes of the farmers toward farm operations as reflected in their monitoring and water management practices.

The first example is Mr. Harjono from Senengsari who is village official in charge of development affairs for which he receives a piece of salary land of 1.75 hectares. More than 1.5 ha of this land is rented out for several years in order to pay off the loan he had to take for slush money to get the job. What he still operates himself is 0.17 ha of which the yield is not enough to feed his family. This does not motivate him to spend much time in his rice field in addition to his daily duties in the village office. Besides, he is not a farmer by profession, but an operator of heavy machinery. In all, he visits his field at most once a week, and hardly ever at the reproductive stage of the crop. Mr. Harjono makes only a two-hour visit to the field before planting for water management. Another eight hours are spent throughout the season on monitoring of pest and water conditions.

The second example is Ms. Suharti from Mulyoagung who operates 0.95 hectares assisted by one of her sons. Her husband is village secretary and so fully occupied with this job that he never looks after the rice fields. In contrast to her husband, Ms. Suharti is a motivated and diligent farmer and so is her son. Throughout the season, they spend 49.5 hours on water management and another 14.5 hours on monitoring of the crop which they visit several times a week, even at the reproductive stage.

5.2.7 *Weed management*

Method and frequency

Weed management is perceived by farmers as a routine operation rather than a pest management measure. Handweeding is most common, applied by 93% of the farmers surveyed (Table III.8). The remaining 7% of farmers, most of whom are from the villages in the subdistrict Sumberagung, in particular the village Sumberagung itself, use a mechanical weeder. Mechanical weeders are not commonly available in most of the study villages, which may partly explain the low utilisation of this tool. Herbicides are not used at all.

The frequency of weeding in the (1989/90) wet season is, on average, 1.8 times (Table 5.3), which implies that the majority of farmers (64%) weed two times (Table III.9). A higher number of families in Plosoksari weed only once. Two possible reasons that can be identified are (1) the shortage of outside labour due to the remote location of the village, and (2) the temporary migration of many men from this village to the town to look for employment resulting in a higher work load for the women at home. In contrast, a very disciplined picture is visible in Jayasari where 85% of the farmers weed twice a season and another 8% even three or more times. There is a tendency to weed more often during the dry season. Although the average frequency is still around two, fewer farmers weed only once, and more farmers three times or more. It is assumed that weeds grow faster in the dry season because of more sunlight. Another reason might be that in the dry season, which immediately follows on the wet season, the soil contains more weed seeds ready to germinate.

Weeding is an activity in the rice field that attracts attention because of the many people working together on a small area. Especially for the larger farmers it is not uncommon to hire a large group of women to do the job. They are paid around Rp 750 (US$ 0.4) and given one meal for half a day work. Smaller farmers usually do it with as many hands from the household as are available and if necessary hire some extra people from outside. Over the eight villages, 61% of the total labour for weeding is done by household members. Among all villages, most family labour is deployed for weeding in Jayasari, and also somewhat more hired labour, which results in a higher weeding frequency in this village. Opposed to

Table 5.3: Average frequency of weeding (in weedings/season) by non-IPM farmers in (1989/90) wet and (1990) dry season.

	Seneng-sari	Sugih-sari	Jaya-sari	Plosok-sari	Kudu-agung	Sumber-agung	Mulyo-agung	Lempung-agung	Total
wet season	1.7	1.8	2.0	1.5	1.7	1.9	1.9	2.0	1.8
N	*66*	*63*	*85*	*85*	*68*	*62*	*85*	*85*	*599*
dry season	2.0		2.1		1.9		1.8		2.0
N	*62*		*85*		*67*		*85*		*299*

Jayasari is the situation in Kuduagung where considerably less family labour is allocated to weeding. The relatively low family involvement in Kuduagung might be caused by the many opportunities for other activities such as trading and education since this village is very close to the subdistrict town and not far from the district town.

Time of weeding
There is good reason to weed just before fertiliser application since all fertiliser is then available for the rice plants only. Farmers, however, tend to weed some days after applying fertiliser, i.e. on average at 21 days after transplanting (DAT) for the first time, and 35 DAT for the second time (Table III.10). Some farmers report that they weed after fertiliser application, because the weeds are easier to pull out. This practice seems uneconomical since part of the fertiliser is absorbed by the weeds. Farmers who push the weeds under into the soil say that there is no loss of fertiliser since the weeds will be decomposed again, a statement which is only partly valid. Others use the weeds as cattle fodder and therefore believe that the fertiliser absorbed by the weeds serves a purpose. A more logical reason seems to lie in the time schedule that most farmers in the study area apply for time of fertilisation which interferes with the time of weeding. In general, Grobogan farmers do not like to give basal fertiliser applications. Thus, they are forced to give the first fertiliser application between one and two weeks after transplanting, otherwise the growth of the seedlings will be disturbed. By that time, there is no need for weeding yet, but with the extra nutrients the weeds will grow fast urging the farmers to start weeding within one or two weeks time. The second cycle of fertilising and weeding will show similar intervals. This application pattern is assumed to contribute to an inefficient use of fertiliser.

5.2.8 Fertilisation
Fertiliser application is almost as important as water management for a good growth of the rice crop. Since the soil can not provide sufficient nutrients in the intensive production system, the necessary elements have to be replenished every season in the form of nitrogen, phosphate and potassium fertilisers. The shortage of organic manure in most rice growing areas in Indonesia forces farmers to use inorganic fertilisers. With the high-yielding rice varieties, urea, a 45% nitrogen fertiliser, and TSP, or triple superphosphate with 46% phosphorus, were introduced. These fertilisers helped considerably to increase rice production. In 1985, two additional inorganic fertilisers were marketed: KCl, a 50% potassium fertiliser, and ZA, or ammonium sulphate with 21% nitrogen, The Agriculture Service in Grobogan gives the following fertiliser recommendations for rice crops, which are equal to the national, blanket recommendations for the whole country:

- 250 kg urea/ha/season;
- 100 kg TSP/ha/season;
- 100 kg KCl/ha/season, and
- 100 kg ZA/ha/season.

These amounts should be given in three applications: one basal and two topdressing applications. Although local variation exists with regard to proportions per application and time of application, general recommendations per hectare include 100 kg TSP, 100 kg ZA and 50 kg of urea at the basal application, and 100 kg KCL plus the rest of urea equally divided over the two topdressings, which should be given at 2-3 respectively 5-6 weeks after transplanting. The first topdressing should coincide with maximum tillering and the second with panicle initiation, which are the stages most effectively influenced by nitrogen (Heenan and Bacon, 1987). Split application can reduce plant height (Wells and Johnston, 1970), and therefore perhaps the degree of lodging, although lodging depends mainly on weather conditions.

Fertilisation is one of the favourite topics conveyed in farmer group extension meetings. It is a clear package message for village extension workers to extend to the farmers, usually not free from personal interest because of the commissions involved in selling fertilisers through the KUD. The message, however, is mostly incomprehensible and inapplicable for farmers since it contains quantities in kilograms per hectare, and does not explain the need for using the right type in the right dosage at the right time. Farmers use all kinds of local area measures and most of them have fields much smaller than one hectare. Conversions from kg/ha to the amount needed in their own fields are not taught or practised together in the routine extension meetings, which are only attended by a selected group of farmers anyway. Furthermore, recommendations are generalising, not considering specific conditions in specific fields. Farmers often complain that the extension worker only tells **what** they have to do, not **why**. It is not clear either what the expected yield is at the recommended dosages. Since there is a strong relation between yield and amount of nutrients extracted from the soil, (recommendations for) fertiliser dosages should increase with increasing yields. For the several reasons mentioned, it is not surprising that farmers' fertilisation practices differ considerably from the recommendations, not only due to ignorance or incapacity to convert dosages, but often because of valid reasons based on own experimentation.

Fertilisation on seedbed

The majority of farmers applied urea (79%) and TSP (61%) to the seedbed during the (1990/91) wet season. KCl and ZA fertilisers are occasionally used on seedbeds. During the (1991) dry season, almost all farmers treated the seedbed with fertiliser: 99% gave urea and 81% TSP. The obvious reason for more intensive fertilisation during the dry season is that the soil is less fertile at the beginning of the dry season right after the wet season rice crop, in comparison with the wet season crop after a fallow period or a less demanding secondary food crop. It should be noted that there are no recommendations for fertilisation of the seedbed in the village extension worker's package, although fertilisation of the seedbed has been commonly observed in other areas of Java (Widodo, 1989).

On average, 19 kg urea and 15 kg TSP was applied to the seedbed in the dry season on an area containing seedlings meant for one hectare of rice field (Table III.11). When these quantities are converted to kg/ha seedbed, appallingly high dosages are encountered, mostly applied in one treatment at around ten days after sowing. For urea the average amount is 1,167 kg/ha, and for TSP 1,054 kg/ha. These figures raise the question whether in some cases the fertiliser, especially urea, does not inhibit the growth of the seedlings. An explanation for the high dosages might be the difficulty farmers have in calculating dosages correctly, especially small quantities as needed for nurseries. Also, there is a belief among many farmers that the more fertiliser is used, the better the crop grows, although some farmers stated that one obtains stronger and healthier seedlings if not too much fertiliser is used.

Basal fertiliser application

As described above, recommendations for fertilisation include one basal application at the moment of planting and two topdressings per season. Farmers in the study area generally do not like the basal application. In the 1990/91 wet season, only 7% of the farmers with irrigation, and none on rainfed lands applied basal fertiliser (Table 5.5). The main reason reported for applying basal fertiliser is to encourage a fast growth of young transplants. Fifty-three percent of all farmers surveyed have ever practised basal fertilisation (Table III.12a), half of which (27%) only for one or two seasons to test the method. The basal fertilisers applied were in most cases (77%) plain TSP, or previously DSP (disuper phosphate), and for the rest mainly urea or urea mixed with TSP. The recommended basal mixture of urea, TSP and ZA was never applied by anybody.

Various reasons are quoted by the respondents for not using basal fertilisation (any more), among others that it is uneconomical regarding expenses on inputs (19%) or on labour (13%), they do not have the extra money needed available (12%, of whom especially many in Jayasari), no additional result was observed in the trials (16%), and many farmers (17%) just do not do it because it is not a common practice in the area (Table III.12b). Various other interesting arguments mentioned include the uselessness of the practice since the fertiliser is trampled too deep into the soil while planting (which is actually favourable to prevent volatilisation of nitrogen), or it is washed away by the rain, it irritates the skin on the feet of the planters, the faster and more fertile growth of the transplants attracts rats, and it encourages the growth of weeds. Farmers' experiences with basal fertiliser application are not consistent. Some say it is favourable for the crop's development, some admit that it is, indeed, but not worth the extra trouble, and others say it does not make any difference.

Although fertilisation experts ascribe advantageous effects to basal fertilisation for initial crop growth (De Datta, 1981), farmers' reasons not to do it seem sensible. Basal fertilisation timely provides phosphate to the young plants since the diffusion of phosphates in the soil is very slow. However, in soils where phosphorous fertilisers have been applied for long periods so that the water contains enough phosphates as a result of desorption from soil colloids, this necessity for basal fertilisation is no longer valid. The need for nitrogen supply through basal fertilisers can also be called doubtful since it is likely that high losses due to volatilisation and washing-away might occur, and the little amount of nitrogen needed by the transplants can probably be supplied by the soil. The argument not to spend extra labour on fertilisation in a period that farmers are already very busy seems to be outweighing the doubtful advantages of basal fertilisation.

Fertilisation of crops

All farmers surveyed over four seasons apply urea on their rice crops, and all but a few TSP (Table 5.4a). Farmers agree on the reason to use urea: urea is needed for the growth of the leaves, for a good green colour, and for good tillering (Table III.13). TSP, the activity of which is less directly observable, is experienced by two-thirds (66%) of the farmers to cause a higher weight of the grains. Thirty-five percent of farmers have noticed that the dark-green colour of the leaves caused by urea is better preserved after a TSP application. Other characteristics ascribed to TSP include the prevention of lodging, a better soil structure, and stronger plant roots.

In contrast with urea and TSP, which were adopted by most Grobogan farmers within only a few seasons time after their introduction in 1969, KCl and ZA are still not widely used even though these two fertilisers have been available since 1985. The availability, however, is limited to the KUD shops and larger kiosks in the subdistrict towns. Supply and demand at the village level seem subject to a vicious circle. KCl is used by 19% of the farmers, with little variation over the four study seasons (Table 5.4a). ZA is even less popular: the number of farmers using this fertiliser decreased from 9% in the baseline season via 6% in the two following dry and wet seasons to 1% users only in the 1991 dry season. What are farmers reasons not to use these fertilisers? No convincing example of the result is mentioned by 31% for KCl and 51% for ZA. Other reasons include the shortage of money, and the belief that KCl and ZA are not needed or not suitable under the prevailing (soil) conditions. The failure of an INSUS demonstration plot

Table 5.4: Type of fertiliser applied by non-IPM farmers, as % farmers reporting.

a. Types used over four seasons.

	1989/90	1990	1990/91	1991
urea	100%	100%	100%	100%
TSP	100%	100%	99%	99%
KCl	19%	22%	19%	17%
ZA	9%	6%	6%	1%
N	*596*	*299*	*549*	*263*

b. Types used per application, 1990/91 wet season (*N* = 549).

	1st topdressing	2nd topdressing	3rd topdressing
urea	96%	96%	25%
TSP	94%	84%	18%
KCl	9%	14%	4%
ZA	4%	4%	1%

in Jayasari, implemented in 1985, induced the belief among the farmers there that KCl and ZA have no additional effect on the yield (Santoso, 1992). Farmers who do use KCl have experienced that the rice plants get heavier grains (11%), and that the crop is more resistant to lodging (5%). Because these characteristics are similar to TSP, several farmers believe that KCl is an alternative for TSP, not an additional fertiliser.

Apparently, the current extension system has not been able to convince rice farmers to use KCl and ZA as additional fertilisers. The need for ZA is, indeed, doubtful since enough, and often abundant, nitrogen is supplied by urea applications. The second component of ZA, sulphate, is most likely available in adequate concentrations in the relatively fertile, volcanic soils of Java. Considering the 21% nitrogen contents in ZA, in contrast with 45% nitrogen in urea, and the fact that the price per kilogram of these two fertilisers is exactly the same, ZA is relatively expensive for farmers. Another unfavourable effect of ZA is that it causes acidification of the soil, faster than urea does. ZA also has a lower efficiency of recovery by rice than urea since ZA nitrogen is stronger absorbed by clay colloids (De Datta, 1987).

Potassium (KCl) is assumed to contribute to further production growth, and, what is more, to a more economic use of TSP, which is urgently required considering the extremely high dosages of TSP applied to most fields in the study area. Potassium is an element of quality. It encourages the growth of cell walls, which can make the plants more resistant to pests and diseases. Potassium is supplied by nature in considerable quantities through floods and precipitation of volcanic ash. In certain places in Java, a maximum annual precipitation of volcanic ash of 3 mm has been measured, which corresponds with applications of 60-900 kg potassium/ha, or 120-1,800 kg KCl/ha (Janssen and Van Beusichem, 1991). This means free potassium fertilisation in those areas. It is not known what the levels in Grobogan are. Some farmers having used KCl obtained better results than without, which suggests that potassium has been deficient before.

An interesting mechanism converting farmers to use KCl was observed in the villages of Jayasari and Mulyoagung during the last study season (1991/92). The rice grains of crops having received KCl had a yellowish rather than a white colour, and seemed to be more resistant against crushing in the electric dehullers (which can be a result of thicker cell walls). Rice traders valued these two characteristics as a higher quality of the rice, and offered higher prices. A higher quality, and thus higher prices motivated many farmers to start using KCl in the next season. The convincing proof to adopt KCl for farmers was, not surprisingly, provided by a market mechanism. An increased demand for KCl in these (remote) villages will hopefully also result in an increased supply through the local kiosks.

Almost all farmers give a mixture of urea and TSP for both the first and the second application (Table 5.4b). TSP which is recommended to be used only for a basal application or for a first topdressing where no basal fertiliser is given, is broadcast at the second application by 84% of the farmers, for 5% of which it is the first TSP application on the crop. Even at a possible third application, 18% of the farmers again apply TSP. According to De Datta (1981), phosphorus

Table 5.5: Type of fertiliser application by non-IPM farmers classified by water supply system in 1990/91 wet season, as % farmers reporting.

	Irrigated	Rainfed
basal + 1 topdressing	2%	0%
basal + 2 topdressings	5%	0%
1 topdressing	1%	0%
2 topdressings	74%	41%
3 topdressings	18%	59%
N	*521*	*41*

applications later than fifteen DAT are not effectively used for grain production, which helps explain the inefficient use of TSP by Grobogan farmers. In the cases where KCl and ZA are used, most of these fertilisers are applied at the second application, but split applications also occur. Fertilisation patterns exactly following the recommendations were not observed.

Frequency and time of application

Regarding frequency and time of fertiliser application, different fertilisation practices apply to irrigated and rainfed crops, considering the unreliable water supply and the possibly higher fertility of the soil on rainfed lands since these are not cultivated continuously. On irrigated fields, the large majority (76%) of farmers give two fertiliser applications, either one basal application and one topdressing or two topdressings, 23% give three applications, and 1% only one application (Table 5.5). Most of the farmers who apply basal fertilisation usually give two topdressings, some only one. On rainfed fields, not less than 59% of the farmers apply fertilisers three times per season, whereas the remaining 41% applies fertiliser twice. No basal applications are given, probably because of the water supply and fertility conditions mentioned above.

Fertiliser is much sooner applied to irrigated than to rainfed crops. On average, the first topdressing on irrigated lands is given at 13 DAT (Table 5.6). The second and possible third applications are applied at 31 and 39 DAT respectively. The minority of farmers who give three applications generally apply the second treatment a bit earlier, and the ones who give two topdressings, a bit later than the average. On rainfed lands, the first application is, on average, given at 24 days,

Table 5.6: Average time of fertiliser application (in DAT) by non-IPM farmers classified by water supply system, 1990/91 wet season.

	Irrigated			Rainfed		
	1st appl.	2nd appl.	3rd appl.	1st appl.	2nd appl.	3rd appl.
average	13	31	39	24	46	96
st. dev.	6	9	12	11	17	13
N (users only)	*521*	*506*	*96*	*41*	*41*	*24*

the second at 46 days, and the third at 63 days after transplanting. The latter application coincides with the flowering stage of the crop which makes its efficiency questionable. There are no differences in the time of fertiliser application between the wet and the dry season.

The reason most frequently mentioned by farmers (35%) to apply the first fertiliser application between one and two weeks after transplanting is that then 'the rice plants and its roots just start to live', which means that the transplants are just well established and start to grow (Table III.14). A second reason, reported by 22%, is that then the plants are most actively tillering for which nutrients are needed. A similar proportion of the farmers (19%) select the appropriate time for the first fertiliser treatment to encourage a fast growth and a green colour of the plants. Not less than 16% of respondents could not mention any reason for their practice. Main reasons for the timing of the second fertiliser application include that it is the right moment for panicle initiation (27%), the right moment for replenishment according to the light green colour of the crop (26%), and it is the time just before weeding so that weeds are more fertile and easier to pull out (17%). A remarkable number of farmers in Senengsari (24%) had to postpone their planned second fertiliser application during the 1991 dry season because of the (temporary) shortage of irrigation water. The main reason reported for the timing of the third application is, again, to encourage panicle initiation.

Quantities of fertiliser

The average total dosage of urea, as reported for the 1989/90 wet season (258 kg/ha), is practically equal to the recommendation of 250 kg/ha (Table 5.7a). Some variation exists among the eight study villages in that Senengsari and Mulyoagung farmers give, on average, higher dosages, and Plosoksari farmers lower. High variation is observed among individual farmers. One third of farmers (34%) give less than 200 kg/ha, 37% within the range of 200-300 kg/ha, and 30% more than 300 kg/ha (Table 5.7b). Regarding TSP, the total average dosage (221 kg/ha) is more than twice as high as the recommended amount of 100 kg/ha. Again, Senengsari and Mulyoagung farmers give higher amounts than this average, and Jayasari farmers lower amounts, although still much more than the recommendation. The number of farmers giving more than 200 kg TSP/ha equals 51% of the respondents.

In all categories of farmers, whether under- or overdosing, the most frequently heard reason for applying a certain quantity of urea (36% of farmers) or TSP (32%) is that they have found out by experience what quantities are most appropriate (Table III.16). In addition, a considerable number of farmers, especially the ones overdosing, take the condition of either the crop, soil, water or weather into consideration in deciding how much urea to apply. Shortage of money is a reason not to give more fertiliser for 16% of the farmers who underdose urea and for 11% who underdose TSP. In general, farmers have a low opinion of the fertiliser recommendations conveyed by the extension worker. They experienced that these recommendations are usually not appropriate under the conditions they face in the field. More important might be that they do not trust the extension

Table 5.7: Fertiliser dosages applied by non-IPM farmers, 1989/90 wet season.

a. Average dosages (in kg/ha) as classified by village (users only).

	Seneng-sari	Sugih-sari	Jaya-sari	Plosok-sari	Kudu-agung	Sumber-agung	Mulyo-agung	Lempung-agung	Total
urea	312	249	232	229	262	264	286	244	258
TSP	271	197	199	200	215	230	250	209	221
KCl	122	135	111	90	97	85	106	80	109
ZA	114	129	117	77	76	81	52	55	95

b. Frequency distribution of dosages, as % farmers reporting (N=596).

	urea	TSP		KCl	ZA
< 200 kg/ha	34%	49%	0 kg/ha	81%	91%
200-299 kg/ha	37%	32%	1-99 kg/ha	9%	4%
300-399 kg/ha	18%	13%	100-150 kg/ha	6%	3%
≥ 400 kg/ha	12%	7%	≥ 150 kg/ha	4%	2%

c. Average dosages (in kg/ha) as classified by farm size.

	< 0.30 ha	0.30-0.49 ha	0.50-0.99 ha	≥ 1.00 ha
urea	273	268	243	231
TSP	241	224	204	193
N	*202*	*161*	*156*	*77*

worker because they think that this person has no practical experience (Santoso, 1992).

By classifying farmers according to farm size, it becomes visible that especially small farmers tend to overdose urea and TSP (Table 5.7c). The smaller the farm size, the higher the average dosage of urea, and the higher also the percentage of farmers applying quantities of more than 300 kg/ha. Farmers operating less than 0.5 hectare apply more than the recommended dosage, while farmers on larger farms than 0.5 hectare tend to underdosage.

Considering the behaviour of phosphate in the soil, and the ability of the rice plants to absorb only the amount of phosphate really needed, mainly at the vegetative stage (De Datta, 1981), the outcome of farmers' experiences seem quite illogical. Too high quantities of TSP have no visible adverse effect on the rice crop since most of the phosphate is adsorbed and bound by colloid compounds in the soil. Since phosphorous fertilisers have been used for more than twenty years in the study area, the soils are expected to be saturated, resulting in a high recovery fractions of phosphate fertilisers. In experiments at one of the Rural Extension Centres in the study area, where soil conditions are comparable to the villages studied, gradually increasing dosages of TSP between 25 and 150 kg/ha were applied. No yield increase was observed above a dosage of 75 kg/ha which is lower than the government recommendation (pers. comm. head of REC). This dosage corresponds with calculations based on the need for TSP of a rice crop yielding 5-6 tons/ha with high recovery fraction of phosphate.

Another explanation for the high quantities of TSP that are applied by farmers in the study area can be found in their perception of the concept of balanced fertilisation which is promoted by the Agriculture Service. The meaning of this concept is that the various nutrients needed by the rice plant, N, P and K, have to be applied in appropriate proportions. As became clear from group discussions and individual interviews, many farmers, however, believe that balanced fertilisation implies equal quantities of both urea and TSP. This belief is put into practice by a large majority, as is shown by the correlation between quantities of urea and TSP depicted in Figure 5.1. The misconception on balanced fertilisation causes great financial loss since expenses on fertiliser are one of the greater budget items for rice farmers.

The few farmers using KCl and ZA generally apply these fertilisers in quantities close to the recommendations (Table 5.7a). KCl and ZA users are either the better informed farmers willing to try out something new, or the farmers taking the KUD package for which the amount is determined by the standard package. Mistakes in calculating the amounts needed will definitely occur considering the inconsistent area units used locally. When KCl is underdosaged, this is mostly based on the notion that KCl only contains complementary elements, not a basic nutrient.

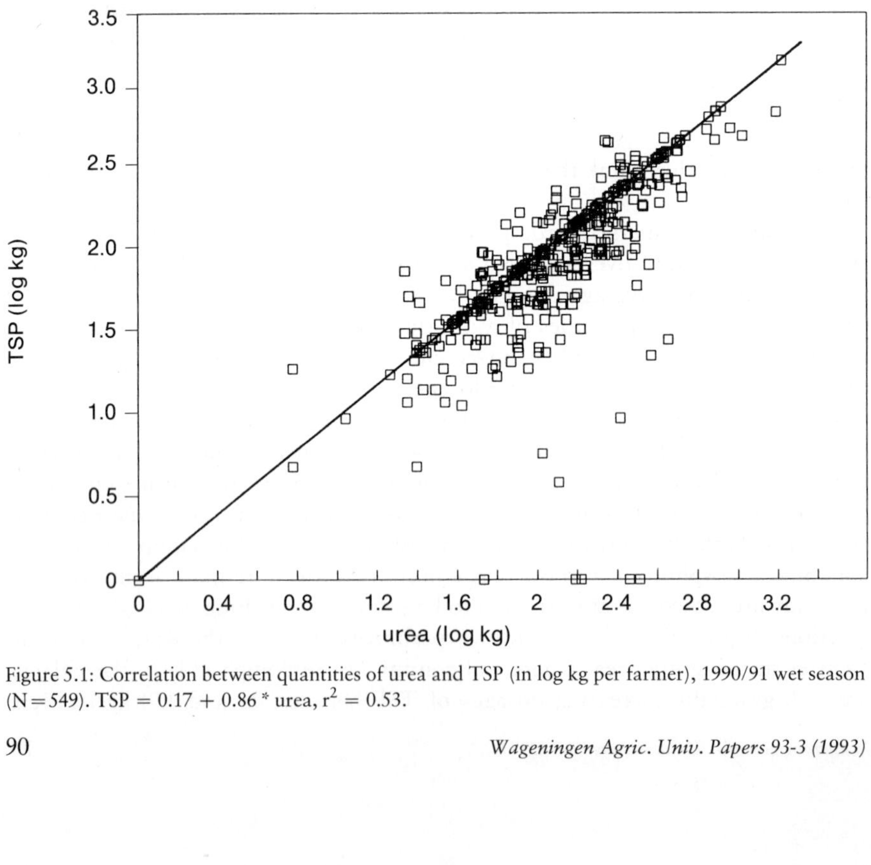

Figure 5.1: Correlation between quantities of urea and TSP (in log kg per farmer), 1990/91 wet season (N=549). TSP = 0.17 + 0.86 * urea, r^2 = 0.53.

An interesting phenomenon observed is that quantities of particularly urea, but to some extent also TSP, applied per hectare seemed to increase over the four seasons studied (Table III.15). This tendency was visible in all eight study villages. The average dosage of urea increased gradually to 304 kg/ha in the 1991 dry season, and of TSP to 248 kg/ha. The expected seasonal variation, in that relatively exhausted soils at the end of the wet season would require higher quantities of fertiliser during the following dry season, is not shown by farmers' fertilisation practices with crops, as was the case with seedbeds. Farmers have experienced that the soil is extremely fertile at the beginning of a wet season crop after a very long dry season, as was the case in 1991. They believe that this is caused by a better aeration made possible by the many cracks in the clayish soils over the long dry period, which is true especially regarding the inorganic nitrogen made available by a faster humus decomposition under aerobic conditions (Birch, 1958). Some say that the soil matures better under long periods of drought, and that decomposed elements are not washed off but stored in the soil.

Considering the fairly high yields of 5-6 tons/ha attained by the farmer respondents, and assuming that the soil (classified as vertisol with a pH of 6-6.5) in which the stubbles and the ash of the burnt straw are ploughed under, also provides some nutrients, calculations can be made on the nutrient supply needed by the crop. Rough estimates resulted in quantities of 200-400 kg urea/ha (when no ZA is used), 50-100 kg TSP/ha, and 50-70 kg KCl/ha. This shows that farmers' high applications of urea are not necessarily overdone. The method of application, however, might be an important indication for altogether inefficient fertiliser use. In all the study villages, farmers apply fertiliser to the crop without draining the field. A level of 5-10 cm of water is usually standing in the field at the moment of fertiliser application. This might result in great losses of nutrients, particularly nitrogen and potassium, which are washed away in diluted form with the ever flowing water, and, in the case of nitrogen, volatise. Farmers could not mention a reason why they leave so much water in the field at the moment of fertilisation, except that it is customary in the area. They do not have problems letting the water in again, with the exception of a few fields on somewhat higher locations, and at the occasional moments when water supply is not reliable. Broadcasting fertiliser into the standing water, which is considered to be a great source of nutrient loss, is not only a habit of Grobogan farmers but of most Asian lowland rice farmers (De Datta, 1987). Broadcasting urea in floodwater resulted in a 30% recovery, on average (Craswell et al., 1981). Higher yields were obtained with deeper placement of nitrogen fertilisers (Humphreys et al., 1987; De Datta, 1987). This method, however, implies a higher and different labour demand, although studies in East Java have shown that it is still economical under the prevailing wage rates (O'Brien et al., 1987).

Foliar fertilisers and manure

Foliar fertilisers containing N, P, K and/or microelements became part of the KUD package for rice intensification several years ago, although the brands change almost every season. The percentage of farmer respondents using these devices was negligible (2%) during the first survey seasons, but increased to 7% in the second seasons, and decreased again to 5% in the last season (Table III.17). Especially during the 1990/91 wet season, a surprisingly high number of farmers in the villages Sumberagung (28%) and Kuduagung (13%) used foliar fertilisers, which corresponds with the fact that relatively more farmers took the KUD credit package in that particular season (see also Box 2.1). The farmers using foliar fertiliser have no explicit opinion about the result of the treatments. Some say that application does not make any difference, others that the growth of the crop was good, but they are not too sure whether that is because of these additives.

Manure is hardly ever used in the study area, a result of the fact that there is not much cattle, mainly due to mechanisation of soil preparation. Farmers with cattle report that it costs too much labour to carry the manure to the rice field. The six (out of 263) farmers who started to use manure (again) in the 1991 dry season were satisfied with the result. In the same season, the KUD in the subdistrict of Sumberagung started an animal husbandry programme by giving loans to farmers to buy young cows to which quite a few farmers responded. These kinds of activities might encourage the use of manure on rice land in the future.

5.2.9 *Harvesting*

As was described in Chapter 4, some radical changes in harvest procedures took place recently after the introduction of the short, modern varieties. A clear distinction is visible between the two subdistricts studied, Sumbersari and Sumberagung, with respect to the type of wage paid for harvesting. A regional distinction is not surprising since harvesters usually also work outside the village boundaries, in which case consistent agreements in a certain region are desired. Since the introduction of the sickle, the wage in Sumbersari is paid in money which varies from Rp 25-35 per kilogram harvested paddy over the study period. In case of a lodging crop or a location far away from the house of the farmer, prices can rise to Rp 40-45/kg. When the harvesting season happens to fall in the fasting month, wages are also considerably higher. In Sumberagung, the wage for harvesting is still paid in kind which is commonly a 10% share. Sometimes, as in Mulyoagung, a lower share (8%) is paid when meals are provided to the harvesters, and higher shares (12%) are paid to labourers from outside the village. In this subdistrict, only traders who buy the standing rice crops ('*penebas*') use money wages to pay harvesters. Paying in kind is generally cheaper for the farmer than paying money, especially at the peak harvest season when rice prices are low. On the other hand, paying money is easier and less subject to fraudulent practices by the harvesters (NODAI, 1982). A farmer who has the facilities to store the yield, can wait to sell his or her paddy until prices are favourable, which is a contributive factor determining return (see Box 5.4).

After the harvest, the straw is mostly burnt in the field, thus providing the soil with phosphorus, potassium and micro-nutrients that remain in the ash of the burnt straw. The ash is usually not distributed over the field, causing that only a part of the field is favoured by this nutrient supply. This is often visible in the field from a locally better growth of transplants before the first fertiliser treatment.

In the last study seasons, an increasing activity was observed by people from other areas who collected straw for cattle fodder, which implies a serious output of nutrients from the area.

5.2.10 *Post-harvest practices*

Part of the harvesting labourers' task is to bring the threshed paddy to the house of the farmers. After that, it is the responsibility of the farmer to process the yield. Many farmers, in need for cash at the end of the growing season, sell a part of the wet paddy right away to traders. This is particularly common for people who have no suitable place to dry and store the paddy, i.e. mostly the poorer people with a small house and yard. Selling wet paddy, however, is usually less profitable especially since prices are low at the peak of the harvesting season. So people who can manage, which means people who have access to a drying floor, storage and labour, will process the greater part of their wet paddy first and wait to sell it until either good prices are offered or money is urgently needed. The practice of selling paddy at the moment money is needed is common in the study area.

Drying of the paddy is in most cases done by the family members themselves, especially the women. Labour needed for this practice is seldom counted and is, indeed, difficult to count since it is scattered over the day. But if an opportunity cost is calculated, as was done for the nine farmers who kept daily records during the 1990/91 season, the labour needed for drying the paddy equals almost 5% of the total costs for rice cultivation.

5.2.11 *Labour distribution*

A general picture of the labour task distribution in the study area includes:

a. family labour, man:
- hoeing and repairing the bunds;
- preparing the seedbed and the field for transplanting;
- pulling the seedlings out of the seedbed and distributing the bundles of seedlings over the field;
- water management;
- monitoring of the crop;
- broadcasting fertiliser;
- pest control, including occasional weeding;
- supervising the harvest;
- selling the harvest;

b. family labour, woman:
- transplanting;
- weeding;
- monitoring of the crop;
- providing food for labourers in the field;
- drying the harvest;
- selling the harvest;

c. hired labour:
- ploughing and levelling the field (tractor operator);
- hoeing;

- transplanting;
- weeding;
- harvesting and threshing.

It is obvious that many variations to this generalised picture are possible. Especially on very small farms, almost all the work is done by husband, wife and children together, where the man helps transplanting and the woman helps cutting the rice with a sickle. No women have been observed in the study area spraying pesticides, which is not uncommon in North and West Sumatra, but they are often in charge of preparing and placing poison bait for rat control.

5.3 Pest management

The fauna of the wet rice land and its transitions through the seasons in a natural situation without the use of chemical inputs were described in detail early this century (Koningsberger, 1915), of which some findings will be described below. It was observed in these undisturbed ecosystems that before the rice plants emerge in the field, a large diversity of insects is already present in water and soil. Mosquito larvae are the first inhabitants after water is led into the field, soon followed by predatory dragonflies, water beetles, water bugs, small crabs and frogs that live initially from the mosquito larvae. By the time the rice is transplanted in the field, attracting rice-eating insects, the population of natural enemies is so large that theoretically insect pest populations have no chance to develop, and will be kept at low levels not causing any damage of importance to the crop. Brown planthopper populations, for instance, breed and multiply very rapidly but they are decimated with the same speed through predation by ladybird beetles (*Coccinellidae* sp.). This strong controlling ability of natural enemies can only be effective when the ecosystem stays in balance, without any external disturbance from, for instance, extreme weather conditions or human activity.

Although human disturbance is inevitable in a highly controlled cultivation system such as irrigated rice, the most fatal disturbance to the natural enemy population in the rice ecosystem began with the introduction of pesticides. In Grobogan, this happened a few years prior to the introduction of the high yielding varieties, fertilisers and pesticides in credit packages. Pesticides like Endrin and DDT had been sprayed on traditional rice varieties in Mulyoagung, causing a brown planthopper outbreak in 1968, which was 'controlled' by aerial spraying. The introduction of the new varieties in 1969 was a relief for the farmers since these varieties were, initially, disliked by the pest. But soon farmers had to cope with many more pest problems than they were used to before they started to use pesticides.

When talking about indigenous pest control knowledge and practices with older rice farmers, they always mention that when there were no pesticides, there were no (serious) pest problems either. The only insect that was considered a pest in those days, and for which traditional control practices existed, was the rice seed bug (*Leptocorisa* sp.). Koningsberger (1915) confirms that the rice seed bug was the only pest of economic importance in traditional rice growing, which he explains by the fact that this stinky bug has hardly any natural enemies. A second pest in traditional rice growing mentioned by both old farmers and the old Dutch literature are rice rats (*Rattus* sp.), in older days reasonably controlled by snakes and birds of prey.

Considering the increased number of important pests and their intensity since the intensification of rice growing, it is obvious that much has changed in farmers' pest management practice. The following sections describe farmers' current perception and knowledge about pests, natural enemies and pest management, and their control practices.

5.3.1 *'Hama jangan sampai merajalela' – Farmers and pests*

Perception

'Don't let pests break out' is an often heard expression from rice farmers in the eight villages studied in Grobogan, which clearly indicates the prevalent attitude towards pests in the rice crop: one pest insect in the crop is the forerunner for many more, if not for an uncontrollable outbreak. This fear for pests is definitely related to the fact that the consequences of pest outbreaks have already been felt several times: brown planthopper in 1968, 1975, 1979, and 1985, gall midge in 1973, 1978 and 1990, stemborer in 1982 and 1990, a list which is not exhaustive and does not include the continuous threat by rats.

In group interviews conducted during the 1990 dry season, farmers ranked rice pests according to their damaging ability and to the difficulty to control. Rats were qualified as the most damaging by far, and the most difficult pest to control. Almost equal qualifiers regarding damaging ability were used for rice seed bugs and brown planthoppers. Farmers' experience with these two pests is totally different in that rice seed bugs cause damage in (almost) every crop but usually at a low level, whereas a brown planthopper attack occurs only occasionally but with disastrous effects. The control of rice seed bugs was considered not as difficult as the control of brown planthoppers. The fourth rice pest of importance as perceived by the farmers was rice stemborer.

Deadhearts, the symptom occurring after stemborer attack of young, vegetative rice plants, was found more damaging and more difficult to control than attack at the reproductive stage of the crop (whiteheads). This statement seems contradictory since whiteheads can not be cured any more, and the yield of the whitehead panicles is zero. On the other hand, before the stemborer outbreak of 1990/91 which caused merely whiteheads, farmers in the study area had never encountered such a heavy whitehead damage. The low level of damage experienced almost every other season is apparently not perceived as causing a loss, which is understandable considering the ability of rice plants to compensate. Lower ranked pests mentioned in the group interviews included caterpillars and mole crickets.

Knowledge

Farmers' knowledge about rice pests was tested in the baseline survey by asking them to describe the appearance and symptoms of four major insect pests (brown planthopper, stemborer, rice seed bug and rice leaffolder) and of one leaf disease (bacterial leaf blight). Three-fourths of the farmers know the symptoms of brown planthopper infestation (hopper burn), and almost all of these farmers (72% of the total) know that this hopper burn is caused by a small insect sucking at the stem of the rice plant (Table III.18). Although the causal insect seldom occurs in observable numbers, the experience of one or two brown planthopper outbreaks in the past has left the farmers with adequate knowledge about this insect, more than they posses about stemborers which occur every season.

Stemborer occurrence is usually addressed by Javanese farmers with the local names for the symptoms of the infestation, the equivalents for 'deadheart' and 'whitehead'. The cause of deadhearts is understood by only 34% of the farmers, whereas 88% can describe the symptoms. Whitehead occurrence seems to be somewhat more comprehensible, maybe because the white panicles can be easily pulled out after which the caterpillar frass is obvious at the lower part of the stem. Almost all farmers (95%) can describe whiteheads, and a small majority (59%) know the causal insect.

The best known pest by rice farmers in the study area is the rice seed bug, known by 99%. Reason for this comprehension is that this insect is clearly observable even from the bunds because of its size and its bad smell which is also described in the Javanese name for the pest. In addition, it is a regular visitor of rice crops especially at the reproductive stage.

Farmers have more problems in describing rice leaffolder and bacterial leaf blight. Only 5% of the farmers surveyed know the symptoms of rice leaffolder infestation, out of which 3% also knows that a leafeating caterpillar causes the damage. Although rice leaffolder is a common pest, the identification problem might be due to the fact that it is not considered an economically important pest. Additionally, farmers have no common name for the symptoms they observe in the field, or for the caterpillar. When talking with farmers about the white-coloured, rolled leaves, most of them know that it is a caterpillar that causes the damage. Seldom, however, do farmers observe the damage more closely to discover that most of the affected leaves do not have caterpillars any more. This leads them to wrong decisions to apply pesticides since so many leaves seem to be affected.

A high percentage of farmers (98%) does not know about bacterial leaf blight. This is not surprising since it is not a very common disease in the area and, if occurring, is seldom of economic importance. Additionally, farmers have no common names for any plant disease other than describing the colour changes it brings about. An exception is the tungro virus that occurred in the past, resulting in the fact that many disease symptoms are called 'tungro'.

From these examples, a strong relation is visible between the experience of farmers and their knowledge about pests. Pests having caused major problems in the past, and easily observable pests score higher than those that are common but less visible, either in terms of effect on the crop, or in appearance. The eight villages surveyed showed some variation in knowledge levels of farmers, which could mostly be traced back to local variation in previous occurrence of pests.

Brown planthopper, for instance, is best known in the villages where outbreaks in 1985-86 were severest (Kuduagung and Lempungagung). The knowledge test was repeated in the 1990/91 wet season survey, not showing any major differences, though (Table III.18).

5.3.2 *'Kinjeng iku mung iseng-iseng wae' – Farmers and natural enemies*

'Dragonflies just hang around' is an expression of some farmers during a group interview in Mulyoagung discussing natural enemies. Another one is about frogs that are thought 'only to be jumping to and fro all the time'. Fortunately, these two expressions do not represent the general opinion of farmers about natural enemies. Almost half of farmers interviewed in the baseline survey (48%) know about animals living in the rice field that eat pest organisms (Table 5.8). Natural enemies known best include snakes (mentioned by 28% of farmers), birds (15%), and frogs (15%). Only a few respondents can name spiders, beetles and dragonflies as predators of pests. Parasitism is a completely unknown and never observed phenomenon for farmers, as is the occurrence of insect pathogens.

In group interviews, in which farmers are encouraged to discuss their field experiences together, more detailed observations on predation of pests are expressed. The best known example is snakes eating rats, although many people are concerned about the current snake population since these animals are wanted objects for hunters. The same holds for frogs which are caught and sold for their meat, whereas some farmers have observed that frogs eat planthoppers, one of the most feared rice pests. In addition, spiders have been observed eating seed bugs and planthoppers, water bugs eating caterpillars, and dragonflies and salamanders chasing all kinds of insects. The greater the animal involved, and therefore the easier to observe, the more farmers are acquainted with its behaviour.

Although quite some farmers, after probing, seem to be aware of different functions for different creatures in the rice field, the fact that pesticides also kill the

Table 5.8: Non-IPM farmers' knowledge about natural enemies, as % farmers reporting.

	Seneng-sari	Sugih-sari	Jaya-sari	Plosok-sari	Kudu-agung	Sumber-agung	Mulyo-agung	Lempung-agung	Total
Mention:									
snakes	42%	22%	48%	9%	18%	40%	12%	35%	28%
birds	14%	6%	25%	5%	16%	16%	9%	25%	15%
frogs	23%	13%	27%	8%	9%	16%	8%	13%	15%
spiders	2%	2%	4%	1%	1%	3%	0%	1%	2%
beetles	5%	2%	1%	0%	0%	0%	0%	0%	1%
dragonflies	2%	0	4%	0%	1%	2%	1%	0%	1%
other[1])	30%	10%	26%	13%	18%	19%	13%	18%	18%
Do not know	33%	67%	31%	73%	57%	44%	72%	39%	52%
N	66	66	66	66	66	66	66	66	66

[1]) 'Other' includes cats (8%), dogs (6%), ducks (2%), salamanders, chickens, mongooses, fish, crabs and crickets.
Totals exceed 100% due to multiple response.

useful animals in the field, and, therefore, encourage the development of pest populations was a new perspective for most of them. The concern that snakes and birds of prey are hunted by (other) people resulting in high rat populations, however, exists. The same has been observed in IPM trainings where the participants initially have some knowledge about predation in the rice field, although often limited to the bigger animals, but never realised the effects of pesticides on the natural enemies and, thus, on the pest populations. The only effect of pesticides thought to be negative for the rice ecosystem and often mentioned by farmers is the death of many earthworms after the application of carbofuran granules.

5.3.3 'Sedia payung sebelum hujan' – Farmers and pest management

Perception

'Have an umbrella ready before it starts raining', although often heard, does not exactly describe the actual perception of many farmers towards (chemical) pest control. A more adequate expression would be: 'Use an umbrella even when it does not rain', explaining the confidence that many farmers have in preventive pesticide use imposed by the rice intensification programmes that farmers were exposed to for over twenty years. This seems to be especially true in the case of granular pesticides (carbofuran and, less frequently, quinalphos) which is mostly applied mixed with fertiliser. Labour can be saved when the pesticide is broadcast at the time of the more or less compulsory fertiliser application, but this implies a granular pesticide application without the assurance that it is needed and timely.

From group interviews with rice farmers discussing chemical pest control, it became clear that two opinions prevail. Some farmers believe that applying pesticides is always necessary even if there are no pests in the field, which is contradicted by others who only spray after having observed pests in the crop. The level of pest intensity on which decisions for chemical control are made, seems to be arbitrary and is generally very low. This is understandable considering farmers' perception (and fear) of pests. Farmers associate pest problems strongly, if not exclusively, with chemical solutions, preventively or curatively. This attitude of strong dependence on pesticides is confirmed by the fact that only 10% of the farmers surveyed choose to save on pesticides in case they have to economise on inputs for rice cultivation, whilst larger proportions prefer to save on labour (33%), borrow money (25%), or save on fertilisers (15%).

Another aspect influencing farmers perception of chemical pest control can be found in their terminology for pesticides. Pesticides are commonly called '*obat*', which literally means 'medicine'. 'Spray medicines' ('*obat semprot*') are distinguished from 'broadcast or soil (granular) medicines' ('*obat tabur/tanah*'), but the function of these poisons is, in principle, not distinguished from medicines for health. More than half of the farmers (57%) believe that pesticides do no harm to people (Table III.19). The high-pesticide users in Sugihsari score highest (75%). Farmers knowing about poisonous effects of pesticides mostly mention headaches and vomitting as observed symptoms. Farmers who apply pesticides more intensively usually consider them as not dangerous to their health (pers. comm. vegetable farmers in North Sumatra; Hussein, 1987). Apparently, poisoning symptoms followed by a quick recovery are not considered harmful, or are accepted as something inevitable related to pesticide use. The danger of pesticides for the environment, in particular their effects on other animals, is realised by only 37% of the farmers interviewed, of whom a low percentage in Plosoksari (20%) and Mulyoagung (24%).

In the case of pest problems, 59% of the farmers seek advice from their fellow farmers, whereas 21% (partly) rely on a village extension worker and 8% on themselves (Table III.20). The proportion of farmers relying on the extension worker is relatively low in Senengsari and Plosoksari, and high in Sugihsari and, not surprisingly, Kuduagung.

Practice

When looking at farmers' pest control practices (excluding rat control), the apparent attitude favouring prophylactic chemical control is not put into practice as often as would be expected from their views expressed in group discussions. In the eight study villages, as surveyed in the baseline season, pesticides are given at an average frequency of 2.2 applications per season (Table III.21a and Figure 5.2). These applications consist of 1.7 pesticide spray, and 0.5 granular pesticide application. The sprays are applied by 74% of the farmers, and granules by 33%, whereas in total 20% of the farmers do not use pesticides at all (Table III.21b). Thirty percent of the farmers give one or more preventive sprays, and 25% one or more preventive granule applications (Table 5.9). Spraying pesticides after pests have been observed in the field is practised by 60% of the farmers, whereas granules are curatively applied by 8%. Some farmers apply both preventively and curatively, and/or both sprays and granules.

In practice, various considerations seem to be decisive not to use pesticides preventively. An often mentioned reason by farmers is the lack of cash money to buy the chemicals. Somehow, when pests occur, the money needed can be made available, usually by selling paddy or borrowing from relatives or neighbours. Another, more likely reason for the contradiction between farmers' confidence in preventive pesticide use and their practice is that what they express in meetings (even when no village extension worker is present) does not correspond with their real opinion, as became clear after many individual discussions with farmers. Probably afraid of losing face in front of their fellow farmers, they express what is supposed to be the public opinion, i.e. what they have heard for many years from the village extension worker and on the radio broadcasts. This is not necessarily, and even not likely, their own opinion, which is openly expressed only in an individual discussion. In their own situation farmers reckon with many more aspects that influence their decisions, above all previous experiences including failures with pesticides.

In an attempt to understand who the high-pesticide users are, farmers giving four pesticide applications or more per season (19% of the farmers during the 1989/90 wet season) were separated from the ones giving less than four applica-

Table 5.9: Insect pest control practices by non-IPM farmers over four seasons, as % farmers reporting.

	Seneng-sari	Sugih-sari	Jaya-sari	Plosok-sari	Kudu-agung	Sumber-agung	Mulyo-agung	Lempung-agung	Total
1989/90 wet season									
Preventive measures:									
pesticide sprays	23%	38%	47%	25%	19%	26%	24%	36%	30%
pesticide granules	9%	8%	41%	21%	9%	24%	28%	47%	25%
Curative measures:									
pesticide sprays	27%	48%	86%	71%	79%	66%	48%	52%	60%
pesticide granules	0%	0%	5%	5%	7%	23%	5%	21%	8%
No control measures	55%	25%	1%	20%	7%	15%	26%	15%	20%
N	*66*	*63*	*85*	*85*	*68*	*62*	*85*	*85*	*599*
1990 dry season									
Preventive measures:									
pesticide sprays	18%		41%		42%		33%		34%
pesticide granules	26%		41%		58%		62%		48%
Curative measures:									
pesticide sprays	32%		66%		70%		52%		56%
pesticide granules	11%		13%		25%		24%		18%
No control measures	35%		6%		3%		9%		11%
N	*62*		*85*		*67*		*85*		*299*
1990/91 wet season									
Preventive measures:									
pesticide sprays	41%	58%	56%	73%	74%	40%	13%	44%	49%
pesticide granules	19%	51%	40%	30%	56%	52%	27%	46%	39%
Curative measures:									
pesticide sprays	12%	33%	35%	34%	28%	41%	32%	54%	34%
pesticide granules	2%	4%	0%	5%	5%	12%	4%	10%	5%
No control measures	41%	7%	21%	15%	8%	17%	48%	19%	23%
N	*59*	*55*	*80*	*74*	*61*	*58*	*82*	*80*	*549*
1991 dry season									
Preventive measures:									
pesticide sprays	2%		0%		0%		0%		0%
pesticide granules	15%		36%		62%		59%		44%
Curative measures:									
pesticide sprays	13%		25%		23%		17%		20%
pesticide granules	4%		1%		2%		26%		9%
No control measures	64%		28%		22%		20%		32%
N	*55*		*72*		*60*		*76*		*263*

Totals exceed due to multiple response.

tions. Data of these two groups were compared with regard to several characteristics and practices. Most of the high-pesticide users live in the villages Jayasari and Lempungagung, and they can hardly be found in the villages Senengsari and Sugihsari. Proportionally more farmers in this group operate large farms, usually consisting of fields owned by themselves and additional fields that are leased. Relatively more high- than low-pesticide users have enjoyed secondary education (also an indication for affluence) and somewhat more of them are aware of natural enemies in the rice field, and of the negative effects of pesticides. The most remarkable difference between the two groups is the contact with the extension worker in that relatively more high-pesticide users, especially in Jayasari, often meet an extension worker, and regularly visit farmer group extension meetings. This is an(other) indication that especially the larger, wealthier farmers are exposed to the messages from the extension service.

In selecting a type of pesticide, many farmers (36%) rely on their own experience (Table III.22). Another 11% of farmers report to be advised by their fellow farmers, whereas 8% is advised by the pesticide salesman and 6% by the village extension worker. Not uncommonly, farmers do not know the type of pesticide they have used, or have their own names for it. In interviews, types are mentioned that have not been available in the market for years, and bottles or packages to check the label are thrown away. Because of this unreliability, types of pesticides used by farmers in the study area are not presented here quantitatively. Several pesticide salesmen admit that farmers often come to buy old brands that are no longer marketed, in which case they sell one of the pesticides allowed for rice by INPRES 3/86. It has become clear, however, from both farmers and pesticide salesmen that a lot of chemicals actually banned for rice are still used.

No other, for instance mechanical or traditional, measures are reported by the Grobogan farmers to control rice pests, except for rat control which is discussed separately below. Regarding pest observations, most farmers (62%) look at the pest problems in their rice field from the bunds. Twenty-one percent of them will step into the field as soon as they detect some symptoms of damage. The group of farmers who have the habit of always closely observing their rice plants in the field is only 36%.

Seasonal variation

Small variations in farmers' chemical control behaviour are visible over the four seasons studied. The average number of pesticide applications slightly increased in the 1990 dry season to an average of 2.5 applications, and decreased in the 1991 dry season to 1.7 applications (Table III.21a), corresponding with increasing respectively decreasing pest populations during these seasons. Twelve percent more farmers did not apply pesticides at all in the last survey season compared to the baseline (Table III.21b). Pesticide use did not increase in the 1990/91 wet season in which a yellow stemborer outbreak occurred, because the damage appeared so suddenly and everybody knew that nothing could be done once whiteheads were there.

The variation over seasons is somewhat more obvious if we look at the various

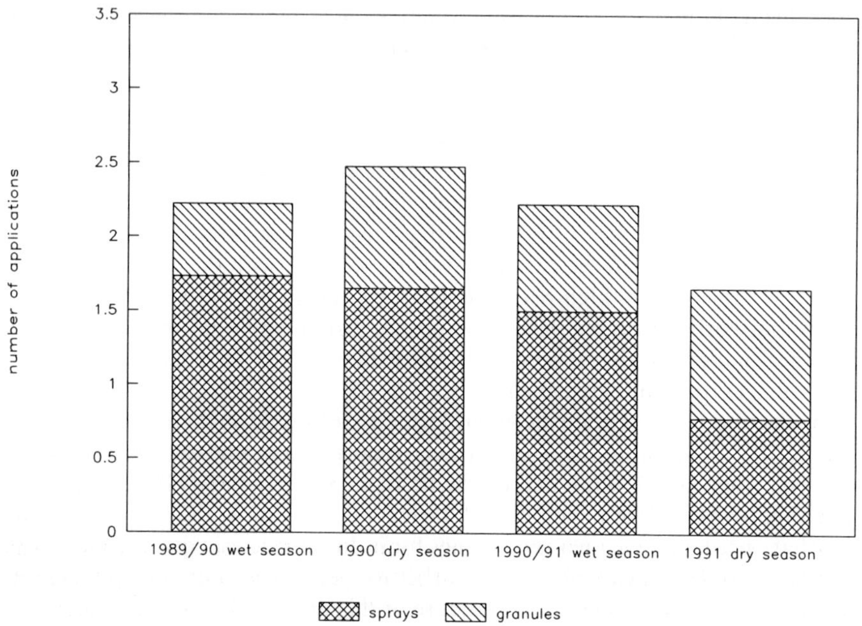

Figure 5.2: Average frequency of pesticide use (in number of applications per farmer per season) over four seasons.

types of pesticide application, distinguishing spray versus granule, and preventive versus curative applications (Figure 5.2). A tendency to reduce pesticide sprays is visible, particularly curative applications (Table 5.9). This trend is especially observed in the intensively studied villages, both those given IPM training in the 1990 dry season and non-IPM, which suggests an influence of the study activities on its outcomes. The application of preventive spraying initially increased to come back to almost the same level in the last survey season. In Plosoksari and Kuduagung not less than 73%, respectively 74%, of the farmers sprayed prophylactically during the stemborer outbreak season, and, whether by coincidence or not, exactly these two villages were the places most seriously affected by this pest.

The use of granular pesticides increased over the four study seasons, from 0.5 applications/season in the 1989/90 wet season, to 0.9 in the 1991 dry season. This increase is mainly due to an increasing number of farmers using granular pesticides, although some consistent seasonal variation is visible in that more farmers apply granules in dry than in wet seasons. More granule applications means more preventive applications, again especially during the two dry seasons (1990 and 1991), practised by 48% respectively 44% of the farmers surveyed. A reason for (preventive) granular pesticide use at the beginning of a dry season is to prevent the development of pests surviving in the field when planting immediately follows the precedent wet season rice crop. The strong increase during the 1990 dry season is most probably caused by more promotion for granular pesticides by the KUD, kiosk and village extension workers during this season.

Variation among villages
A large variation among the eight study villages regarding farmers' pest control behaviour is observed, in which Senengsari and Jayasari form the two extremes. During the baseline season, 45% of the farmers in Senengsari used chemicals in contrast with 99% of the farmers in Jayasari (Table III.21b). The average frequency of pesticide application (Figure 5.3 and Table III.21a) is much higher in Jayasari (3.4 applications/season) than in Senengsari (0.9 applications/season), although this difference is not only due to the larger number of farmers applying pesticides in Jayasari, but also to relatively more farmers applying three or more applications per season rather than only one or two in Senengsari. According to the farmers interviewed, the low pesticide use in Senengsari is partly a result of the difficult financial situation several people in this village always find themselves in, which corresponds with the profile portrayed for Senengsari in Box 4.2.

A remarkable decrease in pesticide frequency was observed in Jayasari over the four study seasons, ending at 1.8 applications in the last survey season. This decrease is assumed to be the result of IPM activities in the area (although no training was organised in Jayasari), and the influence of the activities of this evaluation study, mainly through the village study assistant. The same phenomenon is observed in the intensively studied non-IPM village Mulyoagung, but not in the control non-IPM villages Plosoksari and Lempungagung, which indicates that the study activities were probably the main influence in this respect. In the intensively studied villages, village study assistants who regularly monitored rice fields, might have had some role in disseminating IPM principles. Most likely, farmers who already have the habit of observing their crop before spraying are receptive to the message that pesticide sprays can be reduced.

In Kuduagung and Sumberagung a special case of pesticide promotion is present, the village extension worker who sells pesticides at his house, often repacked without labels explaining the name of the ingredient and user instructions. This person, originating from estate crop extension where an extra income from commissions on the frequently used pesticides is normal, does not hesitate to give an incorrect advice to rice farmers to promote his sales (Box 2.1).

Rat management
Rat management forms a topic by itself, since rats require an other approach than insects and pathogens. Since rats are strongly adaptive animals that can move over long distances, effective rat management should be implemented collectively, continuously and over a large area (Van de Fliert et al., 1993).

In the subdistrict Sumbersari and Sumberagung, as in most of the district Grobogan, rats are a continuous threat to the rice crop. Farmers in all study villages perceived rats as a major pest (Table III.1). Many farmers call rats the most difficult pest to control. This difficulty lies especially in the need for collectivity and continuity of rat management. It is, apparently, more difficult to organise and coordi-

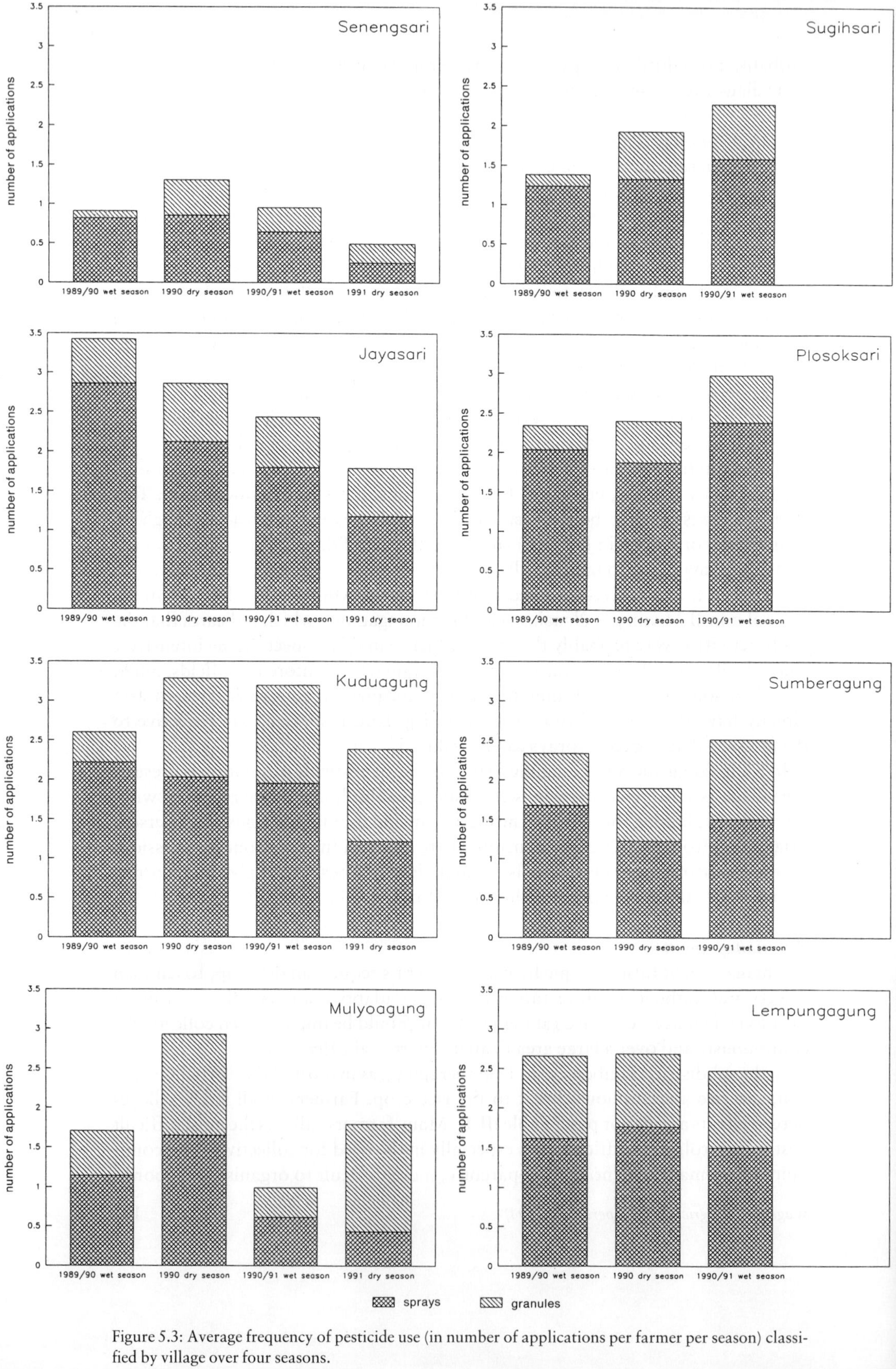

Figure 5.3: Average frequency of pesticide use (in number of applications per farmer per season) classified by village over four seasons.

nate farmers than to kill a rat. Control measures practised by farmers include poison baiting, rat drives, and fumigation using sulphur. Sometimes, seedbeds are protected from rat attack by fencing. Usually, these control measure are taken individually. In some villages, like Senengsari during the 1990/91 wet season and the following dry season, and Jayasari usually at the start of every season, collective rat drives are organised by the village government. In Jayasari, also collective baiting is arranged by the village officials. Rat poison and bait are initially bought from the village treasury after which a contribution proportional to land size from all farmers is required. In Mulyoagung, collective rat drives by using large nets are organised occasionally when rat populations have become too large, a moment when it is usually too late to prevent considerable yield loss.

Most of the rat control measures are applied by farmers after rat injury has been observed. Almost half of the farmers (49%) surveyed in the baseline season go out to hunt for rats after damage occurred, whereas 21% hunts for rats preventively, usually before planting (Table III.23). An empty field, indeed, facilitates the hunt for rats, and rat burrows in the bunds can be dug out when there is no crop yet. The second favourite control measure is poison baiting practised by 46% of the farmers curatively and 15% preventively. The favourite 'rodenticides' are zinc phosphide and aldicarb, two acutely working and very poisonous chemicals. The latter, better known as Temic, is actually a nematicide, sold as a poison to kill wild pigs. The less dangerous, slow-working anti-coagulant rodenticides are hardly ever used by Grobogan farmers, except when donated by the Agriculture Service in cases where high rat populations are reported by the village extension worker. Averaged over the eight study villages, 1.4 rodenticide applications were given in the 1989/90 wet season (Table III.24). High variation, however, is visible among and within villages, an extreme being a few farmers in Senengsari and Jayasari who place poisoned bait in their fields every two or three days throughout the season. Fumigation is practised by only a small number of farmers (10%), mainly in Senengsari and Jayasari, where either a few people or the village own fumigators.

In the 1990 dry season, rat populations were quite high which was an inheritance of the previous wet season. This resulted in a much higher activity of farmers to kill rats. Up to 78% of the farmers went out for rat drives, and even 86% put out poison bait. Average frequency of rodenticide application increased to 6.4 during this season. The 1990/91 wet season started after a four-month rice-free period, during which many rats died of starvation or migrated to other areas. It was a relatively quiet period regarding rat occurrence and, consequently, rat management. An inevitable increase in the next dry season took place, but not as drastic and damaging as the year before. Although most farmers are always on the alert for rats and do something to keep the population low, a response of farmers' rat control practices to actual rat occurrence is obvious.

5.4 Production

Yield

Yields obtained over the four study seasons reflect the description of the seasons given in section 5.1. The 1989/90 wet and the 1990 dry season in which rat infestation occurred, both resulted in an average yield of 5.07 tons wet paddy per hectare (Table 5.10), a level that equals the provincial average for irrigated rice. The 1990/91 wet season, with serious yellow stemborer outbreak later in the season causing whitehead symptoms, resulted in an average yield of only 4.11 tons/ha. The villages most seriously affected during this season (Plosoksari and Kuduagung) did not even obtain an average of 3 tons/ha, and almost one-fifth of the farmers there harvested less than 1 ton/ha (Table III.25). In spite of farmers' fear for another stemborer outbreak, the 1991 dry season had almost no pest damage which, in combination with a timely irrigation water supply and good weather conditions, resulted in relatively high yields of 5.92 tons/ha, on average. This last average is the highest of all seasons. The low standard deviation shows that it was an allround good season with little risk as compared to the previous seasons.

Even better, not to say extraordinary, conditions were experienced in the 1991/92 wet season. This season was preceded by an extremely long dry period which, according to the farmers, always brings about a much more fertile soil. In addition, weather conditions were optimal with not much rain and a lot of sunshine, irrigation water was provided timely and sufficiently, and there was no pest infestation of any importance. Although no survey data are available for this season, a sample of eighty fields in the four intensively studied villages (Senengsari, Jayasari, Kuduagung and Mulyoagung) monitored during this season showed an average yield of not less than 8.29 tons/ha (± 1.35). This has been the best season within living memory of the farmers in Grobogan.

Yield variability

Yield variability depicted by means of a frequency distribution is an important

Table 5.10: Average yields (in tons/ha) by non-IPM farmers over four seasons.

	Senengsari	Sugihsari	Jayasari	Plosoksari	Kuduagung	Sumberagung	Mulyoagung	Lempungagung	Total
1989/90	5.60	5.51	4.09	3.99	4.72	5.78	6.12	5.44	5.07
N	*62*	*63*	*85*	*83*	*67*	*62*	*85*	*85*	*596*
1990	4.99		5.21		4.31		5.59		5.07
N	*62*		*83*		*67*		*85*		*297*
1990/91	3.58	3.90	4.65	2.72	2.66	4.58	5.88	4.36	4.11
N	*58*	*52*	*79*	*74*	*61*	*58*	*80*	*79*	*541*
1991	5.85		5.87		5.52		6.36		5.92
N	*52*		*66*		*57*		*71*		*246*

variable to show risk in rice cultivation. The steeper the curve, and the narrower and higher the peak of the curve, the smaller is the risk during a particular season or in a particular area. Factors determining this risk can include environmental conditions and/or farmers' practices. Yield variability displayed for the four study seasons shows seasonal variation, and for four study villages (the two worst and the two best) during the disastrous stemborer season of 1990/91 shows local variations (Figure 5.4).

A comparative analysis of the yield variability diagrams for the four seasons (Figure 5.4a) leads to the conclusion that the 1991 dry season was the least risky, followed by the 1990 dry season. The two wet seasons show a broader yield distribution with lower peaks standing for higher variability. This high variability is assumed to be mainly due to the variation in pest infestation in combination with farmers' inability to control the pests adequately. This explanation also holds, to some extent, for the 1990 dry season diagram since pest attack, in particular rats, was reported by a large majority of farmers. The 1991 dry season diagram represents the variability that exists when there is no yield loss of importance due to pests. This variability is likely to be caused by differences in local conditions regarding soil composition and water supply, and by differences in farmers' practices.

Variation in yield variability among the eight villages studied over four seasons can, in most cases, be explained by local differences in pest infestation or timeliness of water supply and, thus, of the planting season. Jayasari and Plosoksari, for instance, achieved low yields during the 1989/90 wet season because of relatively high rat or stemborer damage, whereas during the following dry season yield reduction was caused by rats and caterpillars in Kuduagung. A somewhat earlier water supply and consequently earlier planting, although still late, could prevent serious yield loss in Mulyoagung and Lempungagung during the stemborer outbreak season of 1990/91, although less damage was also observed in Jayasari and Sumberagung due to local variation in occurrence of the pest. A comparison of the yield variability diagrams of the two worst and the two best villages for this season (Kuduagung and Plosoksari versus Sumberagung and Mulyoagung, Figure 5.4b), shows that the situation in Plosoksari was subject to highest variability, and riskiest considering the peak of the curve at a low yield level and the broad shape of the curve. The earlier planting in Mulyoagung which resulted in less stemborer damage is reflected in its relatively narrow, steep curve. Sumberagung although with an average yield nearly equal to Mulyoagung, shows a much higher variability, and thus more difference in local conditions mainly caused by stemborer.

Factors determining yield

It is of interest to identify factors that determine the level of productivity in rice growing, and especially to know whether such factors can be influenced by optimising farmers' practices. Adequate quantifiable practices supposed to have an impact on the production level, which can be used in regression statistics, include fertiliser quantities (in particular urea and TSP since KCl and ZA are only

a. Seasonal variation:

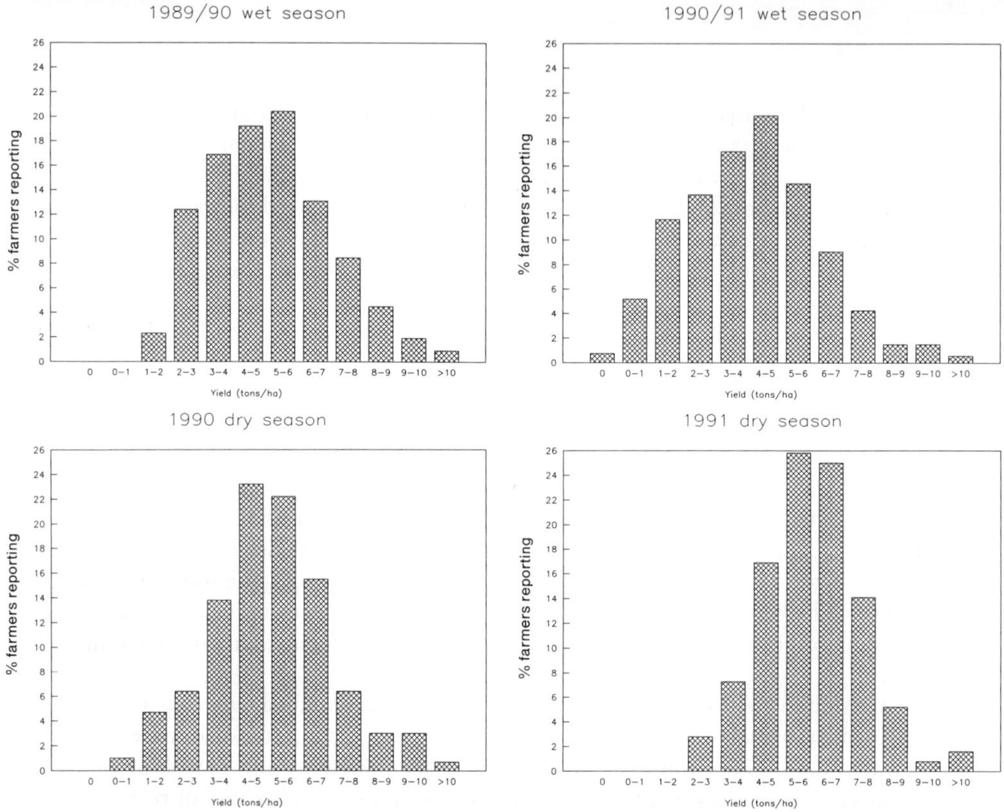

b. Local variation:

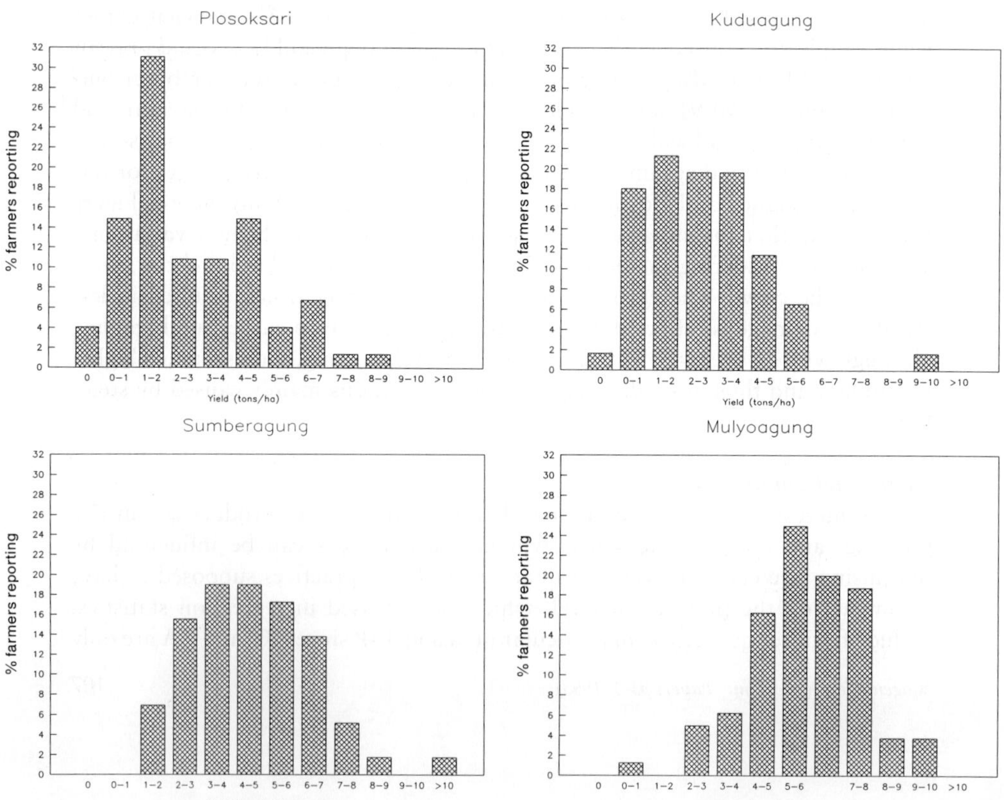

Figure 5.4: Yield variability over four seasons, and over four villages in the 1990/91 wet season. The narrower and steeper the curve, the less risk in rice cultivation.

used by a very small minority), frequency of pesticide application, and frequency of weeding.

Figure 5.5 shows the scatter diagrams for these regressions, using data from the 1989/90 wet season, an average season with some but not extreme pest infestation, over all eight study villages. From these diagrams it can be concluded that there is no relation whatsoever between yield and any of the factors plotted. Although the correlations in the figures for urea and TSP dosages are slightly positive, (Figures 5.5a and b), the diagrams are so scattered that these fertilisers (in the quantities applied) cannot be considered to determine yields. Additionally, the urea figure shows that the highest yields are obtained at applications around the recommended range of 200-300 kg urea/ha, but also much lower dosages of urea (around 50-100 kg/ha) still result in relatively high yields, probably due to overfertilisation of the land for years and influx of nutrients from other fields.

Frequency of pesticide application and weeding are in no way correlated to yield, as can be concluded from the low regression coefficient and the completely scattered patterns in these diagrams (Figures 5.5c/d). It is worth mentioning that the farmers who obtained the highest yields did not apply pesticides or only once.

Considering the conclusion above that there is no relation between yields and quantifiable farmers' practices expected to have some impact on the level of production, an explanation for yield variation has to be sought in other, non-quantifiable factors or practices. The case of Mulyoagung might give indications. In Mulyoagung, significantly higher average yields were achieved over all study seasons than in any of the other villages. Box 5.3 describes what factors possibly determine the exceptional position of Mulyoagung. Summarising, Mulyoagung farmers differ from others in their use of high quality seed, in rowplanting at a distance of 16-17 cm, and in early weeding, but considered more important is their timeliness and diligence which is reflected in accurate water management and crop monitoring, synchronisation of planting, and collaboration among farmers.

When the amount of monitoring and water management time per hectare is related to yield, no relation is visible meaning that yields do not (necessarily) increase with more hours spent in the field. It is probably the quality of water management and the practices based on decisions after monitoring that matter rather than the quantity of time. Farmers (72%) consider pest damage as the main factor determining yields loss, followed by water supply (mentioned by 16%) and weather conditions (13%). A glance at the data reporting farmers' perception of pest occurrence (Table III.1) in relation to the yields obtained confirms the first point, although it has to be realised that farmers' perceptions of pests are, to some extent, subjective. After having obtained a low yield, whether factually due to pest damage or not, a farmer will often perceive the pests observed more seriously than before. Figure 5.5c shows that, on the whole, chemical measures have not been effective in controlling pest problems. Farmers in the study area seem not to have non-chemical measures available that are adequate, except mechanical rat control methods. Again, the case of Mulyoagung gives some explanation in this regard in that early planting, simultaneity and collectivity of the farmers can help prevent serious pest damage, and thus yield loss.

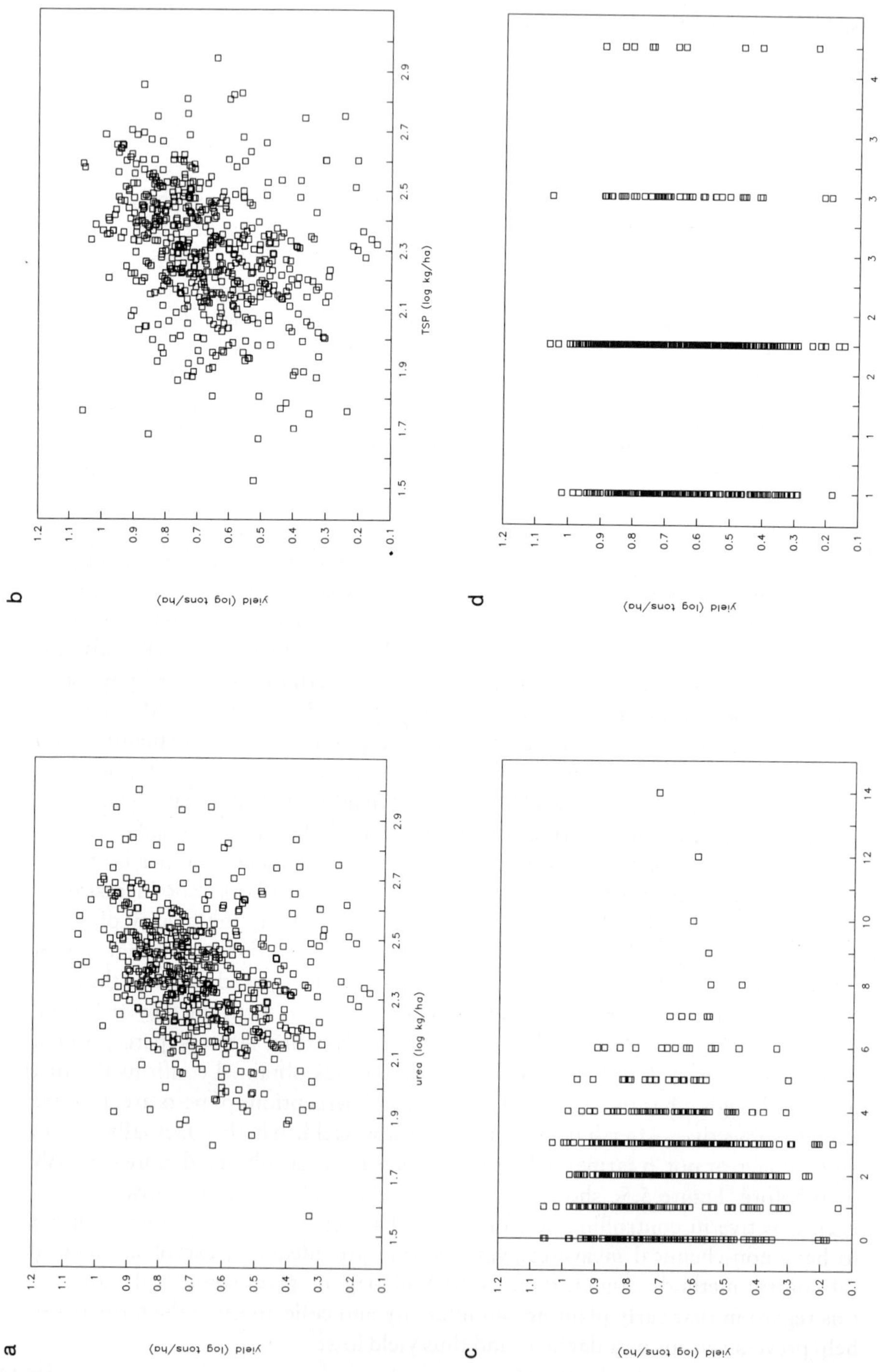

Figure 5.5: Correlations between yield versus urea and TSP dosages, frequency of pesticide application, and frequency of weeding, 1989/90 wet season (N=574).

Box 5.3
High yields in Mulyoagung

Travelling to Mulyoagung is quite an experience. The shortest road to the subdistrict town Sumberagung leads via Kuduagung, the first part being a four-km, slippery clay-soil track full of deep holes, completely impassable after a heavy rain. This is the road through which farmers have to supply their agricultural inputs, although the most commonly used fertilisers (urea and TSP) and pesticides are provided in the local kiosk. Nobody in Mulyoagung owns a car, which is of no use anyway, and only a few people have a motorcycle. The irrigation water originating from the large Kedungombo irrigation scheme has an easier way to reach this village via the recently constructed primary and secondary channels. Irrigation water passes Mulyoagung and the neighbouring village Lempungagung first before proceeding to Sumberagung and Kuduagung further downstream. Consequently, farmers in Mulyoagung and Lempungagung can always plant about one to two weeks earlier than their colleagues in the other two villages. Earlier water supply appears to be advantageous since the irrigation water is often late. Late irrigation was obviously the reason for the rice stemborer outbreak during the 1990/91 wet season. By the time the stemborer population reached the intensity of an outbreak, the older crops in Mulyoagung and Lempungagung had passed the vulnerable stage, and were, therefore, not that severely affected, in contrast with Kuduagung.

An early water supply, however, cannot be the only reason why significantly higher yields are achieved in Mulyoagung over five study seasons, considering the fact that irrigation water conditions in Lempungagung are equal. As described above, Mulyoagung farmers were somewhat different compared to the other villages in a few agronomic practices, including the relatively intensive usage of high quality seed, the habit of rowplanting, earlier weeding, and the application of higher doses of urea and TSP. High fertiliser doses are not assumed to contribute much to the higher yields since the fertiliser levels exceed the recommended rates which is supposedly not efficiently used by the plants. The fertiliser application method in Mulyoagung (broadcasting the fertiliser onto the flooded field) is as wasteful as in the other villages. Therefore, we have to seek a part of the explanation for the higher yields in the other practices mentioned: high quality seed, early weeding and rowplanting.

Discussions on this topic with farmers and village officials yielded some more interesting perspectives. First, it is thought to be advantageous that the whole rice area in the village, which is relatively small, can be planted timely and within a brief period (only 10-14 days), so that crops grow synchronously. In addition, farmers seem to be very cooperative and coherent as a group. Although large-scale collective activities only consist of rat drives once in a while on command of the village head, collaboration among neighbouring farmers appears to be strong. Another fact mentioned is that the remote location and difficult accessibility of the village discourage people to go for other activities or enterprises in the nearby towns. Much time is, therefore, spent in the rice field, and most farmers are qualified as diligent. The great example regarding rice growing for many people are two middle-aged farmers who are always out in the field. They are especially qualified for their diligence in water management. Timeliness and diligence of cultivation practices seem to be important factors to obtain a good crop. The last point favouring a high production in Mulyoagung, as mentioned by some village officials, is that farmers are generally willing to try out and adopt new methods, in spite of difficulties in the transportation of agricultural inputs. This innovative attitude is partly due to a confidential relationship between the people and the village government, which is not a common phenomenon in other places.

The fact that a considerable number of families in Mulyoagung have a (female) member working in the Middle East, regularly sending large sums of money to the relatives at home, might contribute to farmers' good cultivation practices. The availability of (extra) money definitely facilitates the timely purchase of agricultural inputs for the farming families concerned.

5.5 Rice production inputs and outputs

Detailed records of rice cultivation inputs and outputs were made with the survey respondents at the end of the two wet seasons studied (1989/90 and 1990/91). Such records, especially when relying on retrospective interviews, are never totally accurate and complete, but trends can be detected by comparing various categories. Since expenditures and returns of farmers operating under different tenure status are not supposed to be comparable, a distinction is made between owner-operators (44% of the farmers surveyed in the 1989/90 season), tenants with a fixed rent agreement (19%), and tenants with a sharecropping agreement (15%), further referred to as owners, tenants, and sharecroppers. A residual group can be identified consisting of farmers operating on two or more fields with different land tenure status, either owner-operators who lease another piece of land under a fixed rent or sharecropping agreement (19%), or tenants with both fixed rent and sharecropping agreements (2%). These two groups together are referred to as mixed status farmers.

Table 5.11 gives an overview and economic analysis of the input and output data for rice production during the baseline season (1989/90). It contains paid-out and opportunity costs and returns directly related to rice farming. Paid-out costs include hired labour, irrigation fees, and purchased cultivation inputs, whereas owners have additional expenditures on land tax, and tenants on land rent. Opportunity costs include family labour for all categories, and a land opportunity value for owners. Expenditures on credit are not included since loans, especially from banks, are often obtained for several enterprises at a time. Only 24% of the farmers interviewed in the 1989/90 wet season took a loan for which they had to pay an interest or other cost to obtain it. The other farmers were either self-supporting or got informal loans from relatives or neighbours for which no interest was paid. Own capital and non-rice income sources are not considered in the analysis, since responses related to these factors are often incomplete and unreliable. Besides, a complete income analysis would go beyond the purpose of this study.

Money values are mostly rounded off to thousands of rupiahs (Rp) in the text and tables below. No conversions to US dollars are made not to disturb the reader with too many figures. The official exchange rate equalled Rp 1,980 for US $ 1.– in mid 1991, but quick estimates can be made by dividing the figures below by a factor 2,000 to get US dollars.

Gross return

Conform the statement in section 4.3, Table 5.11 shows that the relatively larger and wealthier farmers in the villages are those operating on several fields with mixed tenure status, whereas the sharecroppers are the smaller farmers. Average yields obtained (5.07 tons/ha in the 1989/90 wet season) are not significantly different among the various categories of farmers.

Farmers' gross return from rice production was estimated by as realistic a calculation as possible. Average recovery rates of 80% were used for conversion from wet to dry paddy, and of 66% from dry, unhusked paddy to husked rice. These

Table 5.11: Average inputs and outputs of rice production by non-IPM farmers as classified by tenure status, 1989/90 wet season.

		owners	tenants	share-croppers	mixed status	total average
General						
number of farmers		260	114	91	130	595
average farm size (ha)		0.53	0.42	0.35	0.92	0.56
Output						
yield (tons/ha)		4.91	5.23	5.12	5.23	5.07
gross return (Rp 1,000/ha) [1]		925	957	518	886	860
Costs (Rp 1,000/ha)						
land:	opportunity cost [2]	301	0	0	*	*
	tax	14	0	0	6	7
	rent	0	315	0	157	95
	subtotal	315	315	0	*	*
labour:	family labour [3]	152	146	182	84	141
	hired labour [4]	158	146	166	157	157
	subtotal	310	292	348	241	298
purchased inputs:	seed [5]	12	14	13	11	13
	fertiliser	105	96	85	98	98
	pest control [6]	25	24	28	22	25
	irrigation	4	4	2	3	3
	remaining expenditures [7]	1	0	1	1	1
	subtotal	147	138	129	135	140
Paid-out costs (Rp 1,000/ha)		319	599	295	455	399
Farm business income of rice production (Rp 1,000/ha) of which return to:		606	358	223	431	461
	land	301	0	0	*	*
	family labour	152	146	182	84	141
	capital and management	153	212	41	*	*

1) Where possible, returns are computed using real prices. For stored product and family consumption an opportunity cost of Rp 230/kg dry, unhusked paddy is used. Share for harvest labour is subtracted when wage is paid in kind. Gross return for sharecroppers includes only farmer' share (usually 50%), therefore no cost for land is stated.

2) Owners' opportunity cost for land is equated with the average rent for rice land with land tax subtracted.

3) Family labour at market wage rate.

4) Includes harvest labour only when wages were paid in money, not in kind.

5) Price of purchased seed or own seed at market price.

6) Includes chemicals, rent for sprayer, and rat control implements.

7) Includes contributions for collective activities, and costs for transportation of inputs.

*) Not computable, or computation does not make sense because of mixed tenure status.

(wet season) conversion factors were based on records from millers and rice traders in the villages, and literature references (Mears, 1981; Dir. Food Crops Econ., 1988). Four different ways of rice marketing are common in the study area: selling of the standing crop to a middleman ('*tebasan*'), selling of the wet rice directly after harvesting ('*gabah basah*'), selling of the dry unhusked rice ('*gabah kering*'), and selling of the dry husked rice ('*beras*'). A part of the harvest is generally used for family consumption. With each respondent a record was made of how much rice was sold for what price. The remaining part of the yield not sold by the time of the interview or used for family consumption was multiplied by an opportunity value representing the average price for the season. The several ways of rice marketing each have their favourable and unfavourable characteristics with regard to risk, labour, and storage. Labour aspects were included in the calculations, but the use of opportunity values for risk and storage factors was not feasible within the context of this study.

Gross return from rice production as computed for the baseline season is, on average, Rp 860,000/ha. This average implies somewhat higher gross returns for owners and tenants, but a much lower level (Rp 518,000/ha) for sharecroppers who paid a 50% share of the yield to the landlord. The gross return for sharecroppers, however, was more than half the amount obtained by owners and tenants, which is a result of favourable marketing by sharecroppers. Especially selling the standing crop to a middleman, which is practised by 5-9% of the respondents, is considered disadvantageous by farmers. Those farmers who have no opportunity to arrange the harvest themselves or to store the harvested product have no other choice. Sharecroppers generally do not face such problems, due to the small farm size, resulting in better marketing opportunities. Box 5.4 describes a case on how gross return can be influenced by marketing opportunities.

Expenditures

Expenditures on rice production are divided into three main categories: land, labour and purchased inputs. Paid-out costs for land by owners consist of only tax, but a land opportunity value for these farmers needs to be considered which is equated to the current average rent with land tax subtracted. Thus, the total land costs for owners equal that for tenants. The average rent paid by the 114 tenants is Rp 315,000 which is 51% of their total paid-out costs for rice production. The rent agreement still looks more profitable than the sharecropping agreement. The 50% share paid for the land can be equated to the gross return for sharecroppers in Table 5.11 (Rp 518,000). On the other hand, rents vary a lot depending upon interactions between the landowner and the tenant, and influenced by the fertility of the soil and the shortage of land in a certain area. Eight percent of the tenant farmers paid more rent than the value of the sharecroppers' share. For sharecroppers no land value is stated in the table since the share is accounted for in the figure for gross return.

Costs of labour are divided into real wages paid to hired labour, and opportunity cost for family labour which is calculated using the average prevailing market wage rates for men and women. On average, farmers spend a little bit more on

Box 5.4
Return and prices

Seasonal records kept by two farmers in Mulyoagung during the 1990/91 season show an interesting difference in gross return obtained mainly due to a difference in the moment of selling the crop. It is a comparison between Ms. Suharti, operating on 0.95 ha, and Mr. Joko, having 0.67 ha, both owner-operators of their fields. Both farmers planted the modern variety IR 64. Fertilisation practices were comparable since two fertiliser applications were given, a basal application and one topdressing at 2.5 weeks after transplanting, complemented by one (Ms. Suharti) or two (Mr. Joko) foliar fertiliser sprays. Total dosages of fertiliser were somewhat higher on Ms. Suharti's field than on Mr. Joko's: 502 kg/ha (urea, TSP and KCl) versus 442 kg/ha (urea, TSP, KCl and ZA). Storage facilities were equal for the two farmers. The following outputs were obtained:

	Ms. Suharti	Mr. Joko
Yield	6.86 tons	4.60 tons
Yield per hectare	7.19 tons/ha	6.90 tons/ha
Harvesters wage	686 kg	460 kg
Sold wet	700 kg @ Rp 185 = Rp 129,500	none
Sold dry	800 kg @ Rp 230 = Rp 184,000	2,000 kg @ Rp 270 = Rp 540,000
	2,800 kg @ Rp 250 = Rp 700,000	1,000 kg @ Rp 275 = Rp 275,000
	600 kg @ Rp 270 = Rp 162,000	
Consumption	500 kg @ Rp 270 = Rp 135,000	475 kg @ Rp 270 = Rp 128,250
Gross return	Rp 1,310,500	Rp 943,250
Gross return/ton	Rp 191,035/ton	Rp 205,054/ton

Mr. Joko harvested later and sold a part of the harvest at a time when better prices were offered. Therefore, he obtained a 7% higher gross return per ton harvested product than Ms. Suharti. Although price fluctuations now are minor compared to those previously when there were no price interventions by the government, the difference is noticeable within the usually tight budget of Javanese rice farmers.

hired than on family labour (Rp 157,000/ha versus Rp 141,000/ha), a difference which is mainly influenced by the farmers with mixed tenure status. These farmers, who are the relatively larger and richer farmers, tend to deploy less family labour. Sharecroppers are the category of farmers who use most labour, especially family labour, on their relatively small farms. This is conform the statements made in section 5.2.6 that small farms are operated more intensively.

Expenditures on purchased inputs include all paid-out costs except those for land and hired labour. These inputs account for 46% of the total paid-out costs by owners, 23% by tenants and 44% by sharecroppers. The larger part (70%) of the costs for purchased inputs is spent on fertiliser, on average Rp 98,000/ha. Owners tend to spend most on fertiliser, and sharecroppers least. Despite the observation that smaller farms on average use higher fertiliser dosages (section 5.2.8), the statement above indicates that sharecroppers apply less fertiliser on their small fields, probably to save costs since they receive only half of the harvest. Similar observations were reported from India where farmers with both owned and sharecropped lands manage the latter less intensively (Lipton and Longhurst, 1989). Corresponding with the high variation applied among respondents in fertiliser quantities, expenditures on fertiliser also vary greatly, with 38% of the farmers spending more than the average. Expenditures on pest control inputs, on average Rp 25,000/ha, account for 18% of the total costs on purchased inputs, or 6% of the total paid-out costs, thus a minor cost item. No significant differences are visible among the four categories of farmers when pest control costs are concerned. More detailed data on pest management expenditures are described in a separate section below.

When expenditures are classified by village, no outstanding differences among the eight study villages are observable. Land tenure proved to be the main determining factor in which, obviously, tenants paying a fixed rent accounting for 51% of their total paid-out costs have a completely different expenditure pattern than the other categories. The lack of tenure security for sharecroppers is another factor that seems to influence expenditure pattern.

Farm business income of rice production

Farm business income is calculated by subtracting the rice cultivation costs from the gross return. The average farm business income of rice production for all farmers equals Rp 462,000/ha, which corresponds with other studies done in Central Java (Sudaryanto et al., 1988). As expected, high variation among the various land tenure categories exists. Obviously, owners have the highest income (Rp 606,000/ha) but half of this amount consists of the opportunity cost for land. After subtracting return to land and family labour opportunity cost from the farm business income, tenants appear to have the highest return to capital and management (Rp 212,000/ha versus Rp 153,000/ha by owners), which implies a more efficient farm management.

Sharecroppers' farm business income from rice is lowest of all tenure status categories, on average only Rp 222,000/ha. When this income, obtained in four months' time, is converted to an opportunity value for daily wage, sharecroppers' 'earn' about Rp 1,850/day. This amount is still higher than the wage rate for (male) agricultural labourers, on average Rp 1,500/day, who probably will not be able to find work every day. In this regard, farming, even under sharecropping agreements, seems still more profitable than working as a hired labourer, especially since it can be combined with part-time, off-farm employment. Subtracting the opportunity cost for family labour, sharecroppers' return to capital and management is only Rp 40,000/ha. Considering their average farm size of 0.35 ha, this means only Rp 14,000, on average, which contrasts sharply with the return to capital and management by owners and tenants. A detailed case comparing returns obtained from three fields with different tenure status operated by one farmer is described in Box 5.5.

When looking at the individual records of the respondents, a negative farm business income for rice is not an exception. Seven percent of all farmers, of whom many tenants (14%) and sharecroppers (11%), literally lose money by growing rice (Table III.26), mostly as a result of low yields due to pest damage. This state-

Box 5.5
Return and land tenure

Mr. Budi in Jayasari grows rice on three different fields under three different tenurial arrangements. The first plot is 0.73 ha large and owned by himself, the second 0.71 ha and is leased under a sharecropping agreement, and the third is 0.70 ha for which he pays a rent of Rp 203,000 per season. The three fields are managed in comparable ways, except that the rented field receives more fertiliser because he thinks that the soil here is less fertile than on the other two plots. Yields and returns obtained (at a standard price of Rp 200 per kg wet rice) per plot are:

	owned	sharecropped	rented
Yield	2.70 tons	2.90 tons	2.70 tons
Yield per hectare	3.69 tons/ha	4.11 tons/ha	3.86 tons/ha
Share for landlord	–	1.45 tons	–
Gross return	Rp 540,000	Rp 290,000	Rp 540,000
Purchased inputs + labour	Rp 305,000	Rp 282,000	Rp 277,000
Rent	–	–	Rp 203,000
Net return	Rp 235,000	Rp 8,000	Rp 60,000
Net return per ha	Rp 322,000/ha	Rp 11,000/ha	Rp 86,000/ha

Although yields were relatively low during this season due to stemborer damage, the ranking of the three examples shows that particularly sharecropping, but also operation on rented land are relatively low-profit enterprises.

ment is confirmed by a classification of farm business income by village which shows that especially in Plosoksari and Jayasari low net returns were obtained, as a result of relatively low gross returns but normal costs. Farmers in these villages reported serious rat damage as the cause for low yields. The frequency distribution of farm business income for rice production (Table III.26) shows that highest variation exists among the tenants, mainly due to a large variation in rents.

Seasonal variation

The same analysis of farmers' rice production inputs and outputs was made for the 1990/91 wet season (Table III.27). The different picture obtained for this season is mainly influenced by two events: (1) the outbreak of yellow stemborer causing severe damage to the rice crops, and (2) the price increase of land rents and fertiliser.

The stemborer outbreak, obviously, resulted in low yields and, thus low gross returns, on average Rp 758,000/ha. Yield (and gross return) variability was large among all categories of farmers as a result of local differences in rice stemborer attack, indicating high risk during this season. Average land rents increased with Rp 80,000/ha (or 25%) within one year, giving land an opportunity value of around Rp 380,000/ha.

A shift in labour allocation is visible in the 1990/91 season in that, over all categories, more hired labour is used and less family labour. In total, less labour is deployed compared to the previous season. Labour wages increased slightly in some but not all villages, which can explain only a small part of the higher costs on hired labour. A more likely reason is that more people, especially men, were observed in the later survey seasons to migrate to the towns for employment after crop establishment to return before harvesting, thus requiring more hired labour for weeding and maintenance work, although no quantitative data are available to substantiate this statement. Other studies in Java have shown that off-farm jobs have become increasingly available during the 1980s, attracting many people from rural areas where job loss through mechanisation appears to be overtaking job creation through increased land utilisation (Collier et al., 1988).

Costs on purchased inputs increased with 32% in one year, reaching a level of, on average, Rp 185,000/ha in the 1990/91 season. This increase was mainly caused by higher expenditures on fertiliser, a result of both higher dosages and higher prices because of reduction of fertiliser subsidies. Apparently, the higher prices did not influence farmers' fertilisation practices negatively. Their conviction that more fertiliser was needed, was stronger, although a lot of complaints could be heard in the villages about the rising fertiliser prices. All in all, total paid-out costs increased with 23% over the two wet seasons surveyed, be it more for tenants and less for sharecroppers.

Considering the substantial increase in costs while low gross returns were obtained, it is not hard to imagine that the farm business income for this rice season was very low: on average Rp 267,000/ha, but for the tenants, for instance, only Rp 26,000/ha. Under these disastrous conditions of pest attack, sharecroppers appeared to be relatively better-off than tenants, although their incomes were similarly low. All three categories of farmers had to face a negative balance if they took the opportunity values for land and labour into account. The farmers in Senengsari, Plosoksari and Kuduagung, the villages affected most by stemborer attack, suffered most.

Pest management

As stated above, average expenditures on pest control inputs during the 1989/90 wet season totalled Rp 25,000/ha (or 6% of the total paid-out costs), and one year later (1990/91) Rp 22,000/ha (4% of paid-out costs). When labour on pest control measures is taken into account, the averages are Rp 31,000/ha respectively 28,000/ha. Pest control labour includes both hired labour, mainly for spraying, and family labour, mainly for rat control.

More detailed records of farmers' pest management costs were obtained during the 1991 dry season in which only four villages were surveyed. Pest pressure happened to be very low in this season. Table 5.12 shows the expenditures for pest management (rounded off to hundreds of rupiahs), distinguishing three main types of measures: spray applications, granule applications and rat control. A classification by village is made here since pest management practice is strongly influenced by local variation in pest occurrence. The observation stated above that no significant differences were observed among the four tenure categories, justifies this classification.

Table 5.12: Average expenditures on pest management in Rp/ha, 1991 dry season.

	Senengsari	Jayasari	Kuduagung	Mulyoagung	Total
Spray applications:					
chemicals	2,300	5,700	9,800	4,800	5,700
labour	600	2,400	2,500	900	1,600
subtotal [1)]	3,100	8,300	13,200	6,000	7,600
Granule applications:					
chemicals	1,800	7,900	19,600	24,200	14,000
labour	0	0	0	100	0
subtotal	1,800	7,900	19,600	24,200	14,000
Rat control:					
poison baiting	4,500	5,300	4,900	5,300	5,000
rat drives	5,800	5,600	3,800	7,000	5,600
subtotal [2)]	10,300	12,400	9,000	14,100	11,700
Total	15,200	28,600	41,800	44,300	33,300
N	*55*	*72*	*60*	*76*	*263*

1) Subtotal includes also rent for sprayer.
2) Subtotal includes also fumigation and plastic fencing.

Average costs on spray applications were Rp 5,700/ha for chemicals and Rp 1,600/ha for labour, which, in addition to a possible rent for a sprayer, adds up to Rp 7,600/ha. Since only 42% of the farmers applied pesticide sprays, the actual amount spent by the users equals Rp 18,100/ha. Farmers in Kuduagung tended to spend most on sprays. Granular pesticide application costed about twice the spray applications (Rp 14,000/ha). This amount was almost exclusively spent on chemicals since labour for broadcasting is not counted when the granules are mixed with fertiliser, a common practice. Since granules are used by 52% of the farmers studied, the actual amount spent by these users equals Rp 26,900/ha. On a per hectare base, this amount is lower than one application of the recommended dosage (17 kg carbofuran/ha), costing around Rp 35,000/ha, which indicates that farmers underdosage granules. Because of the high price of granular pesticides, in comparison to sprays, farmers hardly ever use the recommended dosage. Average costs for granular application are high in Mulyoagung and Kuduagung compared to the other two villages due to the higher number of users here.

Expenditures on rat control vary considerably with the rat populations, thus with the seasons. More farmers will go out for rat drives at the beginning of the dry season than of the wet season. The 1991 dry season had relatively few rats compared to the previous dry season, but rat control still costed the farmers, on average, Rp 11,700/ha of which almost equal parts were spent on either poison baiting or rat drives. High variation in expenditures on rat control was observed, due to varying levels of rat attack. Since rats were not around everywhere during this season, mainly farmers close to rat sources, such as roads or irrigation channels, suffered from damage. Fifty-six percent of the farmers did not do anything against rats, and another 36% spent less than Rp 5,000/ha, indicating that only a small proportion of farmers were plagued by this pest and forced to spend relatively much on control measures.

5.6 Adoption of Green Revolution technology

The introduction of the Green Revolution technology in Indonesia is a typical example of the 'Transfer of Technology' model for extension. In such a model, a linear communication process is prevalent in which technology is generated at International Agricultural Research Centres, locally adapted at national research centres, translated for extension purposes into technical recommendations by subject matter specialists, and passed to contact farmers by village extension workers (Röling, 1988). Through natural diffusion processes, the recommendations should trickle down to the other farmers. Farmers are called adopters when they implement the recommendations. As was described before, the Indonesian rice intensification programmes consisted of input packages containing hybrid seed, chemical fertilisers and pesticides to be obtained on credit at the regional village unit cooperatives or KUD, and promoted through the Training-and-Visit extension system. Although initially, during the first, so-called 'BIMAS Gotong Royong' programme, farmers were forced to take the package, later programmes emphasised extension and demonstration, with participation in the KUD on a voluntary base. Farmers who do not want to take the credit packages are allowed to buy inputs for cash and on a need base. Participation is only compulsory in special intensification programmes implemented in potential areas that are carefully selected. These areas are, then, to serve as examples for others. The latest intensification programme, SUPRA-INSUS, still operates in this way.

To assess the current level of adoption of the rice intensification technology among rice farmers in Grobogan, we should look a bit closer at the actual implementation of the government programmes in the study area. The 'BIMAS Gotong Royong' programme started in 1969. Village heads received a letter from the District Agriculture Service informing them about the programme, and commanding them to organise a farmer meeting. The subdistrict agricultural officer ('*mantri tani*'), at that time serving a role as extension worker), and/or a representative from the bank providing the credit attended the meeting to convince the farmers to accept the package. Implementation was the responsibility of the village governments. Comparing the eight study villages, various strategies can be identified with various results, of which six are described in Box 5.6. Summarising these examples, three strategies prevailed:

1. Advantage was taken of the situation in which brown planthopper had attacked the rice crop for several seasons. New, resistant varieties were a welcome innovation for farmers which was easily adopted;
2. Demonstration crops of high yielding varieties were cultivated by either officials or pioneer farmers as an example to other farmers;
3. Farmers were threatened by coercive measures and sanctions by village officials to adopt the new technology.

Box 5.6
The introduction of 'BIMAS Gotong Royong' in six villages

In group interviews in the four intensively studied villages, farmers (mainly older participants in the groups) rendered the transition from traditional to modern rice varieties in a role play, after which a discussion on this topic was opened with the whole group. The results per village are described below. For two other villages (Sugihsari and Plosoksari) information on this topic was obtained from interviews with village officials.

Senengsari

To introduce the BIMAS package in Senengsari, the then village government created an example group of fifty farmers growing high yielding varieties, who were guided by the '*mantri tani*' and the Agriculture Service. At first, there was a lot of resistance among the farming community, but after seeing the success of this pioneer group, i.e. much higher yields than had ever been obtained using the traditional varieties, bit by bit other farmers followed the example. Total adoption by all farmers took somewhat longer in Senengsari than in other villages in the area, since no coercive measures had been used by the village government.

Sugihsari

This was different in the neighbouring village Sugihsari, according to the village officials. The '*mantri polisi*' (a subdistrict police officer) in this village commanded that everybody had to take an input package. Seed was supplied by the subdistrict, and fertilisers and pesticides by a local irrigation project. Initially, many farmers were reluctant, but they were forced to buy the new seeds, fertilisers and chemicals. The village officials watched the rice fields during the planting season (one month). The farmers were threatened that the irrigation water would be closed, and that their fields would be harrowed over if they tried to plant traditional varieties. This happened during one season. After experiencing the new technology, the farmers felt confident and continued planting the new varieties themselves, but also started selling some of the inputs that were obligatory in the loan packages.

Jayasari

In the first year of BIMAS, farmers in Jayasari were obliged by the Agriculture Service via the village leader and his officials to register at the village centre for an input package and credit, initially without any sanctions. During the previous season, the '*mantri tani*' and the village head had already planted a demonstration plot with the new varieties to give an example to the farmers. Still, there was reluctance among farmers to try out themselves based on the belief that the rice of the new variety was not as tasty as the traditional ones. The second reason was that they would encounter problems in harvesting and threshing the crops. The rice plants of the new variety were much shorter which made it difficult to find labour willing to harvest it with the traditional rice knife ('*ani-ani*'). The village leader solved this harvest problem right away by contracting labourers from a neighbouring district willing to do the job. The majority of farmers (estimated at 70%) finally took the BIMAS package, attracted by the high yielding and pest resistant characteristics of the new variety, and by the guarantee of the loan with low interest. Although a proportion of farmers refused to follow during the first season, it was said that the transition to high yielding varieties was realised for 100% during the second season.

Kuduagung

Farmers in Kuduagung first heard about high yielding varieties and accompanying inputs through the radio and the '*mantri tani*'. All farmers received a form to apply for a BIMAS input package on credit. The '*mantri tani*' strongly recommended the farmers to participate, and apparently the majority of farmers was willing to do so because there was credit involved.

In addition, one farmer in the village had already experimented with the new varieties, and tried to convince farmers with the good results he had obtained. Some farmers, however, refused, and continued to use traditional varieties and manure. These farmers were forced by the village head to participate with the sanction that they would be detained in the subdistrict office. With this coercion, everybody followed. One method practised by farmers in order to get away with the heavy burden of the loan they had to pay back, was the sale of inputs acquired from the package. But after a few seasons, everybody was convinced that with the new varieties and fertilisers they were better off. Most of the farmers who attended the group interviews expressed that they had rather not taken the loan. Having a loan is considered a burden, especially in the case of a risky enterprise like rice farming: 'Who knows, the crop might die'. Contracting a loan is something one only does when forced to.

Mulyoagung

BIMAS was first introduced to the people in Mulyoagung in a routine group meeting, attended by all farmers from one neighbourhood and the '*mantri tani*'. The '*mantri tani*' recommended that all farmers use the new varieties. Reasons included the duration of the new variety which was short and made it possible to grow rice twice a year and, thus, increase income, and the yield which was supposed to be much higher than the traditional varieties. In addition, the new variety was resistant to brown planthopper, which was the most convincing reason for farmers in Mulyoagung at that moment, because the rice crops in the village had been attacked by brown planthopper for several seasons successively. Although at first there was some reluctance and suspicion among a few farmers in the village, within two seasons the new technology was adopted by everybody. There had not been any need for coercion by the village government to have farmers take the BIMAS package, because the pest pressure on the rice crop itself forced farmers to change their practices. It appeared to be a welcome way-out of the unfortunate situation they were in. Pesticides included in the package were seen as a useful prevention for another brown planthopper attack.

Plosoksari

A similar situation as in Mulyoagung occurred in Plosoksari. Because of severe brown planthopper attacks during a couple of seasons, farmers were eager to try out the new varieties supposedly more resistant to brown planthopper. The village government did not apply any coercive measures. The adoption of fertilisers and pesticides, however, took much longer, about five years. During the first years of the BIMAS programme, many of the farmers who had taken the package, sold (part of) the fertilisers and chemicals. It was compulsory to take the whole package, but farmers were not convinced yet of the use of the chemicals. Logically, the ones who did not use fertilisers on their crops achieved very low yields. After a while they started following the example of farmers who applied the complete package. Until 1976, many farmers in Plosoksari took credit packages. After that year, the interest in credit rapidly decreased because farmers felt they had enough capital to be independent.

The first and third strategy seemed to be fastest to change farmers' practice, but were not the most effective. Although the traditional varieties were replaced by the new, high yielding varieties, the other inputs were not, or not optimally, used in many cases. Not uncommonly, they were sold to lighten the heavy burden of the loan, resulting in low yields. Only after being convinced by evidence of successful practices by fellow farmers were the other inputs also adopted.

After this initial, rather obtrusive acquaintance with modern cultivation technologies, most farmers in the study area had no further, direct experience with rice intensification programmes. The later BND and INSUS programmes had a more voluntary and individual character. Farmers who were in no need for a loan had actually nothing to do with the KUD and the credit packages, and could buy inputs according to their needs. Promotion for participation in the KUD depended strongly on initiatives of the village head or the village extension worker. Demonstration fields with INSUS technology were once implemented in Jayasari and Kuduagung. Farmer group activities through the SUPRA-INSUS programme were limited to high potential areas. Except Sugihsari, none of the study villages were ever selected for participation in these programmes.

Changes in farmers' practices after the introduction through 'BIMAS Gotong Royong', such as the use of the ever-changing new varieties and pesticides, as well as adaptation of fertiliser dosages, seem mainly to have taken place through natural processes of marketing, experimentation, and exchange of experiences among farmers. The additional fertilisers KCl and ZA, as well as foliar fertilisers were available in the market around 1985. They became part of the KUD package, and were further promoted to the farmers through the regular extension system. Since only a small proportion of farmers in the study villages makes use of the services of the KUD, and meetings of extension workers with (contact) farmers are mostly limited to one pre-season meeting, usually with the same group of people who take the KUD package, these fertiliser are, until now, not widely used. Farmers' reasons for not adopting KCl and ZA were described above (section 5.2.8). In addition, many farmers report in informal discussions not to know what these fertilisers are, and what for.

Looking from a qualitative point of view, the most important facets of the Green Revolution technology (improved seed, urea, TSP and pesticides) were soon adopted by Grobogan rice farmers, and are now widely implemented, in contrast to many other provinces of Indonesia (Jatileksono, 1987). The additional fertilisers introduced later are, however, not in great demand yet.

Considerable differences were observed between the amounts of seed and fertilisers used by farmers and the recommendations. Many farmers apply certain practices based on their own experience over the years, and their own experimentation. It is not surprising that the national recommendations do not hold for the specific, local conditions of every farmer. Experimentation by farmers, although a proof of independence, is not always done rationally in that sometimes extremes are tested, or incomparable conditions (seasons, fields, dosages) are compared. Another difficulty encountered in this connection is the conversion of dosages into local measures. Regarding the common practice of overdosing TSP, a relation could be detected between the amounts of urea and TSP applied by farmers, based on a misinterpretation of the term 'balanced fertilisation', which shows a shortcoming in current extension messages only explaining 'what' and not 'why' or 'how'.

5.7 A good farmer, a good crop

Farmers seem to have a clear perception about the distinctive features of a prototype good farmer. In group interviews with rice farmers in four of the study villages during the 1991/92 wet season, the participants listed and discussed the characteristics of a good farmer. In all groups, more or less the same characteristics emerged. According to the respondents, a good farmer is someone who:

- is diligent;
- frequently visits the field, especially when the crop is still young;
- applies proper fertilisation;
- applies good soil preparation, including ploughing, levelling and trampling the soil;
- monitors the field regularly;
- manages the water well;
- weeds the crop properly;
- selects good seed;
- possesses knowledge and experience about agriculture.

Most of these points deal with diligence and regular field work, not necessarily with advanced technology. This became obvious when in one of the groups a farmer suggested to include in the list 'applying fertiliser according to the recommended dosage of the extension worker', which was ruthlessly rejected by the others. It is interesting that the possession of knowledge emerged as well. Farmers stressed that this also includes indigenous knowledge and beliefs relating to rice cultivation. Some occasionally mentioned characteristics showing interesting perspectives, such as clearing of the bunds, being able-bodied, having self-confidence, being able to calculate costs and returns, planting synchronously with the neighbour farmers, giving guidelines to other farmers, using a pedal thresher, and knowing the recommendations from the government. All groups could mention one or more people in their village who were seen as the prototype farmer, usually ordinary farmers without any function as farmer group leader or village official, but characterised by their diligence. These people often had an informal role as information source for others.

From the analysis of rice cultivation practices in the study area as described in the sections above, strong indications emerged that, within the range of current practices, yields are not determined by urea or TSP quantities, number of pesticides applications, or weeding frequency. Although it is likely that yields are mainly determined by the quality of the soil and water supply, and the level of pest attack, some influence of farmers' practices could be identified, primarily in the sense of **quality** of cultivation methods and maintenance of the crop. Possibly important aspects in this respect are timeliness (of weeding, fertilising, pest management), good water management, seed selection, and collectivity and coherence of farmers.

Small farms tend to be operated more intensively in the sense that more labour is employed, and higher fertiliser quantities are used. Especially farmers cultivating on miniature fields (smaller than 0.1 ha) apply extremely large quantities of fertiliser. Calculation of the dosage for a small field is, indeed, subject to larger mistakes. A more intensive cultivation, however, did not lead to higher yields, as shown by the non-significant difference among the yields obtained by various farm size categories. Apparently, more intensive cultivation does not necessarily mean a better quality of cultivation. An additional reason can be sought in the fact that the best rice fields are usually operated by the better-off in a village (Hüsken, 1989).

It is assumed that farmers in the study area can economise on inputs and, therefore, cultivate more cost-effectively. This holds especially for fertilisation practices. Inefficient fertiliser use is observed in the timing of fertilisation before weeding, and in the method of application by broadcasting the fertiliser in flooded fields. Both practices are supposed to cause high losses. Better application methods and better timing might improve fertiliser efficiency considerably, and therefore increase the returns for farmers. Other supposedly inefficient practices are the preventive use of pesticides, still practised by a considerable number of farmers, and the use of extremely high seed quantities.

Net return of rice production appeared to be very small for tenants and sharecroppers, due to the high costs involved for the land. Especially during a season with heavy pest pressure, many farm families faced a negative net income. Even in reasonably good seasons, the income stands in no relation to the needs of the family. The rural people in Java, however, continue growing rice even for a low return, as also observed in Nepal where most farm households were indebted but still wanted to continue farming (Bhuktan and Denning, 1990). Farming is, therefore, more a way of life for the small farmer households than a major source of income. Currently, a trend is visible suggesting that small sharecroppers are deemed to vanish. But in general, Javanese rice farmers attempt to continue, through ever balancing opportunities and constraints within the antagonistic environmental and social forces they are facing.

6 IPM farmer field school portrayed

6.1 This chapter

In early May 1990, after a pilot season in Central Java, the first implementation cycle of the National IPM Programme's farmer training started with about 1,800 farmer groups, 450 pest observers, and 900 village extension workers in six provinces in Indonesia. The pest observers had just finished their rice IPM training at the regional Field Training Facilities (FTFs), and were given the task to replicate what they had learned with the farmers in their home areas in season-long IPM farmer field schools.

As one of the main activities of the evaluation study in Grobogan, two field schools with different trainers in the villages Senengsari and Kuduagung were closely studied throughout the 1990 dry season. Most training sessions were observed, individual and group interviews were conducted with field school participants and untrained farmers, and interviews were held with pest observers, village extension workers, and training supervisors at the Rural Extension Centres (RECs), subdistrict agricultural offices, and the Yogyakarta FTF. Two other training groups of the same trainers in the villages Sugihsari and Sumberagung were reviewed only through individual interviews with the participants, farmer group leaders, and trainers, and a group discussion one year after training, serving as a comparison without the influence of the study team's presence. The farmers selected for and participating in the IPM farmer field schools are further called 'IPM farmers', whereas 'non-IPM farmers' are those not involved in training.

Since the four schools evaluated form only a small proportion of the 1,800 trainings given during the 1990 dry season which was only the programme's first large-scale implementation cycle, the results described here are not representative for the whole programme. The process evaluation of these four field schools serves as a case describing how the IPM training model can be implemented at the village level. At the same time, it provides a point of reference for the effects of IPM training measured in the study area in the seasons afterwards. As a comparison, various characteristics of the training groups will be compared with those of all farmer groups trained from the same FTF during the same season. In addition, various comparative observations were done in field schools in other areas, in other seasons, and under different training models.

6.2 Organisation

The design and overall organisation of the National IPM Programme in Indonesia was described in detail in section 2.4. The first large scale implementation cycle of IPM training started in January 1990 with the four-month rice IPM training for pest observers in nine regional Field Training Facilities (FTFs). The two pest observers from Grobogan selected for this evaluation study were sent to the Yogyakarta FTF. During the last month of this training, in consultation with the heads of the RECs, two village extension workers per pest observer were chosen to become apprentice co-trainers in the following extension season. These extension workers joined the pest observers for one week at the FTF to become acquainted with the principles of IPM and IPM training, and to make a work plan together with the pest observers for the implementation at the village level during the extension season. First, two villages per trainer team (which implies four villages per pest observer), and next, one farmer group per village were selected to participate in the training (Figure 6.1). Back home, the extension workers approached the heads of the selected villages to discuss participation, and to determine the selection of participating farmers.

Before the field implementation began, training materials were distributed from the Yogyakarta FTF to the RECs, consisting of large plywood boards, a bundle of newsprints, crayons, clothes-pegs, stationery, printed leaflets with colour pictures of rice pests and natural enemies, and various materials for the special topics. The materials were distributed to the participating villages and stored in the village centres. Because of the field-oriented nature of the IPM training, the most important training 'material' was a piece of rice land of about 0.1 hectare that had to be made available by the village for observations and experiments during the season. The owner of the field received a compensation for possible yield loss as a result of trampling and of pest damage in case wrong control decisions would be taken. This field was divided into two plots, of which one plot received a preventive carbofuran granule application, standing for the current local technology package, and the other received IPM treatment, meaning pest control measures after group decisions based on observations. Rice variety, fertilisation, weeding and water management of the two plots were exactly the same. Other training

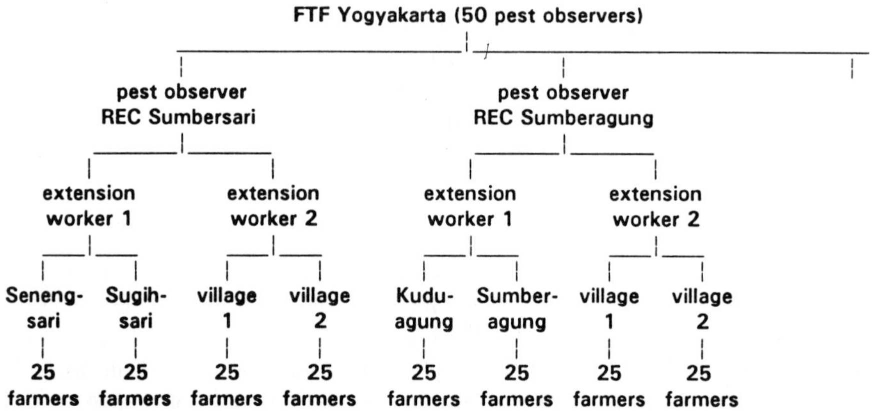

Figure 6.1: Organisation of the IPM extension season.

activities took place close to the observation plot. In Senengsari farmers sat down on the field road under a row of trees, and in Kuduagung a small house close to the field was used for the field school.

The opening and first session of the IPM farmer field schools in Senengsari and Kuduagung coincided with the start of the growing season. In Senengsari, a part of the rice area was still being planted, whereas in Kuduagung the crop was, on average, two weeks old. The training consisted of ten schooldays held once a week, which were all realised in both villages. During this period of ten weeks, all stages of the rice crop could be examined. In both groups, the farmers together with the trainers decided to have the field school in the morning, starting at 8:00 a.m., which is the best time for field monitoring. On the last field school session, a field day was organised to present the result of the training to the surrounding community in order to create interest and obtain support for follow-up activities. Each field school participant received Rp 1,000 (US$ 0.5) per session for attending the training to compensate possible loss of income. In Senengsari, the pest observer suggested that half of this money be used for a study tour at the end of the season. In the end, they could not get the trip organised, mainly because it was difficult to find a suitable day for everybody, so the money was reimbursed to the farmers. In Kuduagung, a similar initiative arose, and at the end of the season a study tour to tourist and agricultural sights in Central Java was made together with the three other field school groups of this pest observer.

Throughout the extension season, the RECs, in the person of the head of the REC and/or the senior extension officer (PPM) in charge of food crops, served as coordinators and supervisors. Intensive supervision from the FTF was done by the Field Leaders I and II, who had to cover large distances on their motorcycles to visit each pest observer-extension worker team at least once during the season. To support the activities in the field, all pest observers returned to the FTF for a workshop of one week per month in order to exchange experiences, and get feedback from fellow trainees and supervisors. In those weeks, the village extension workers were responsible for conducting the farmer field schools. The two pest observers usually prepared the materials for the field school at home, but briefed their extension workers every week at the REC. The school sessions at which the pest observers could not be present because of the workshops at the FTF, were thoroughly prepared together with the extension workers.

6.3 Group selection and composition

Selection procedure and criteria

According to the guidelines for pest observers from the programme, the procedure for the selection of IPM farmer field school participants for this first large-scale implementation season consisted of three steps: (1) the selection of two village extension workers, usually based on personal criteria like motivation and suitability for collaboration with the pest observer, which was done by the pest observer in consultation with the head of the REC, (2) the selection of two villages or farmer groups within the work area of each selected extension worker, and (3) the selection of twenty-five farmers per village or farmer group. The selection of villages or farmer groups, made by the pest observer-extension worker teams, was based on the following criteria:

- availability of rice land with irrigation;
- easy accessibility from the REC;
- presence of an active farmer group;
- synchronisation of the planting season with the start of the field school.

With these criteria it is not likely that average villages or average farmer groups are selected for participation – a common syndrome in extension programmes (Röling, 1988) –, but the easily accessible villages with farmers who, in general, have been more exposed to extension messages and promotion campaigns. As described before, active farmer groups are rare in the current extension system, but active individuals able to motivate and mobilise others are present in almost every village. In most farmer field school groups observed, such an active individual who already had regular contacts with the extension worker could be found. The characteristic that most IPM villages had in common appeared to be location of the village on, or close to, the main road (also Winarto, 1992). To a certain extent, these two characteristics are related because it is likely that active farmers living close to the extension worker (who seldom lives in a remote village) more often meet him or her, in contrast with people living far away. On the whole, people living close to the main road have easier access to all kinds of information sources and facilities. The other way round, the extension worker visits nearby villages more frequently, and, therefore, knows the people there better than those in remote places. In selecting a village for participation in a programme, the better known places are obviously chosen. In addition, the farmer field school implementation was part of the pest observers own training, so they wanted to be assured of success.

The last step in the selection procedure, i.e. the selection of field school participants, was primarily the responsibility of the village extension worker. In both Senengsari and Kuduagung, the extension workers left de selection of farmers to the village officials. In Senengsari, the village secretary assisted by a farmer group leader selected the trainees. They organised a meeting with the candidates to explain what was expected from them, and to register the ones willing to participate in the field school. In Kuduagung it was the so-called farmer group coordinator, a typical example of an active individual without an active farmer group, who selected the trainees. The criteria used for the selection of field school participants for both groups included (1) the ability to read and write, (2) the possibility to attend the training regularly, and (3) the ability to disseminate what is learned to other farmers. After initial selection, there were a few mutations in the composition of the groups. Of the trainees who participated in the ballot box pre-test, two farmers in Senengsari, and two in Kuduagung quit after the first or second week, because they did not feel capable, and were replaced.

Over all 200 groups trained from the Yogyakarta FTF during the same season, the selection of participants was left completely to the villages by 58% of the extension workers. In other cases, the extension worker usually consulted with people in the village. The farmer group leader was involved in 85% of the cases, and the village head or other village officials in 50%. Despite the programme's guidelines to select all twenty-five participants per training group out of only one farmer group (usually farmers living in the same neighbourhood), the trainees in Senengsari and Kuduagung, as in many other groups throughout the country, were selected from **all the farmer groups** and from **all the neighbourhoods** in the village. The explanation given by the village officials was that faster spread of the IPM technology could be expected over the village.

Composition

A comparison of some characteristics of IPM farmers with non-IPM farmers in the four IPM villages studied, as collected in the baseline survey prior to training, shows that the field school groups were composed of relatively more younger farmers with higher education, operating on larger rice lands than the average farmers in the villages (Table 6.1). Proportionally more of the IPM farmers are owner cultivators, especially those operating on extra leased fields (the group identified in Chapter 2 as the richer farmers), and more part-time farmers. Average frequency of pesticide application in the season before IPM training was somewhat higher among the IPM farmers compared to the non-IPM farmers, with the exception of Kuduagung. High variation in pesticide use is visible among the villages,

Table 6.1: Characteristics of IPM farmers in comparison with non-IPM farmers in the four IPM villages. Data refer to the (1989/90) baseline season prior to training.

	Senengsari		Sugihsari		Kuduagung		Sumberagung	
	IPM	non-IPM	IPM	non-IPM	IPM	non-IPM	IPM	non-IPM
Age								
average (years)	40	46	41	42	45	48	38	44
less than 30 years	16%	8%	4%	16%	4%	7%	20%	13%
30-49 years	60%	44%	76%	54%	60%	44%	60%	47%
50 years or more	24%	49%	20%	30%	36%	49%	20%	40%
Formal education								
average (years)	6.6	4.2	7.2	4.3	7.5	5.0	6.3	4.5
no education	4%	17%	4%	13%	0%	15%	0%	13%
1-3 years (SD)	12%	26%	8%	27%	4%	12%	12%	31%
4-6 years (SD)	48%	47%	44%	52%	60%	59%	64%	42%
7-9 years (SMP)	16%	9%	20%	6%	16%	10%	24%	11%
10 years or more	16%	2%	24%	2%	20%	4%	0%	4%
Farm size								
average (ha)	0.91	0.66	0.95	0.41	0.98	0.68	0.78	0.69
less than 0.30 ha	17%	33%	16%	43%	8%	16%	12%	15%
0.30-0.49 ha	17%	26%	40%	35%	12%	19%	16%	29%
0.50-0.99 ha	17%	23%	20%	16%	52%	49%	44%	44%
1.00 ha or more	48%	18%	24%	6%	28%	15%	28%	13%
Land status								
owners	29%	38%	56%	41%	64%	62%	48%	34%
tenants	8%	9%	0%	14%	16%	22%	4%	16%
sharecroppers	8%	30%	24%	30%	0%	9%	12%	21%
mixed owner + t./sh.	48%	20%	16%	8%	20%	7%	28%	27%
mixed tenant + sh.	4%	3%	4%	6%	0%	0%	8%	2%
non-cultivators	8%	0%	0%	0%	0%	0%	0%	0%
Frequency of pesticide use								
average (times/season)	1.1	0.9	2.1	1.4	2.5	2.6	2.5	2.3
% of farmers not using pesticides	48%	55%	8%	25%	4%	7%	20%	15%
% of women	0%	8%	0%	4%	0%	3%	0%	2%
% of part-time farmers	70%	53%	40%	48%	64%	40%	52%	61%
N	*25*	*66*	*25*	*63*	*25*	*68*	*25*	*62*

in that Senengsari farmers (both IPM and non-IPM) are relatively low pesticide users, and Kuduagung and Sumberagung high users.

Especially the three characteristics concerning farm size, land tenure, and farm occupation indicate that the farmers selected for participation in the IPM field school are the more affluent people in the village, having access to more land and to more enterprises, again a common phenomenon in extension programmes.

The same train of thought probably holds for the selection of participants as for the villages, in that success has to be guaranteed which is more likely through the involvement of the better informed and more affluent people. Another reason often mentioned is that the message will be spread better through people with some status in the village than through an average person. One of the results of this last thought is that relatively many village officials and neighbourhood heads were involved in training. An exception is Sumberagung where the village head purposely selected a group of 'real' farmers. Neighbourhood heads are, indeed, the obvious medium for dissemination of any message in a village, but village officials, on the contrary, often have a distant relationship with the majority of villagers. It is striking that no women at all were involved in the IPM trainings in the villages in Grobogan.

The composition of the four field school groups studied is neither exceptional with respect to age, education, farm size, and tenure status, compared to other groups trained throughout Indonesia, nor with respect to the biased representation of certain layers of the farming community in the training groups (Source: Data base IPM Secretariat, Jakarta; Anonymous, 1991). Table 6.2 gives some details on the involvement of women and village officials in 200 field school groups from the Yogyakarta FTF during the 1990 dry season, based on reports from the pest observers in a written questionnaire at the end of the season. The table shows that, overall, very few women participated in the IPM training (only 3% of the total number of trainees), and relatively many village officials (11%). The actual percentage of village officials is expected (and in some cases noticed) to be higher since the pest observers, who reported on this, usually do not know all village officials. Box 6.1 provides more detail on the problem of women involvement in IPM training describing high variation among provinces, among others due to cultural differences.

Table 6.2: Involvement of women and village officials in IPM farmer field schools of the Yogyakarta FTF during the 1990 dry season ($N = 200$).

Participation of women:	
Average number of women per group	0.8 ± 1.2
Percentage of women from total	3%
Groups with no women	77%
1-2 women	9%
3-4 women	9%
> 4 women	6%
Participation of village officials:	
Average number per group	2.7 ± 1.7
Percentage of officials from total	11%
Groups with no village officials	9%
1-2 village officials	42%
3-4 village officials	32%
> 4 village officials	17%

Box 6.1
Women participation in IPM farmer field schools

The important role played by women in rice cultivation is widely recognised, but often neglected (e.g. IRRI, 1985; Boserup, 1984; Kodiran dan Hudayana, 1990; Siwi et al., 1990). Women are involved in all stages of rice production and processing, and often have an important voice in decision making relating to agricultural activities. Agricultural extension, however, is rarely directed to women.

Practising IPM would fit well in the tasks of the women observed in the study area, since they are, in general, actively involved in routine crop nursing, monitoring and decision making in rice growing. This is especially true for the increasing number of farm families where the husbands leave the village after crop establishment to find work in the town, forced by the insufficient income obtained from the small rice lands. In the four IPM villages studied in Grobogan, not a single woman participated in the farmer field schools. The comment of the village extension workers on the question why no women farmers were selected to attend the IPM training, was: 'Oh, we never thought about it!', indicating that women are never invited for extension meetings. Women, indeed, are seldom members of farmer groups. Nevertheless, the extension workers, as well as the village officials, did not see any major objections to involve women in training. Although men are seen as the ones who should represent the family in official meetings in the villages (unfortunately still including agricultural extension), no main cultural barriers except this habit could be detected that would inhibit the involvement of women.

The IPM programme actually welcomed more women in training, but had neither a directly encouraging nor inhibiting policy in the selection procedure. Indirectly, however, women were screened out of training since trainees are selected from among membership of the farmer groups. Also by leaving the selection of participants completely to the village extension workers who, in turn, often left it to the village officials, it can be expected that habitual patterns are followed, resulting in the selecting of mainly men. The only exception observed in another subdistrict of Grobogan, where three women were invited to participate in the IPM training, was a special case of women with higher education.

The situation described above holds mainly for most parts of Java. Great variation relating to women involvement in field schools could be detected among provinces. Whereas during the first training cycle in Java only 2% of the farmer field school participants were women, in North Sumatra 14% of the trainees were women farmers, a proportion that increased to 23% during the second training cycle (Source: IPM Secretariat, Jakarta). Also West Sumatra, which has a matrilineal culture, scored high (17%), but, in contrast, South Sulawesi and Bali scored low with only 1% women. These figures seem to confirm that cultural habits strongly influence the selection of women for training.

According to the records of interviews with trainees and trainers of some groups with relatively high proportions of women participants in North and West Sumatra, the involvement of women had a positive effect on the training. Women performed as well as men, and they were actively involved in the group exercises. Many men reported that women were more accurate in identifying insects which forced the men to have a better look, too. In mixed groups, women seldom gave presentations in front of the group, again, as a result of cultural habits, except for the ones with higher education. The women trainees mentioned that they benefitted much from the IPM training which allowed them to make better pest (and crop) management decisions and save expenditures.

Structural changes in the programme's trainee selection procedure are expected to have great impact on the involvement of women farmers in IPM training, which definitely will benefit an important group in the farming community. Such a procedure, however, should guarantee that active involvement in rice farming is a major criterion, and not, for instance, the level of education (Sumayao, 1986).

6.4 Ten weeks field school

Field school activities

The standard time schedule for a farmer field school day, as proposed by the IPM programme, takes 4.5 hours, preferably starting early morning:

07:30 – 08:30	**Field observation:** the participants sample the observation field in five small subgroups (photo 6.1);
08:30 – 09:45	**Agroecosystem analysis:** the subgroups make drawings of what they have found in the field; each drawing contains pictures of the rice plants, pests and diseases, natural enemies, and weather, soil and water conditions, both for the IPM plot and the local package plot (photo 6.2);
09:45 – 10:00	**Presentation and discussion:** the subgroups present their drawings, discuss the field situation, and all participants take a decision together whether a pest control measure should be taken or not, and if so what measure;
10:00 – 10:15	**Break** with tea and snack;
10:15 – 10:30	**Group dynamics exercises;**
10:30 – 12:00	**Special topic:** experiments, lessons, exercises, and discussions on special topics dealing with problems that the field school farmers are facing at that particular moment (photo 6.3).

Photo 6.1: Field monitoring by farmers.

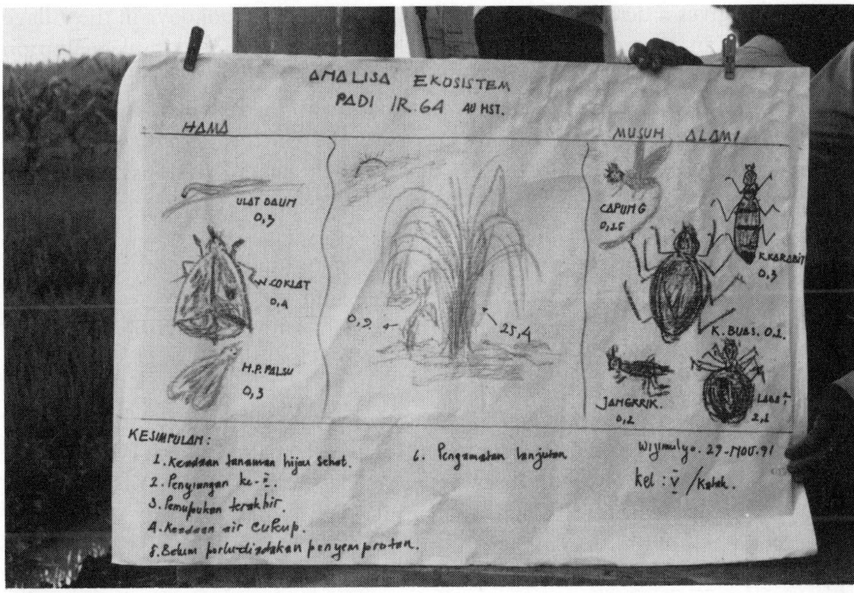

Photo 6.2: Drawings of an agroecosystem analysis.

Photo 6.3: Farmers do the experiment of the special topic 'Carbofuran, monocrotophos and natural enemies'.

Appendix II gives a detailed description of the ten field school days in the village Senengsari as an example of how a field school can be run. The implementation of the field school in Kuduagung did not differ considerably from Senengsari. The sessions in both villages followed the schedule mentioned above with regard to the sequence of the activities. Time spent on the various activities, however, departed from the proposed schedule. In most field school sessions, more time was spent on the presentation of the drawings and discussions, and less on special topics, field observation, and agroecosystem analysis.

The average amount of time spent per field school day during the eight core sessions of the training (not including opening session and field day) was 3.6 hours per school day ($\pm$ 0.5 hours) for Senengsari, and 4.1 hours ($\pm$ 0.7) for Kuduagung. Usually, no time was assigned particularly for a break. Tea and snacks were consumed while doing the group exercises. The school lasted slightly shorter on the days when the pest observers attended the workshops at the FTF and the programme was conducted by the village extension workers alone (week IV and VIII). For the sessions where the pest observers were present, the amount of time spent per day was quite constant in Senengsari, but slightly decreased in Kuduagung in the second half of the training.

It was expected that in the course of the season less time is spent on field observations and agroecosystem analysis because farmers become more skilled in monitoring and identification, and more time on special topics because they learn to identify and solve their problems. These trends, however, were not observed (Figure 6.2). In Senengsari, somewhat less time was spent on field observations in the second half of the season, but no distinct trend in time allocation for the other activities was visible. In Kuduagung, the time for field observation also decreased after the fifth week, even more than in Senengsari, but the spare time was not filled with more special topics, consequently reducing the total time per day. Time spent on agroecosystem analysis was fairly constant in both groups. In Kuduagung more time was spent on presentation of the agroecosystem and discussion than in Senengsari, but high variation existed over the whole season. The discussions in Kuduagung were mainly held by a small number of intellectuals in the group, and did not contribute much to the understanding and knowledge construction of the larger part of the group. Time spent on special topics varied over the season, depending on the topic presented.

The main pest problem in the rice crop in both Senengsari and Kuduagung was rats during the 1990 dry season. At the young stage of the rice crop, Kuduagung also had to cope with bacterial leaf streak infestation, but this was not very serious since the plants compensated rapidly by making new tillers that were not infested. Special topics in the IPM field school should link up with rice cultivation problems of the participating farmers. The special topics presented in the schools in Senengsari and Kuduagung are listed in table 6.3. The topics for week IV and VIII, when the pest observers attended the FTF workshops and the field school was conducted by the village extension workers, were preset. Almost all other topics done in the two villages belong to the 'top ten' of special topics presented in all farmer field schools from the Yogyakarta FTF trainers in that season. Relatively much time

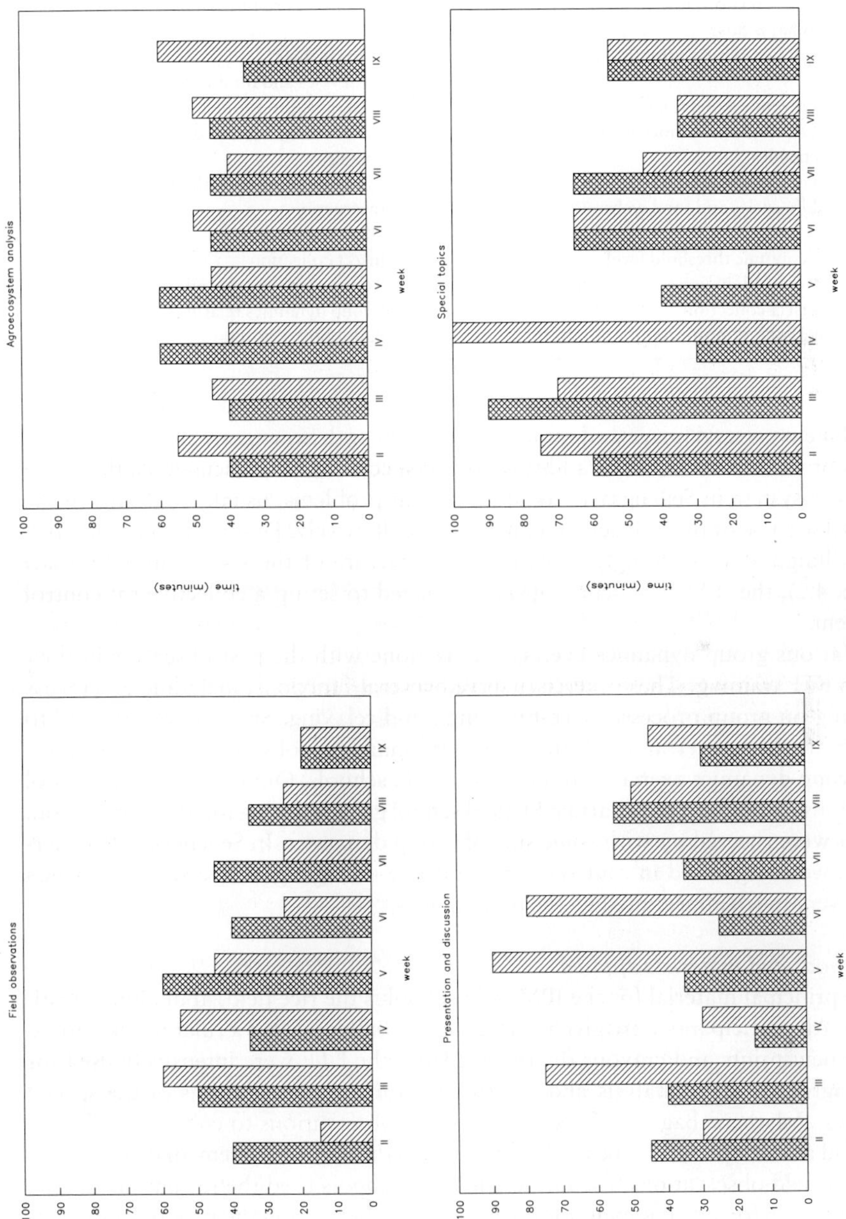

Figure 6.2: Time allocation for the different field school activities throughout the season in Senengsari (left bars) and Kuduagung (right bars)

Tabel 6.3: Special topics presented in the IPM field schools in Senengsari and Kuduagung.

Week	Senengsari	Kuduagung
I	Ballot box pre-test What is this?	Ballot box pre-test
II	Rat population growth	What is this? Life cycles and food webs
III	Prevention of damage	Economic threshold level
IV	Carbofuran, monocrotophos and natural enemies	Carbofuran, monocrotophos and natural enemies
V	Prevention of rat damage	Rat population growth
VI	Life cycles and food webs	Insect zoo What is a predator?
VII	Economic threshold level	Insect collection
VIII	Roots, plant vessels and systemic pesticides	Roots, plant vessels and systemic pesticides
IX	Insect collection	Group dynamics relating to rat control
X	Ballot box post-test	Ballot box post-test

was spent on rat damage and rat management in both Senengsari and Kuduagung: two of the ten sessions. In addition to the special topics, discussions about the agroecosystem in Senengsari often touched rat problems, usually focusing on the unwillingness of the villagers to control rats collectively. Despite the fact that this unwillingness was strongly rooted in the structure of the society in Senengsari (Box 4.2), the IPM field school group managed to set up a collective rat control system.

Various group dynamics exercises were done with the pest observers in their own FTF training. These exercises served several functions, including awareness raising on group processes, 'ice-breaking', and relaxing. Since they were told to replicate their own training with the farmers, many pest observers organised several group dynamics exercises in the farmer field schools. Out of all the field school sessions in the 200 Yogyakarta FTF field school groups during the 1990 dry season, 54% were cheered up with some sort of group dynamics. In Senengsari, the exercises were organised in four sessions, and in Kuduagung in five sessions, which was very much appreciated by the participants.

Methods and materials applied

The principal material for the IPM field school is the rice field, abundantly available. All participants were given a notebook, a ballpoint and a ruler to take notes. The newsprints and crayons distributed from the FTF were intensively used for the agroecosystem analysis and mostly also for the presentations of the special topics. Polythene bags were used for the field observations to collect insects that served as examples for making the drawings in the agroecosystem analysis.

For field observations, the farmers in both schools used the format of the pest observation form that is commonly used by pest observers in their routine monitoring. The field school participants were given a copy of the form on one of the first sessions, and were told to copy the format in their notebooks for the following weeks. The format, quite complicated and distinguishing various categories of pests and levels of infestation, was only used in the first weeks. After a while, the field school farmers just noted down everything they encountered in the field during the observation, not following the format any more. In fact, for observations in their own fields they never noted down anything. This practice, and not the use of complicated forms, is actually expected by the IPM Programme since farmers do not go to the field with a piece of paper and a pencil. Nevertheless, making observation records in the field school appeared to be a good exercise for farmers to promote accuracy in monitoring.

In the agroecosystem analysis, a formula taught by the pest observer was used in both groups to calculate percentages of damage, pest and natural enemy occurrence, again a practice not expected by the programme. The results of the calculations were often incorrect and confusing. Insect counts presented as numbers per hill often represented all insects in the environment of that hill. Percentages were sometimes calculated per hill, sometimes per tiller. It was not always clear whether the farmers used scores or percentages, and sometimes scores were called percentages. The main use of the scores calculated was to compare pest and natural enemy populations which was practised well. An absolute number of insects or damage symptoms found in the whole field, however, would have been more logical.

The concept of economic threshold level (ETL) was often used, and misused. The special topic about ETL was presented reasonably well both in Senengsari and Kuduagung, but the explanations stayed at a very abstract level. The pest observers were acquainted with the constraints of the ETL concept during their FTF training. They were explained the many factors that have to be considered in the economic analysis of a farm, and not only the number of pest insects found in the field, and the complicated and hardly applicable use of the ETL concept at the farm level. The pest observers did not seem to be comfortable yet with these restrictions of the ETL concept, and fled back to their old, straightforward knowledge when they had to explain it to the farmers. Examples of damage threshold levels of the major rice pests as applied by the Agriculture Service were given in the IPM field schools. The majority of farmers did not grasp the concept, which holds especially for Kuduagung, where the ETL was presented in the third week, before the farmers had a good understanding of the agroecosystem. Despite this incomprehension, many farmers loved to use the term ETL in their presentations, but usually not adequately and without any relevance to the factors that determine the ETL.

Performance of trainers

The pest observer in Senengsari showed a very committed and enthusiastic attitude, especially during the first few weeks. Later on he seemed to be less self-assured. He had high expectations of the training, and became easily disappointed when things did not turn out the way he had planned. His relationship with the farmers was excellent. He always used Javanese in his speech, and was often humorous, which was appreciated by the farmers.

The village extension worker was well known in the village, and already had a good relationship with the farmers. At the meetings conducted by the pest observer, he was silent but supportive, e.g. in giving useful, additional examples. He enjoyed to organise and conduct the field school on his own on the days when the pest observer attended the FTF workshops. The pest observer and the extension worker collaborated well, and both were satisfied about it. The pest observer was not very happy with his second co-trainer who did not show much initiative and had too many other activities.

In Kuduagung, the pest observer did not originate from the REC Sumberagung but from another REC in the district of Grobogan, and was only stationed there for the IPM field school season. He did not know the area well, and he was not known either. At the opening session of the field school, the pest observer was nervous. He jumped from one topic to another, using a lot of scientific terms which he did not explain. From the third week onwards, he showed a more self-assured behaviour. He was motivated and visibly enjoyed conducting the training. The last few weeks, though, his performance changed. He did not actively guide the farmers any more, was sometimes not at the location when exercises were done, and did not participate as actively as before in the discussions. He admitted afterwards that he was not very committed to become a trainer and would rather do field monitoring.

The village extension worker, in contrast, was well known in Kuduagung. He lived in the neighbouring village where he sold pesticides at his house – an inheritance from his past assignment as estate crops extension worker during which he became used to the extra income from pesticide sales –, a service used by many farmers in Kuduagung. Relatively many farmers in his work area, and especially among the farmers selected for the IPM field school, often contracted credit from the KUD (the Village Unit Cooperative) for which the village extension worker was a mediator.

Collaboration between pest observer and village extension worker in Kuduagung was not as smooth as in Senengsari. The extension worker was in a quiet support role, and facilitation was merely done by the pest observer. Actually, the village extension worker wished to do more, but felt that he was not given the opportunity by the pest observer. He reported later that he had found the pest observer rather 'bossy', as if he was the one who owned the programme and did not want to share it with anybody else. The pest observer's comments on the functioning of his two extension workers was that they were 'quite helpful', but that they had too many other activities to be fully committed to the IPM field school.

Both the pest observer and the extension worker in the field school in Kuduagung used only Indonesian when talking to the farmers, which appeared to be the language usually used in farmer group meetings. Some farmers admitted that they had problems following the lessons but said that it was not because of the Indonesian, but because of the many foreign (scientific) words that were used by the trainers. In other farmer meetings Indonesian was used, indeed, in cases where the meeting was chaired by certain people, e.g. the farmer group coordinator being a school teacher, or the village head. In other cases, Javanese was predominantly used, and obviously appreciated by the majority of farmers. In the IPM field school, the use of Indonesian visibly inhibited about half of the trainees from participating actively in the discussions.

Performance of trainees
The group of field school participants in Senengsari was quite heterogeneous, but it was interesting to see how they changed as a group. After about four weeks of training, they had formed into a coherent group. There was an open atmosphere allowing everybody to express their views, and they made a lot of fun together. One farmer, an ordinary farmer who had not been farmer group leader before and apparently a very committed and humorous person, was appointed to become the IPM group leader. Javanese, the local language, was commonly used, encouraged by the trainers' example. During the later weeks of the training, some farmers, mainly the village officials in the group, used more and more Indonesian, the national and official language, in the presentations. But others who felt less comfortable to express themselves in Indonesian were not discouraged to react in Javanese.

In general, there was a good cooperation in the small subgroups. Most members participated in making the drawings and conclusions of the agroecosystem analysis. More complicated group exercises, like counting rat population growth, determining economic threshold level, or executing an experiment with insecticides, were usually carried out by only two or three people per subgroup. The presentations were often performed by the same people, each group having about two representatives. At first, there was a lot of response to the presentations, and discussions were lively and useful. From the fifth week onwards, however, response decreased, and during the last weeks the farmers seemed in a hurry to finish the programme. Apparently, as was admitted by some farmers, they already understood the material, and would have liked to get additional lessons or exercises. However, they did not show an increased interest in the special topics during the last weeks, which might have been a result of tiredness or saturation by the time the special topics were done.

The group of participants in Kuduagung was a relatively select group, unrepresentative for the rice farmer population in this village, especially with regard to education and occupation. There were seven village officials and two school teachers in the group. These people dominated the exercises and discussions. One of the teachers did most of the exercises for the agroecosystem analysis and special topics on his own, neglecting his other subgroup members. He was the one who always did the presentations. Other subgroups also had only one or two representatives during the whole season to do the presentations. Neither the pest observer nor the extension worker ever dared to give a warning to the dominant people in the group. The pest observer admitted that he did not feel comfortable to rebuke someone older than himself. On the whole, the farmers were very participatory and eager to learn. Because of the relatively high intellectual level of the group, the discussions were of a good standard, and usually lasted long (on average almost one hour, compared to 35 minutes in Senengsari). In the course of the training, the analyses of the ecosystem improved significantly. More and more ecosystem factors were included in the presentations. The discussions, however, were strongly dominated by a part of the group only, partly due to the language used. Only a few farmers in this group gained communication skills because of the training programme. The others already either had them, or were not given the opportunity.

It took about four field school weeks before all subgroups in Kuduagung stopped giving recommendations for chemical control on the IPM field school observation plot. In Senengsari, on the contrary, the farmers did not talk about pesticides any more after the first week. This difference is definitely influenced by their previous practices and attitude regarding chemical control, Kuduagung farmers being high pesticide users, and Senengsari farmers low users.

6.5 Evaluation

6.5.1 *About the trainers*

As described above, both pest observers conducted the IPM field school with enthusiasm and commitment, at least during the greater part of the training. The Field Leader I, the direct supervisor from the FTF, valued the pest observer in Kuduagung for his technical and facilitation skills, although it was noticed that he often could not adapt his explanations to the knowledge level of the majority of the participants. The pest observer in Senengsari was valued for his social skills. His facilitation behaviour was more natural because of these social skills, although he sometimes showed an insecure attitude.

The farmers in Senengsari were very satisfied with both trainers (pest observer and village extension worker). They had noticed, and appreciated that the trainers observed and listened first, and only interrupted the discussions when mistakes were made. They found the explanations clear, and the language understandable. They also noticed that there was a good collaboration between pest observer and village extension worker. In Kuduagung, the farmers' opinion about the trainers was quite positive, but they expressed that they had difficulties with the language used by the pest observer. It was not only the fact that he spoke in Indonesian, but also that he used too many scientific terms which were not explained. A less close relationship was established here between trainers and trainees than in Senengsari.

6.5.2 *From the trainers*

The pest observer in Senengsari was not completely satisfied with the result of his IPM field schools. He found the group in Senengsari not coherent enough, and the farmers in Sugihsari had too many other activities. Actually, his expectations had been too high, and he was disappointed that he had not been able to meet them fully. This might explain his less self-assured attitude during the last part of the training. He complained about the supervision given by the REC Sumbersari. He did not feel supported by the REC head and the PPM Food Crops, and he felt that responsibility for the IPM field school had been fully on his shoulders. He was aware of the fact that the REC staff did not feel involved in the programme, and that they were not given a clear task, as was reported by the PPM Food Crops. This officer perceived the IPM programme as a training programme for pest observers with a theory part at the FTF, and a practical part in the field with farmers. He regretted that the REC staff and village extension workers (except the two that had been assigned as apprentice co-trainers) were still completely ignorant of IPM. This made him sceptical about the follow-up of the programme in the field, although he was positive about the impact of the training sofar. He found the training model attractive and effective.

The village extension worker was not as negative as the pest observer about the result of the field school in Senengsari. He knew his clients, and seemed convinced that something was initiated that could develop by itself. He believed that the follow-up activities would have a clear impact. His more positive attitude was definitely influenced by the fact that he would stay in the area and could still have a role in organising activities, whereas the pest observer had to return to the FTF for the coming half year. The extension worker was not completely satisfied about his own functioning in the field school. Because he was also involved in another extension programme, he had found it hard to divide his time well. Although the training of one week at the FTF was far too short, according to him, he had learned much from the field school with farmers, and the preparations of the field school sessions with the pest observer. He felt convinced that he could conduct a similar training on his own. In his opinion, the shortcomings of the FTF training lay specifically in technical aspects of IPM, such as knowledge on pests and natural enemies and on control measures.

In Kuduagung, the pest observer was very satisfied with the result of the field school season, and expected that the IPM technology would be disseminated in the villages that were given training. He was especially impressed by the field day in Kuduagung because of the high intellectual level of (a part of) the participants. He was also positive about the role of his supervisors at the REC and subdistrict office (in person of the '*mantri tani*') who had been very supportive according to him. The '*mantri tani*', indeed, visited a great part of the field school sessions, although she did not have any role at all, even did not have a short discussion afterwards with the trainers. The difference between the two pest observers in perceiving support from the REC, which was actually more or less the same, probably lies in different expectations.

Of his four field school groups, the pest observer found the group in Kuduagung most responsive, but the one in Sumberagung most coherent. According to him, the farmers in the latter group, which consisted of real farmers only without village officials, were creative, though the level of discussions was lower than in other groups. One of his other groups had three women participants who played a very active role in the field school. All three women had a high school diploma, and, unfortunately, left the village after the IPM season to find work in the city.

The village extension worker considered the IPM field schools in Kuduagung and Sumberagung successful. He was especially pleased with the response of the farmers in Kuduagung, but concerned about a follow-up. He was not satisfied with his own role in the field school. He felt that the pest observer had not given him the opportunity for any inputs to the training, and had wanted to keep all responsibility to himself. Communication had not been good between the two trainers. The FTF training was not sufficient to get prepared for the field school, according to this extension worker. Even after following the farmer field schools for a whole season, he estimated that he understood only about two-thirds of the materials on IPM. According to him, village extension workers need to be given more training, especially on group dynamics. This opinion contrasts with that of the extension worker in Senengsari who needed more technical training. Their needs correspond with their personal weaknesses. Obviously, the two extension workers are completely different personalities, the one in Senengsari having strong social skills, and the one in Kuduagung being technically stronger, but socially weaker.

6.5.3 *About the trainees*

The majority of trainees (68% in Senengsari, and 72% in Kuduagung) attended all ten session of the field school. Nobody attended less than six times. During the two last weeks of the training in Senengsari, when quite a few farmers were unable to attend because of other occupations, they were replaced by their sons or neighbours. During the field school sessions, which took place primarily in the open air, many bypassing farmers stopped and had a look what was happening. No one, however, attended a whole session or returned regularly, as reported for many other field schools throughout the country.

The participants were tested on their knowledge by means of a ballot box test, consisting of a pre- and a post-test. Results of the tests of the four field school groups are given in Table 6.4. On average, scores are higher for the pre-tests in Senengsari and Sugihsari than in Kuduagung and Sumberagung. The opposite is true for the post-test. As the tests were constructed by the two pest observers, and no significant differences are visible between the two villages from one pest observer, the scores are obviously influenced by the way the tests were formulated. This makes it difficult to compare groups. It is worth mentioning that the tests

Tabel 6.4: Results of the ballot box test in Senengsari and Kuduagung, as % of participants (N = 25 farmers for each village).

Score (range 0-100)	Senengsari		Sugihsari		Kuduagung		Sumberagung	
	pre	post	pre	post	pre	post	pre	post
< 30	0%	0%	0%	0%	28%	0%	24%	0%
30-49	48%	0%	60%	0%	48%	0%	52%	0%
50-69	48%	56%	40%	80%	24%	36%	24%	28%
70-89	4%	44%	0%	20%	0%	60%	0%	72%
≥ 90	0%	0%	0%	0%	0%	4%	0%	0%
Average	52	66	50	63	40	75	41	74
Learning gain	27%		26%		88%		80%	

in Sumbersari were in Javanese, and those in Sumberagung in Indonesian.

A more legitimate way to compare ballot box scores of groups is to look at the gain of knowledge between pre- and post-test, not at the actual scores. Learning gain is calculated by the difference between pre- and post-test scores as a percentage of the pre-test score. Groups with the same trainer showed similar gains. Considering the different compositions of the groups in Senengsari versus Sugihsari, and Kuduagung versus Sumberagung, this finding indicates that the field school training model is suitable for various groups. An analysis of ballot box scores obtained in a large number of IPM field schools throughout Indonesia during the 1990 dry season, classifying trainees by age, education, sex and farm size, also showed that no significant differences existed in the performance of farmers from various categories (Setti, 1990). If the scores of the groups of the two trainers are compared, the groups in Kuduagung and Sumberagung seemingly gained far more knowledge than the ones in Senengsari and Sugihsari. This difference, however, was not at all perceptible when the direct observations on the trainees are compared, on the contrary. The finding confirms the statement that the results of the tests depend on the way the tests are formulated. The pre- and post-tests made by the pest observer in Kuduagung and Sumberagung differed most likely in level of difficulty.

In the course of the ten weeks training, a remarkable change in social interaction was visible in the two groups, especially in Senengsari. The trainees learned rapidly to express themselves in an open way in front of the group. Where discussions were formal and led by a few people during the early weeks, more people (in Senengsari almost all) freely participated during the second half of the training.

6.5.4 *From the trainees*

Senengsari

All the IPM trainees in Senengsari considered the field school useful. Strong points of the training programme, according to these farmers, were that:

- they obtained knowledge about pests, natural enemies, economic threshold levels, and the correct methods and moments to control rats;
- IPM can reduce their expenditures;
- they learned to calculate their expenditures, and not use chemicals indiscriminately;
- agriculture can be improved with IPM;
- they learned to discuss and express their views in a group.

As weak points they mentioned that:

- the group of participants was not functioning well, especially in organising collective pest control activities; and
- IPM had not reached farmers outside the field school yet.

Some other issues that they called weak points were actually conditional constraints reflecting their own weaknesses, such as:

- farmers usually do not have enough money to buy all inputs according to the recommendations;
- they find it difficult to extend the field school materials to other farmers;
- they often could not come to an agreement in decision making during the discussions in the field school.

The majority of farmers enjoyed to learn in the field, and not in the classroom. As most interesting topics they mentioned the field observation/agroecosystem analysis, and the topics about rat control and rat population growth. The group differed in opinion about the topic of economic threshold level. Some farmers found it the most interesting, others the least interesting topic. Although most of the farmers reported that there was nothing that they disliked, a few people did not like the special topics on insect collection, and on 'carbofuran, monocrotophos and natural enemies'. Everybody had enjoyed the group dynamics exercises, and found them useful.

In the course of ten weeks field school, the trainees experienced many improvements of their own performance, especially in their field observations which became more accurate, and in their ability to identify insects, to estimate damage levels, and to take a decision. In the end, they felt much more comfortable to express their views in the group, and not just to agree with the general opinion, as compared to the first weeks of the training. They appreciated the way the pest observer and village extension worker had guided them. In general, they had no problems in understanding the trainers' explanations. The language was suitable, and they appreciated that Javanese was spoken. They were satisfied with the training materials.

Most of the farmers in Senengsari (85%) found the programme at the field school too relaxed. This finding corresponds with the observation that, especially during the last weeks, the participants seemed in a hurry to finish and go home. The reason is probably that many of them had other jobs or activities in addition to farming. On the other hand, they also expressed that they would have liked to spend more time in the field school, but with a more varied programme.

After the training, 60% of the graduates were convinced that they could identify pests. Even more farmers (84%) were familiar with natural enemies, and 92% could give a correct example of the impact of natural enemies on pest populations. Although the field school graduates could not identify all insects in the rice fields by name, they were always able to identify them by function. They generally had more problems in identifying diseases: only 48% felt knowledgeable. Most farmers reported that they knew what control measure they should take in case there are too many pests (84%). They still had problems with the concept of economic threshold level: 18 of the 25 trainees had no or partial understanding of ETL, but 40% could give a correct example of the application of an economic threshold level in pest control decision making, in the way it had been presented in the field school.

Kuduagung
Also in Kuduagung, IPM trainees without exception found the field school useful. They reported the following strong points:

- participants gained knowledge and understanding about pests, diseases and natural enemies in the rice crop;
- if IPM is implemented, expenditures and labour can be reduced;
- the field school focused on practice, and not on theory, so the material was easy to understand;
- it taught the farmers to always observe the crop first before taking a pest control decision;
- farmers had to be diligent, disciplined, and creative in the training programme;
- they learned how to calculate the expenditures for chemical control measures.

As weak points were mentioned:

- the period of ten weeks was too short, the field school should last until harvesting;
- IPM reached only a small group of farmers in the village, and had not been extended yet;
- the training covered rice IPM only, and did not include secondary food crops;
- there was no information about the success of IPM in other areas;
- the trainers often used foreign (scientific, Indonesianised English) terms that were not understood by the farmers.

Topics taught in the field school were perceived to correspond with the problems they encountered in the field. The most interesting topics for the trainees in Kuduagung had been the field observations and agroecosystem analysis, and the calculation of rat population growth. Farmers felt confident to calculate the percentages of damage and the economic threshold levels, but still found it hard to identify and draw stemborers and various sorts of caterpillars. In the individual interviews, 80% of the farmers reported that they fully or partly understood the concept of economic threshold level, and 72% good give a correct example about an application of the ETL in the way it had been presented in the field school.

The time spend on each topic was long enough, according to the farmers in Kuduagung. They reported that field observations only took five to ten minutes, although they spent, on average, half an hour on observing the crop during the field school sessions. They expressed that during the last weeks they got somewhat bored with doing the agroecosystem analysis, but the drawings became better and the discussions more lively. Most of the trainees admitted that the presentations of the agroecosystem should have been done in turns rather than by the same people all the time. But no one dared to take over from the few dominant people in the group.

The majority of IPM field school graduates in Kuduagung felt convinced that they could identify pests (76%), natural enemies (80%), and diseases (68%) after

Table 6.5: Yields of farmer field school observation plots (in tons/ha).

Location	Variety	IPM plot	local package plot
Senengsari	IR 64	6.1	6.0
Sugihsari	IR 64	7.9	7.8
Kuduagung	Semeru	7.1	7.8
Sumberagung	Cisedane	6.5	6.9
Grobogan (average of 16 locations)	various	7.1	7.2

having followed the field school, and that they knew what control measures to take (80%). Most farmers (80%) could give a correct example of the impact of natural enemies on pest populations.

6.5.5 *IPM field school observation plots*

The actual difference in treatment of the IPM observation plot and the one with the local technology package was one application of carbofuran granules at the time of the first fertiliser application, and in Kuduagung and Sumberagung also one application of foliar fertiliser.

According to the season-long observations executed by the field school participants, and reported in the agroecosystem analyses, pest and natural enemy populations did not differ much between the two treatments. At harvest, the yields of both plots were weighed. Yields obtained in the four IPM schools, in comparison with the average of all the sixteen field schools in the district of Grobogan, showed no significant difference between the two treatments (Table 6.5). In the two schools in the subdistrict Sumberagung, the yields of the plots with the carbofuran and foliar fertiliser applications were slightly higher than of the IPM plots not treated with chemicals. This is more likely a result of the foliar fertiliser than of the carbofuran since pest populations were observed to be similar. Local variation in the soil or water conditions of the two plots, however, is expected to be the most important factor determining this difference, since the effects of foliar fertiliser have, until now, been doubtful.

The results in the field schools in Grobogan are consistent with the results of twenty-two yield samples that were taken all over Central Java during the pilot extension season in 1989/90, not showing any significant differences in yields (6.1 versus 6.3 tons/ha for IR 64, and 6.8 versus 6.5 tons/ha for Cisedane) between IPM and local package plots (Salac and Peralta, 1990).

6.6 Follow-up activities

The IPM Programme provides an introduction to IPM through the farmer field school but expects that activities will not stop there. Therefore, follow-up activities in trained farmer groups are encouraged, through follow-up meetings, farmer-to-farmer training, and an IPM newsletter for, and partly by, farmer groups and extension staff ('*Laba-laba*' which means 'spider'). During the FTF training on IPM in secondary food crops, which follows the farmer field school season, pest observers are given two opportunities to return to their field schools for follow-up activities.

Sumbersari

In the subdistrict Sumbersari, the pest observer had already organised meetings with his groups immediately after the closing of the IPM field schools to initiate follow-up activities. Both groups in Senengsari and Sugihsari had agreed that they should get together with all the participants on a fixed day once in five weeks ('*selapan*', which is the time interval of thirty-five days on which a certain normal week day and a certain Javanese market day (five in one Javanese week) fall together). The pest observer could not attend these meetings himself, because it interfered with his FTF programme.

The group in Senengsari started immediately after the closing of the IPM field school. The meetings, held in the Village Centre, were generally attended by about 50% of the IPM farmers. To keep the group together, a revolving lottery fund ('*arisan*'), in which the group members contribute a certain amount of money (Rp 500), and take turns in winning the aggregate sum reduced by an amount for tea and snacks at the meeting, was organised every meeting. In addition to the '*arisan*', the meetings usually contained a talk by the IPM group leader, and sometimes by the village head. The village extension worker was present whenever he had time, and usually conveyed the latest extension messages from the REC to the farmers.

After four months, the IPM group leader together with the village head reformed the existing five farmer groups in the village that had been inactive for about eight years. They gave the IPM group a name, and appointed it as the coordinating organ for the seven new farmer groups. The IPM farmers were split up over the seven groups, and all of them were given a task as group leader, secretary, treasurer, or contact farmer (following the model of the Extension Service). Formation and tasks of the new farmer groups were discussed in the '*selapan*' meetings. A letter from the village head was sent to farmers outside the IPM group who were also assigned as contact farmers. The first activity of the new farmer groups was to organise rat drives, which went quite successfully. In spite of the many complaints during the field school about the unwillingness of the villagers, they could mobilise a fair number of people. The farmer group leaders were also made responsible for collecting a contribution from their members for collective rat control activities. In the first season, complaints were heard that hardly any other villagers outside the IPM group joined the organised rat drives, but later the system functioned fairly well. For the first time since years, rats could be effectively controlled in Senengsari. The regular meetings of the IPM group stopped after two seasons.

The IPM group in Sugihsari held its '*selapan*' meetings, also with an '*arisan*', in their farmers' shack located in the field. Before the meeting, which was scheduled in the morning, the members did observations in their own fields, the results of which were reported to the group leader. In the meetings, they usually discussed pest problems, and developments and problems in the farmer groups. The group seemed active and motivated to continue their meetings, field observations, and discussions, even though there were no serious pest problems. Although the activities were officially continued until the end of the study, active participation of the individual members decreased in the course of the seasons due to lack of motivation.

Sumberagung

In the subdistrict Sumberagung, the first follow-up meetings in the IPM villages Senengsari and Sumberagung were organised by the pest observer on the special return days during his FTF secondary food crop training. In Kuduagung, the farmers agreed to have meetings once in two weeks, combined with an '*arisan*'. They also formed a cadre of IPM farmers coordinating the normal farmer groups, not active hitherto. Each IPM farmer should have five followers. The general farmer group coordinator had the intention to integrate the ordinary farmer groups with the IPM group and the irrigation groups. The village extension worker was never present at the meetings, but not specifically invited either.

The first meeting took place mid January, six months after the field school was closed. Only nine farmers were present at this meeting. The following meetings were held irregularly and occasionally, attended by only a small number of field school graduates. The topics discussed had nothing to do with a follow-up of the IPM field school, but focused on the long-term planning of how to get the farmer and irrigation groups active again, coordinated by the IPM farmers. Two people (being the two teachers in the group) had been drafting this plan, and were very authoritarian in bending the other members to their will. Apparently, most of the IPM farmers just wanted to focus on pest problems within the IPM group meetings, but this was easily rejected with the remark that there were no pest problems, so they should focus on other problems like irrigation water. They should take advantage of the situation that finally something had been established to get farmers together (the IPM group). Within one season, the IPM group fell apart. The meetings had not been fruitful and motivating because they dealt with the interests of a few people only.

In Sumberagung, the IPM farmers also agreed to meet every '*selapan*', in combination with the planned regular farmer group meetings. The meetings were never realised, as had been the case with the 'regular' farmer group meetings for years.

The village extension worker in these two villages never attended any IPM follow-up activities, not unlikely because IPM interfered with his own interests since he was one of the main pesticide suppliers for the farmers in Kuduagung and Sumberagung. His personal 'follow-up activities' on the IPM field school were rather more aggressive promotion campaigns for pesticides and other agrochemicals. The conflict between IPM and the conventional rice intensification technology package, with which his generation of extension workers was brought up, seems to express itself most at the level of the village extension worker. The case in Sumberagung (Box 6.2) is a typical example in this respect, but fortunately not too common.

Box 6.2
IPM and the extension worker: a conflict at the village level

'Do you know what war we get after the Gulf War?', asked one of the IPM graduates in Kuduagung when I met him just after he came home from a meeting with the village extension worker and a seed company promoting hybrid corn. 'We, farmers, are going to have an Information War! In addition to our own experiences and trials, we get information from the extension worker, from the IPM field school, from salesmen who visit our village, from the village officials, and the more from traders in the market. And it is all not consistent, so we really get confused, and just go our own way.'

This is not the only farmer in this village who complains that the information received from various sources is contradictory. The village seems to be target for all kinds of projects and promotion efforts since an extension worker originating from a high-pesticide-use estate crops area was designated to the subdistrict. In 1988, a part of the rice area was determined by the district government to participate in the INSUS programme, obliging the farmers in that area to take the complete input package on credit from the KUD. During the 1990 dry rice season, the village received IPM training. The following intermediate season, 15 hectares fell under the new soybean intensification programme. Forced by the failure of the soybean crop, many farmers took credit packages from the KUD during the 1990/91 wet season for which purchase of pesticides and foliar fertilisers was made compulsory by the extension worker, leaving them with even greater debts because the rice crop was heavily attacked by stemborer. A demonstration plot for urea briquets was implemented by the farmer group leader in cooperation with the extension worker in the same season. During the 1991 dry season, the extension worker promoted a new type of insecticide (one of the types abolished by INPRES 3/86!) to some key farmers in the village by distributing free samples to be sprayed on the rice crop. Before the 1991 intermediate season started, several meetings with farmers were organised by the extension worker in cooperation with seed, fertiliser, and pesticide companies to promote hybrid corn, watermelon, the previously promoted insecticide, NPK and micro-nutrient fertilisers. All these activities were carefully planned by the extension worker without the involvement of the IPM group coordinator in the village, although several IPM graduates, like the confused farmer group leader mentioned above, were seduced to start using pesticides again by giving them free samples.

During the IPM farmer field school, this extension worker had performed a passive role by only opening and closing the meetings, and being in charge of domestic arrangements. He had not contributed much to the facilitation of IPM activities like agroecosystem analysis and special topics. In one field school session, when a part of the rice crop in the village was infested by bacterial leaf streak, the extension worker, before the pest observer was present, advised the farmers to spray the crop with fungicides. He recommended various types, informing the farmers casually that he could provide these chemicals. When he was questioned afterwards about the use of spraying fungicides against a bacterial disease, he expressed that it actually did not matter much but that farmers were not satisfied if they did not use pesticides, since they always ask for advice on chemicals. Several farmers in the village were, indeed, observed to spray the fungicides in the following weeks.

Considering his background in high-pesticide-use crops, this extension worker was used to the extra income through pesticides sales and commissions from formulators. The low income of extension workers, as of most civil servants, actually forces them to arrange an extra income source. Some have a photo studio, or grow crops, but others sell agrochemicals, a not illogical choice for an agricultural extension worker. The introduction of IPM, likely to reduce the pesticide use by farmers drastically, is an obvious threat for this last income source. Instead of looking for other sources, the extension worker in Sumberagung preferred to deliberately disadvantage the farmers in order to safeguard his own income, as long as possible. In 1992, he was transferred, back to the estate crops.

In many other places, extension workers soon realised that it is not their personal interest that should come first, but the farmers'. Several of them became committed to the IPM principles and the IPM training approach, accepting that this approach will change their role as extension worker, by making farmers more independent.

6.7 Comparison and trends

Some more schools

Although each IPM farmer field school has its own atmosphere, its direction of success, and its typical happenings, the schools observed in Grobogan were not really different from those in other places. According to the FTF supervisor, both groups scored fairly well among all groups during the same season, although they were not outstanding.

Many other groups in Central Java, North Sumatra and South Sulawesi were visited during one or more of their field school sessions in order to compare. After the first training implementation cycle, an interesting phenomenon occurred, the self-supporting IPM farmer field schools. These are field schools organised on the initiative of the people from the village, and supported with money from the village itself, not uncommonly from the farmers. Such initiative usually arises after having seen the example of a farmer field school in the neighbourhood. Mostly, the pest observer and the extension worker are invited to facilitate the training, but in other cases field school graduates (farmers) become the trainers (Anonymous, 1991). If a request is submitted, these field schools are supplied with materials from the IPM programme, but many groups conduct the training on their own resources.

One example of self-supporting field schools was reported by the village official for development affairs, a woman who is also farmer group leader, in a village in the province of Yogyakarta. Her village was selected for IPM training during the first implementation cycle (1990 dry season), for which she, in consultation with her farmer group members, selected thirty participants from four different hamlets. She was so enthusiastic about the training approach, especially because it empowered farmers to express their views, that she decided to organise IPM field schools for a couple of seasons until all thirteen hamlets in the village were trained. She was able to get money for notebooks and pens for the trainees from the village funds, and arranged tea and snacks from the farmer group funds. Other training materials were supplied by the REC, which appeared to be the leftovers of previous trainings. A clear change in perception of pest management was soon visible among the farmers in this village. Previously being 'pesticide-minded', farmers began to observe their crops thoroughly and reduced their pesticide use (and expenditures) considerably.

Boxes 6.3 and 6.4 describe two other examples of self-supporting IPM farmer field schools that were observed throughout their trainings. One was extremely successful, mainly due to the motivation provided by the farmer group leader, who was the initiator of the field school. The other can be called a failure compared

Box 6.3
Giripeni: an example of success

After a very successful IPM training had been conducted in one hamlet in the village Giripeni during the pilot cycle of the IPM programme, the farmer group leader in the neighbouring hamlet became interested to revive his farmer group with a similar training. This initiative received full agreement and support from the village government which donated some money to realise the idea. The two trainers and a representation of the farmer group members (including four women) were called together, and the field school started a few weeks after the rice was planted. A volunteer contribution from the participants was requested to arrange tea and snacks during the sessions, and a uniform T-shirt with the name of the group was made for each member from the donations of the village government and a rich villager.

The group members afterwards expressed that they initially participated in the IPM field school only because they had received an invitation from the hamlet head without having any idea what the field school would involve. After attending the first field school session, they became interested in the material, and wanted to learn more about IPM. Within one season, all of them became motivated IPM practitioners, actively monitoring their own crop, and hardly ever using pesticides any more. They valued highly what they had learned, both with respect to pest management, and to group processes.

After having harvested the rice crop, most farmers in the village planned to plant soybeans. The farmer group leader assumed that the same principles of IPM would hold for soybeans as for rice, and, thus, suggested to organise another IPM field school in the soybean crop. This gained enthusiastic response from the group members. Also the trainers were willing again to facilitate the group, and the village government to give a donation. Although the soybean field school lasted only three weeks, it had a clear impact on the farmers. They paid more attention to this usually neglected crop, they learned to identify pests and natural enemies (at least on the young crop), and to recognise the importance of weeding. They started experimenting with fertilisation, and they adopted the habit of doing weekly observations. Since they continued the observations after the field school had stopped, they learned and identified the problems at the later stages of the soybean crop by themselves. The farmer group leader assured that this would not be the last IPM field school in the village. He was planning to train new groups of farmers during every season until everybody in the village who wanted to learn about IPM had had his or her opportunity.

A story like this must have its reasons for success. The situation of being self-supporting and supported by own people and village government appeared to reinforce the group. The farmers were proud to be independent, which increased their motivation to continue or initiate new activities. Also, the physical as well as mental support from the village government was a positive influence in the success of this group. Most important, however, was the enthusiastic guidance of the group leader. He managed to be informal within the group (a common problem in farmer groups when the leader is too concerned about the formalities demanded by the Extension Service, therewith losing the interest of the members), but he was formal to the outside world which resulted in much attention to the group. This attention, in turn, encouraged the group by feeding their feeling of pride. On the question what is necessary for farmer groups to become active and self-supporting, the members answered that it is a mix of awareness, deliberation among the members on all topics of concern for the group, guidance by respected persons in the village, and support by the village government.

to the average achievement in farmer field schools, although a small group of farmers benefitted from the training. This failure was mainly due to the fact that the initiative came from the village secretary who did not involve any farmers in the preparation and selection of the participants. Most of the farmers appointed

Box 6.4
Giriselo: an example of failure

At the start of the 1991/92 wet season, an IPM farmers field school was opened in the village Giriselo on the initiative of the village itself. Giriselo is a fairly small village, with a rice land area so small that the average farm size per family is less than 0.2 hectares. Most farmers are owner cultivators but with such small lands that farmers are forced to have other enterprises. For extra income, many families also raise chicken. Another consequence of the small area of arable land in the village is that the village officials receive very small salary lands. Therefore, all village officials, including the village leader and secretary, need to have other employment as well. The two farmer groups in the village have been active in the past, mainly thanks to the support of the rice mill owner who funded farmer group activities for some years. This support stopped after the election of the present village head. This village head enforced some rules on the use of rice mills, which reduced the spirit for collective activities amongst farmers. The present farmer group leaders are not able to get their farmers together. According to the farmers themselves they are not interested, admitting that rice cultivation is not a major concern.

The initiative to organise the IPM farmer field school came from the village secretary. He had heard about IPM training on the radio and in the subdistrict office from people of the neighbouring village where two IPM field schools had been successfully implemented. He became interested, approached the REC where in consultation with the REC head, the pest observer, and the village extension worker a plan was made soon. The village secretary listed thirty-five candidates, and the pest observer requested for training materials from the IPM Secretariat in Yogyakarta. This was all done within a few days to be able to open the field school the following week since the crop was already two to four weeks after transplanting.

The candidates listed by the village secretary included people (all men) who had previously been active in the farmer groups. The village officials involved were the village secretary himself, and his son, the hamlet head. No preliminary meeting was organised but trainees were told by their neighbourhood heads to be present at the farmer group meeting place, a shed in the field, for an extension meeting.

The overall impression of the field school season in Giriselo is one of a chaotic organisation and implementation, due to the following factors:

- the composition of the group changed every week, involving bypassing farmers that were not originally listed, except for a core group of about eight people; average number of attendants was only fourteen people;
- it was difficult to get the participants (taking care of their chicken) together in the morning, causing a late start of the field school;
- there was no fixed time schedule of activities per session, so farmers did not know what to expect for the next week;
- no routine field monitoring was done because of the late start in the morning and the changing composition of the group;
- agroecosystem analysis was not done during the first five sessions because there were no plywood boards;
- there was no communication between pest observer and extension worker, causing weak connection of topics for successive weeks when either of them was not present.

The pest observer had a hard time to get this group to the core of the IPM field school: routine field monitoring and analysis, because of the changing composition of the group. No group dynamic exercises were presented by the pest observer, losing an attractive element of the training, and special topics were limited because of the late start, and thus short duration of the field school. After 11:00 a.m., the farmers wanted to go home because they were hungry and thirsty since no tea and snacks were served during the sessions.

Compared to other groups, most farmers in this group found it very hard to express themselves, both in drawing what they have observed and by presenting it. The pest observer had to push them hard. Except for the village officials, the farmers hardly ever wrote in their notebooks. There were a few farmers in the group who could not write. Very good attention, however, was paid to lectures, and the trainees participated actively in the field and in discussions. Although the average level of education was relatively low in this group, it does not have to be a constraint in the field school implementation. The contents and activities, however, should be adapted to the abilities of the participants which was not always done by this pest observer. The group is an obvious example of a second choice group, not fulfilling the criteria commonly used for field school group selection.

Despite the less successful course of the field school, at the end if the ten weeks (the core group of) trainees reported that they had achieved the following from the field school:

- gained knowledge on pests and natural enemies;
- are now able to distinguish pests and natural enemies;
- learned a good way to grow rice;
- learned not to use pesticides indiscriminately, because the natural enemies will die otherwise;
- stopped or reduced using pesticides.

A main constraint in the field school implementation in Giriselo lies in the initiation of the training which was on the initiative of the village secretary. Even selection of participants was done without any involvement of farmers. Good preparation at the village level is considered to be very important before the field school starts. In addition, no active support from the village government was displayed, neither in organisation nor in materials (e.g. tea and snacks for the farmers). The village secretary and the farmer group leader attended the field school only twice. Farmers expressed that they missed support from the village government, which they ascribed to the fact that the village officials have very small salary lands, and, therefore, are not really interested in these duties. Other farm enterprises than rice growing also seemed to hamper the implementation of the field school. One factor is time allocation (farmers had to feed their chicken first in the morning), another factor is concern (rice is not their main concern). Good communication and deliberation between pest observer and extension worker is found important. Especially if they can not be present together every session, they should meet to ensure good connection of topics.

to attend the training, did not have the interest and time for the IPM field school since they operated on miniature rice fields, and, therefore, depended more on other enterprises, like chicken raising.

Factors determining success

Reviewing all IPM farmer field schools visited in various areas and under various training models, there was one thing that all groups had obviously in common: commitment to do the field school, from both farmers and trainers, in particular the pest observers. The field school model applying field-based and experiential learning was, in principle, positively received by all farmers met, regardless of location, group composition, or trainers. The level of success, in the sense of individual learning, group learning and implementing, varied per group and per individual. Various factors determining this success could be identified based on the (be it limited number of) observations.

A factor strongly determining success was found to be the relationship between the trainers and the trainees. Good relationships were established through open interaction between trainers and farmers. The facilitation approach of training encourages, and actually requires, an open interaction. Although the ability to become a good facilitator, able to establish an open atmosphere in the farmer group, depends to some extent on the personality of a trainer, most pest observers, including those with less natural charm, have become good facilitators. Thoroughly experiencing the rewards of such a training approach themselves during the FTF training is likely to have contributed much to make them good facilitators. Only a few examples of pest observers feeling uncomfortable with the facilitation approach were observed.

A second factor identified to be important for success is the composition of field school groups. A favourable composition is not determined so much by given characteristics of the members, but rather by the presence of one or more enthusiastic, motivating persons in the group, and by mutual respect among the participants. Related to respect is active involvement in farming (non-operators were not considered serious discussion partners), and non-dominance in the group. Village officials were observed to have both inhibiting and encouraging roles in farmer field schools. Inhibiting in the sense that there were too many of them or that they were too dominant, and encouraging in that they can assume a leading role in taking the farmers' problems to a higher level of decision making.

Group dynamics exercises were observed to enhance the coherence and collaboration of the field school groups, which is an implicit objective of the IPM Programme. It was noticed, that these group dynamics exercises tended to be omitted by the pest observers in the trainings they gave after their graduation (such as the self-supporting groups). Two reasons were mentioned by several pest observers not to continue with this attractive part of the training, (1) because there was no budget to buy materials for the exercises (although the costs involved were very low), and (2) because they faced lack of time for both preparation and implementation during the field school. Some pest observers expressed that they feel awkward to do the 'games' with the farmers who are, on average, quite old, although they admitted that response of farmers to the exercises and effects on group processes were positive. Continuation of the group dynamics exercises in IPM training by graduated trainers should be encouraged by the programme.

The importance of rice farming for the participants is identified as a third factor determining success. The more important the rice crop is for the subsistence of a farm family, the more likely a farmer is motivated to attend the field school and to implement IPM. A smaller impact of the IPM field school was noticeable in the case of farmers with other occupations or enterprises. In some cases, the compensation money received from the programme by the trainees was an extra motivating factor to attend the field school for farmers with other occupations. In some self-supporting groups not receiving the compensation, with many busy farmers having other enterprises, a lower attendance of trainees was visible.

Related to the first factor mentioned, an interesting observation was made about the pest observer from Senengsari. This pest observer was visited one year after he had finished his complete IPM training, and had begun to train other farmer groups in his REC area. It was interesting to notice how this pest observer had undergone a change, showing much self-confidence now without having lost his commitment. He had turned into a 100% facilitator, but firmer in leading the group, excited about what he had learned, and thrilled to help the farmers to learn the same. A better collaboration between the pest observer and the extension worker was visible. The extension worker had a more definite role in the field school, and tasks were more equally distributed between the two trainers, mainly because the pest observer could leave more responsibility to his mate. The same improved relation between the two trainers was observed in other places where the pest observer had graduated as the REC's IPM trainer.

6.8 Conclusions

This chapter covered only a few cases out of many IPM farmer field schools conducted sofar. The conclusions below, and their relevance to the entire programme, should be interpreted in this context.

Success

The two IPM farmer field schools observed in the villages Senengsari and Kuduagung, as almost all field schools, were successful. The trainers performed as committed facilitators, being non-directive, and encouraging farmers to discover and learn from their own experiences. There was a good attendance and active participation of the trainees throughout the season. All farmers gained knowledge if results from pre- and post-test are compared. More important, the farmers themselves expressed that they learned much the during ten weeks field school, especially regarding identification of pests and natural enemies, rat control, and analysis of costs and benefits in pest control.

The IPM demonstration plots did not need any chemical treatment during the field school season, according to the trainees' observations and analyses. The yields were not significantly different from plots with prophylactic carbofuran treatment. In other field schools throughout the country pesticide applications were not necessary either (Oka, 1990b).

Success of the IPM farmer field school implementation at the village level is determined by several factors, including the relationship between trainers and trainees, the ability to facilitate rather than teach by the trainers, the composition of the group, the support by leading persons in the village, and the importance of rice farming for the trainees.

Composition of field school groups and representativeness

The composition of the field school groups observed was not representative for the farmer population in the villages. The groups contained many village officials, and farmers with relatively high education, large (owned) fields, and off-farm jobs, and no women at all. The groups, especially the one in Kuduagung, were often dominated by village officials and people with higher education (teachers). Selection of field school participants was done by the village officials and/or farmer group leaders without any criteria given by the village extension worker.

A possible role that the programme can have in obtaining more representative, and socially stronger groups is by supplying stricter guidelines for the selection procedure favouring sofar neglected groups. Women participation in training should be encouraged.

In addition to non-representativeness of field school participants, villages were not representative either. The villages selected for training were those located on, or close to, the main road, where people have more access to information and inputs than in more remote villages. This implies that the IPM farmer field school model as applied during the first implementation season has been tested on an non-representative group of farmers. This does not necessarily mean that the model would work differently on the 'average' farmer in Central Java, if existing. It is not unlikely that a training based on experiential learning is even better received by the groups of farmers underrepresented sofar, as farmers in remote places, small farmers, sharecroppers, women farmers, in general the ones who are fully occupied with rice cultivation, and whose subsistence is more dependent upon a good crop. However, graduated pest observers need to possess the ability to adapt to situations and groups of farmers other than they experienced during their first training season, on which they should be prepared during their FTF training.

Field school activities

Activities in the field schools in Grobogan, and in other areas visited, looked very similar, and similar to the activities in the pest observer training at the FTF. Obviously, the pest observers succeeded in fulfilling their task to replicate in the field what they have learned themselves at the FTF, as also observed in other areas (Pontius, 1990).

Farmers enjoyed being trained in the field instead of in the class-room. They expressed that the exercise of drawing the agroecosystem had been very useful, especially during the first weeks to learn and identify the pests and natural enemies. Both groups, however, reported that they got bored to do it during the later weeks, because they thought to have mastered the material, and would have liked to learn about other topics. From the training point of view, it is likely that making the drawings is still useful until the end, but it could take less time. The pest observers should adaptively compose the activities in the field school sessions in accordance with the progress that the respective groups make.

For the field sampling in the IPM field schools, farmers began by using the observation forms, but they skipped this later and just noted down what they observed in the field. They were given a formula to calculate percentages of damage and insect occurrence. Calculations were often incorrect and confusing. In the observations in their own fields, none of the farmers uses forms to note down scores, nor calculates percentages. A lot of time in the agroecosystem analysis and the discussions was spent (and wasted) on these scores and percentages, which are of no practical use to farmers in the field.

The term economic threshold level (ETL) was often used to justify a decision not to apply a chemical control method, but without any relevance to the factors that determine an ETL. Many farmers reported that they did not understand the concept well. The trainers did not seem familiar yet with the new approach of the ETL either. Considering the use of observation forms, percentage calculations, and preset ETLs, pest observers apparently are not free yet from the old IPM concepts, which appeared not to be suitable for implementation by farmers in their own fields.

Trainers

All pest observers considered were able to perform as facilitators in the training, rather than being teachers. This role sometimes seemed to inhibit them to steer the groups, probably afraid of being too dominant. Steering of the groups was necessary in managing the time schedule, interfering in discussions that lasted too long, and in cases where certain participants were dominant. Farmers were often waiting for the trainers to determine the activities, which resulted in situations with activities lasting too long without progress. It is expected that a smoother time management will benefit the active involvement of the trainees. Minimising the dominance of people is necessary to give all participants a chance to execute tasks, e.g. presentations, experiments, drawing, and, therewith, to learn. More training in group leadership with emphasis on managing time schedules, interfering in discussion, and minimising the dominance of participants might improve the facilitation skills of the pest observers.

The trainers who used the local language in their speech were better understood by the farmers than the ones who used the national language, Indonesian. In some cases, scientific terms used by the trainers were not understood by the farmers. Pest observer and village extension workers should be encouraged to use a language that is understandable for farmers, without scientific terms, and preferably in the local language.

The two village extension workers in Senengsari and Kuduagung were not satisfied with their functioning in the IPM field school. They often did not know what was expected of them. This feeling is probably influenced by the fact that they are usually the trainers, and now they had become apprentice co-trainers. This notion is supported by the observation that they did very well on their own on the days that the pest observers were at the FTF. The preliminary FTF training of one week was found insufficient to get prepared for the field school programme, but apparently enough was learned during their apprenticeship at the field. A clearer task description for village extension worker in the IPM field school might improve their performance.

For pest observers to become the future IPM trainers at the REC, a change in their job description is needed stating that they should be given the opportunity to train instead of only doing their regular pest surveillance.

Trainees

All trainees observed made progress in the ballot-box test. The level of the gain varied with the trainers, depending on the way the questions were formulated. This hampered group comparison. If the aim of the ballot box test is to assess the impact of the programme, preset questions should be used to eliminate the trainer's bias.

More important, farmers themselves perceived the IPM training to be extremely useful and interesting. They were excited about their improved ability to identify pests and natural enemies, allowing them to become better, independent decision makers.

Follow-up activities

In three of the four field school groups observed, follow-up activities were organised, including (more or less) regular meetings, reformation of farmer groups, and collective rat control activities (Senengsari). The role of the pest observer and/or village extension worker in initiating these activities was important. Farmers became aware that the programme is devolved upon them. Follow-up visits of the village extension worker appeared to be stimulating, because the farmers felt that they were given attention. Village extension workers should be encouraged to follow-up on activities in the field, but to leave the responsibility to the groups themselves.

The follow-up activities stopped or decreased after a few seasons, because collective activities became less necessary considering the pest status in the area, or because the group leaders showed an interest in the continuation of the group different from the other members. Having a common problem seemed to be very important for the coherence of a group. Groups observed in other areas which succeeded to continue had a concrete activity plan, e.g. training more farmers in the village, or experimenting with rice-fish culture. Collective goals and activities appear to be of paramount importance for the functioning of groups.

Consistency of the training model and its field implementation

As was suggested in Chapter 3, a training model should be characterised by a consistent configuration of (1) the training contents, (2) the expected farmer behaviour, and (3) the nature of the intervention (Röling, 1992c). In the IPM farmer field school model, IPM principles are promoted through training methods based on experiential learning, with the expectation that farmers take better (informed) decisions. This model is, indeed, consistent (Röling and Van de Fliert, 1991). The implementation of the model at the village level was observed to be for the greater part consistent, too. The trainers presented the IPM principles and only occasionally gave preset recommendations, training was done mainly through facilitation, and trainees performed as independent decision makers in the field school. The training process highly allowed its contents to be applied. The cases where trainers gave recommendations, taught in the form of lecturing, and where farmers applied standard recommendation without informed decision making felt considered to be exceptions in situations where trainers or trainees were insecure with the new approaches, and fell back to practices or patterns induced by previous education.

7 Farmer field school graduates as IPM implementors

7.1 This chapter

Evaluating effects of IPM training at the farmer level includes many aspects of perception, practice, and knowledge relating to crop cultivation in general, and pest management in particular. Whereas Chapter 6 gave a description of the training processes and direct achievements of the IPM farmer field schools in the villages studied in Grobogan, this chapter deals with the effects of the training on the implementation of IPM principles by rice farmers in their own fields. Behaviours of IPM field school graduates are related to the objectives of the IPM programme, and to farmers' own criteria for a good IPM implementor.

The term 'IPM village' is used for the villages studied where the IPM farmer field school was conducted during the 1990 dry season (Senengsari, Sugihsari, Kuduagung and Sumberagung), in contrast to non-IPM villages without training (Jayasari, Plosoksari, Mulyoagung and Lempungagung; see overview of villages and seasons on the flipchart in the back of this book). The term 'IPM farmer' is used for farmers who are, or would be (when referring to the baseline season before training), selected to participate in the farmer field school, in contrast to 'non-IPM farmers' for untrained farmers in both IPM and non-IPM villages. The term 'field school graduates' is used for the IPM farmers who graduated from the IPM training.

In most tables and figures of this chapter, IPM farmers are compared with non-IPM farmers surveyed in the IPM villages only. Since large diversity was measured among the eight study villages, it seems more sensible to relate the behaviours of trained farmers with those of non-trained farmers operating their farms under comparable circumstances. The most important tables belonging to this chapter are presented in the text (indicated by Table 7.*), whereas additional, mainly descriptive tables can be found in Appendix V (indicated by Table V.*).

7.2 IPM competency objectives

The Indonesian National IPM Programme defined as its main training objective the ability of rice farmers to take informed pest management decisions, which stands in contrast to applying preset recommendations, such as prophylactic pesticide use. In other words, farmers are expected to become independent managers of the rice agroecosystem, expressing thus the ecological approach of IPM. The four principles of IPM as applied in the Indonesian programme: (1) grow a healthy crop, (2) observe the field weekly, (3) conserve natural enemies, and (4) farmers become IPM experts and trainers, lead to the programme's competency objectives, or desired practice of field school graduates, which are defined as follows (Program Nasional PHT, 1990):

ad 1:
- choose a variety that is resistant or tolerant to the local disease and insect complex, which yields well under local soil and micro-climatic conditions;
- prepare a seedbed and grow seedlings that recover quickly after transplanting;
- apply correct amounts and combinations of organic manure and chemical fertilisers (N, P, K and Zn), according to soil conditions, at transplanting and panicle initiation;
- correctly identify the panicle initiation stage of the crop;
- irrigate the fields for weed management;
- remove weeds during the second and fourth week after transplanting and before harvesting;
- drain the field one week before harvest for more uniform maturation;
- determine correct harvest time for maximum production and quality after milling;
- dry grains for storage purposes and for better prices;

ad 2:
- recognise phytophagous insects, diseases, and rat damage in the field;
- know that 'pests' are defined by their population density (not by occurrence only);
- estimate population densities of insects, diseases, weeds and rats in the field in relation to economically damaging levels;
- analyse the population density of insects and natural enemies taking into consideration crop health, potential yield, water supply and other factors affecting yield, leading to a field management decision including agronomic and pest control practices;
- take early season action against rats when previous season damage was high, including community organising;

- when insect densities must be reduced by insecticides, choose the proper type for the specific insect, apply the insecticide with minimum exposure to self and non-target species, and with proper dosage and delivery;
- when diseases or insect populations are high, adjust varietal choice for the following season;

ad 3:
- recognise natural enemies in the field and know their prey;
- explain the effects of pesticides on natural enemies;
- promote survivorship of predators by managing habitats for their benefit (short grass on the bunds, straw stacks after harvest, etc.);

ad 4:
- implement IPM independently in own fields;
- disseminate IPM principles to the surrounding farming community.

These objectives stress skills and knowledge of farmers, although a changed perception is often necessary prior to practising the knowledge and skills. Knowledgeable and more skilled farmers are supposed to make better decisions, which in turn is supposed to lead to a better (more efficient, more profitable) farm output. In addition to competency objectives relating to the programme's main objective of improved decision making, the farmer field school model bears another, more implicit, objective, i.e. farmer group reinforcement. Stronger, more coherent farmer groups are expected to result in collective action of farmers, which is important for the effectiveness of IPM implementation. This relates directly to ecosystem conditions: if all farmers in an area practice IPM, a more balanced ecosystem is obtained which, in turn, reinforces IPM implementation. Other areas that require collective action are rat control, irrigation management, and the conservation of large predators (for instance, by preventing poaching). Farmers trained in the IPM field school are expected to become organisers of collective activities in the village. An even higher goal, implicit in the fourth IPM principle, is that farmers become IPM trainers.

Farmers themselves also have a clear perception about the distinctive characteristics of a good IPM farmer. In group interviews with field school graduates in Grobogan, almost two years after the IPM field school was conducted, the following characteristics of an IPM implementing farmer were unanimously listed by the participants:

- monitor the crop frequently;
- do not use pesticides indiscriminately;
- can identify pests and natural enemies;
- give advice to other farmers.

In line with the programme's main objective, farmers emphasise monitoring of the crop and informed pest management decision making. Extra qualifiers of good monitoring were given by the farmers, including thoroughness, diligence, and creativity. In addition, they expect themselves to disseminate what they have learned to other farmers, which connects with the programme's objective of collective action. Several other interesting characteristics of a good IPM farmer were expressed, such as being always alert on pest occurrence, being patient in observing the development of pest populations, knowing the life cycles of pests, weighing expected yield loss against control costs before acting, informing other farmers when pests occur, meeting regularly with the farmer group, and applying timely and balanced fertilisation. These characteristics can be categorised into three groups emphasising practices including decision making, knowledge, and dissemination to other farmers.

In the following sections, practices and perceptions of IPM field school graduates are discussed following the sequence of the competency objectives above, but wherever possible related to the programme's and farmers' higher objectives of IPM training.

7.3 The four IPM principles in practice

7.3.1 *Grow a healthy crop*

Growing a healthy crop actually implies a whole set of agronomic practices and knowledge on plant physiology. As stated in the competency objectives above, most relevant to pest management are the use of resistant varieties, crop establishment through transplanting, good and timely fertilisation, water management, timely weeding, timely harvesting, and good post-harvest processing.

Rice variety

Rice varieties planted by IPM farmers are not different from those planted by other farmers in the study villages, since the selection is more or less limited to short duration varieties in view of the supply of irrigation water and the prevailing cropping pattern. Among IPM farmers, a shift is visible over the four study seasons to a strong preference for IR 64, a brown planthopper tolerant variety, as among non-IPM farmers. In general, this IRRI variety has become popular throughout Java because of its high productivity and good taste (Winarto, 1992), and it spreads more rapidly and widely than other high-yielding varieties (Fox, 1991).

A slight change in IPM farmers' reasons for selecting a rice variety was observed, in that around 10% more farmers mentioned pest tolerance as a factor in their decision making for varietal selection after having followed the IPM field school (Table V.1). In addition to pest tolerance, the high-yielding ability of a variety remained an important characteristic for field school graduates.

Crop establishment

In the study area, as in most parts of Indonesia, rice crops are traditionally established by transplanting. Careful seedbed preparation is, therefore, an inherent practice to began the growing season. Although transplanting is a common practice, its inclusion in the list of IPM competency objectives can be seen as a recommendation to prevent the adoption of broadcasting techniques which generally bring along great weed problems.

Using high quality seed is the basis for a good quality crop. Chapter 5 already

Table 7.1: Seed generation used in 1990/91 season, as % farmers reporting.

	IPM	Non-IPM
certified seed	29%	20%
1st progeny	39%	40%
2nd progeny	40%	23%
3rd progeny	2%	8%
later progenies	4%	12%
do not know	4%	8%
N	*93*	*233*

Totals exceed 100% due to multiple response.

stated that farmers in the study area often use self-grown, progeny seed that might have lost (part of) its germination capacity and in-bred characteristics. In comparison to the non-IPM farmers, field school graduates use fairly high quality seed. Only 6% of IPM farmers used seed of third or later progenies for the 1990/91 wet season crop, in contrast to 20% of non-IPM farmers (Table 7.1). The proportion of IPM farmers using certified seed is particularly high in Sugihsari and Kuduagung (46% respectively 39%), not unlikely because many of them took an input package from the KUD in that season, including obligatory seed.

Fertilisation

No significant differences were observed in fertilisation practices between IPM and non-IPM farmers over the four study seasons (Table 7.2; for more complete comparison see Tables V.2 and III.15). The same average dosages are applied, and the same variation among the villages could be observed in that Senengsari farmers give the highest quantities. The same trend is visible that dosages of urea and TSP increased in the course of the four study seasons. Reasons mentioned by IPM farmers to use urea and TSP are similar to those by non-IPM farmers. Regarding frequency and timing of fertilisation, including basal application and reasons for practices, no differences were observed, either, between IPM and non-IPM farmers.

In the baseline season, relatively more IPM than non-IPM farmers applied KCl and ZA fertilisers to their rice crops indicating once more that these selected trainees are the more innovative farmers in the villages. In later seasons, fewer IPM farmers continued to use these fertilisers, especially ZA, and specifically in the

Table 7.2: Average urea and TSP dosages in kg/ha over four seasons.

	IPM				Non-IPM			
	1989/90	1990	1990/91	1991	1989/90	1990	1990/91	1991
urea	255	260	290	312	272	287	300	323
TSP	220	210	235	243	229	230	245	263
N	*94*	*48*	*93*	*42*	*258*	*129*	*233*	*115*

villages Senengsari and Kuduagung. Three main reasons, similarly to those of non-IPM farmers, are given not to use KCl and ZA, including no necessity considering the fertility of the soil, shortage of money, and doubts about the result. Some farmers believe that KCl can be substituted by TSP, and ZA by urea, only the latter being true. The farmers who still use KCl have experienced that crops lodge less and grains are heavier.

The only fertilisation topic discussed in the IPM farmer field schools was balanced fertilisation, presented by the village extension worker in the same, ineffective, way as is usually done in extension meetings. Quantities in kg/ha, without explaining the way to convert them to local area measures, were used. Little explanation was given about the reasons to use the various nutrients, and about the relation between fertilisation and pest management. No attention was paid to farmers' ability to determine the correct time for fertiliser application through identifying the panicle initiation stage of the crop. For these reasons, it was only to be expected that no differences in the fertilisation practices of IPM farmers were observed after training, other than those following trends in the village. Considering the apparent inefficiencies in fertiliser application, especially TSP, much could have been improved through the field school. The topic seems very relevant in view of the ecological and economic aspects of IPM. The training approach of the field school leaves much room for experimentation and discussion.

One remarkable difference with respect to fertilisation was found in the number of IPM farmers using foliar fertilisers on the wet season crop after having received training. Whereas previously only one or two IPM farmers used foliar fertilisers, in the post-training, 1990/91 wet season, this number increased considerably in Sugihsari (29% of IPM farmers), Kuduagung (35%) and Sumberagung (39%) (Table 7.3). The use of foliar fertiliser also increased among the non-IPM farmers in Kuduagung and Sumberagung, but not as much as among their IPM-trained colleagues. Considering that in the IPM field school no attention was paid to foliar fertilisers, of which the effects had not been verified, and that application of these substances was mainly a once-only action, the sudden use of foliar fertilisers by a part of the field school graduates can not be related to the IPM training. The increase in foliar fertiliser use can be explained by the fact that more farmers, especially in Kuduagung and Sumberagung, took an input package, including obli-

Table 7.3: Application of foliar fertilisers over four seasons in the four IPM villages, as % farmers reporting.

	IPM					Non-IPM				
	Senengsari	Sugihsari	Kuduagung	Sumberagung	*Total N*	Senengsari	Sugihsari	Kuduagung	Sumberagung	*Total N*
1989/90	4%	4%	0%	8%	*94*	0%	0%	1%	8%	*259*
1990	9%		8%		*48*	3%		9%		*129*
1990/91	0%	29%	35%	39%	*93*	0%	4%	13%	28%	*233*
1991	0%		14%		*42*	0%		11%		*118*

gatory foliar fertilisers, on credit from the KUD in this season, forced by the failure of the soybean programme in the preceding season. Farmers in the two villages mentioned were told (and forced) to take the whole package regardless of their needs. The low use in Senengsari, where hardly any affiliation with the KUD exists, confirms this relation between KUD packages and foliar fertilisers. Likewise, low use during the following (1991) season can be related to the fact that farmers with debts at the KUD (which were, after the stemborer outbreak, most of those who had taken a loan before) were not allowed to apply for new credit from the KUD.

Weeding, water management and post-harvest practices

IPM farmers have the same habits of weeding the crop as non-IPM farmers, which is somewhat later than the time suggested in the competency objectives (two and four weeks after transplanting). As explained in Chapter 5, farmers' weeding schedule depends on their fertilisation schedule which is delayed, compared to the recommendations, since basal application is generally disliked. Therefore, weeding is done after fertilisation, presumably leading to inefficient use of fertiliser.

Water management is an art traditionally mastered by rice farmers themselves. Although they are dependent upon water supply from the irrigation schemes, farmers know exactly when the field needs to be flooded or drained. This topic was, therefore, not touched in the IPM field school, except in relation to prevention of the spread of bacterial leaf streak infection which occurred during the training season. IPM farmers' water management practices did not change after the field school. The same holds for harvesting and post-harvesting practices.

Maintenance of the village irrigation system is the collective responsibility of the farmers. Problems with irrigation due to bad maintenance are often a result of difficulties in organising farmers to fulfil this responsibility. In many IPM groups throughout the country, collective farmer activities relating to water management were organised through the farmer field schools. The field school model provides suitable conditions to organise these activities, since farmers are encouraged to discuss their problems openly together and to collaborate. When water problems occurred in Senengsari during the IPM training season, these problems were discussed at length in the field school (Appendix IV). The farmers mainly blamed the village officials for not taking action against the subdistrict's irrigation authorities, but did not organise anything themselves. Nevertheless, the result was that some village officials took action and, whether due to this intervention or not, water problems were solved within a week. In Kuduagung, an attempt was made by the farmer group leader to organise irrigation maintenance through the follow-up of the IPM field school (section 6.6). This initiative failed and even caused the IPM group to fall apart due to no interest of the group members. The authoritarian plan of the group leader, not in line with the experiences on group work that the members had gained in the field school, worked adversely.

7.3.2 *Observe the field weekly*

Habit of field monitoring

The amount of time spent in the rice field for field monitoring and water management outside the labour peak periods was estimated with the farmer respondents during the baseline survey. IPM farmers were not found to have a different behaviour in visiting the crop from non-IPM farmers (Table III.8), despite the fact that IPM farmers, on average, operate larger fields. The estimates of time spent on monitoring were not appropriate to serve a function of measuring change in the field monitoring behaviour of field school graduates. The figures represent the habitual pattern of field visits: go to the field at early morning, check the water and the crop, do some necessary work, and chat with fellow farmers until it becomes too warm. What matters more in IPM implementation is the quality of monitoring. IPM field school graduates are expected to look at natural enemies, and consider pest densities rather than pest occurrence, in relation to the development stage of the crop. In order to get information about farmers' field monitoring behaviour, the topic was informally discussed in both group and individual interviews, field school graduates were observed in the field, and they qualified each other's field monitoring activities in a focus group interview almost two years after the training.

Field school graduates were observed to stick to the same field visiting behaviour as before IPM training, including frequent visits (at least once a week) at the vegetative stage of the crop, and less frequent visits at the reproductive stage. During field visits, the crop is monitored for pests and natural enemies by most of them, although usually not in the systematic way learned during the field school. Some farmers inspect many hills from bottom to top, others only a few, and some farmers admit that they only step into the field to have a close look if they discover some symptoms while walking around on the bund. The number of farmers doing field observations more thoroughly in comparison with before training noticeably increased. In Senengsari, the village study assistant, a young farmer and field school graduate himself, estimated that sixteen of the twenty-five field school participants (64%) monitor their crop at least once a week (during the vegetative stage). Six of them monitor incidentally or not very thoroughly because they have other activities, whereas three always leave the village after crop establishment to work in the town, and never do any observations. In Kuduagung, about the same proportions of more and less diligent field observers were identified, supplemented by two field school graduates who do not operate a farm at all (a teacher and someone who could no longer afford to rent land). The farmers less regularly observing the field are mainly the village officials in this group. Several farmers in the four villages admitted in individual interviews that they are often not up to monitoring the field regularly and thoroughly, without giving a special reason.

Identification and perception of pests

In individual interviews at the end of the 1989/90 wet season (before training) and of the 1990/91 wet season (after training), IPM and non-IPM farmers were

Table 7.4: Knowledge about stemborers and leaffolders, before (1989/90) and after (1990/91) IPM training, as % farmers reporting.

	IPM		Non-IPM	
	before	after	before	after
Stemborer deadheart:				
know symptoms	91%	91%	88%	85%
know appearance	44%	58%	35%	35%
do not know	9%	5%	11%	15%
Stemborer whitehead:				
know symptoms	95%	94%	92%	97%
know appearance	73%	90%	57%	69%
do not know	5%	2%	7%	2%
Rice leaffolder:				
know symptoms	16%	43%	6%	9%
know appearance	10%	31%	4%	12%
do not know	84%	52%	94%	86%
N	*94*	*93*	*259*	*233*

asked to give a description of appearance and symptoms of six common rice pests. Table 7.4 shows the (change in) cognition of IPM and non-IPM farmers regarding the two pests (stemborer and leaffolder) on which field school farmers gained most knowledge, whereas Table V.3 gives a complete overview of the results on this topic for the four training groups studied.

Before training, IPM farmers already possessed somewhat better knowledge about the pests tested than non-IPM farmers, probably a result of their participation in previous extension meetings. Non-IPM farmers' knowledge in one year's time stayed at about the same level, except for a small increase of stemborer whitehead knowledge which is definitely a result of experiencing a stemborer outbreak during that season. IPM farmers, on the contrary, performed better than before after they followed the field school. Their knowledge gain was highest for the previously barely known pest, the rice leaffolder. Although a commonly occurring pest, the leaffolder was not commonly known because farmers had no name for it other than describing the whitish leaves. In the field school, they learned the common name for this pest, and discovered that the white patches are caused by leafeating caterpillars that hide in the rolled leaves. One-third of the field school graduates in the four study villages learned to recognise rice leaffolders and/or its symptoms. The second pest on which trained farmers scored better than before, and better than untrained farmers, is stemborer, a regularly occurring pest in the region. Symptoms of this pest (deadhearts at the vegetative stage of the rice crop and whiteheads at the reproductive stage) were already widely known before training, whereas the caterpillar feeding in the stem was only recognised as the cause of these symptoms by 44% of IPM farmers in the case of deadhearts, and 73% in the case of whiteheads. These percentages increased by 14% respectively 17%

after training, primarily with farmers in the villages Senengsari, Sugihsari and Kuduagung. Surprising is the apparent decrease in knowledge about stemborer by Sumberagung farmers, indicating that the field school had rather been confusing for these farmers. The effect cannot be explained since this group has not been studied intensively.

Despite the knowledge gain reported above, the IPM farmers in Senengsari and Kuduagung were not able to foretell and prevent the severe stemborer outbreak in the 1990/91 wet season. This could have been foretold by noticing flights of stemborer moths, usually visible around lamps at night, and thorough inspection of the rice crop for stemborer egg masses. A stemborer outbreak in Grobogan and surrounding districts was actually forecast on television and radio by the provincial Crop Protection Centre. In the IPM farmer field schools, attention was paid to the rice stemborer life cycle, and to identification and prevention of damage by picking egg masses. This knowledge, however, was not put into practice by the field school graduates the season after. They all reported that the outbreak had come suddenly during several days of rain when they did not visit the crop. Weekly field observations in twenty fields per village by the village study assistants did not show proportionate amounts of stemborer egg masses in the weeks before the whiteheads occurred. Low parasitism and predation rates, maybe due to climatic factors, might be a reason for the 'unexpected' outbreak. After this experience, many IPM farmers did not feel confident that they could handle another outbreak in a better way, indicating that they could not translate the lessons learned in the field school into their own practices. Supposedly, these lessons were too theoretical since during the training season no stemborer damage of importance occurred.

Change in farmers' knowledge about the other pests tested in the interviews was minor. The brown planthopper was already widely known before training, probably as a result of the outbreaks experienced before. Since brown planthoppers hardly occurred during the training season, there had not been many opportunities for learning. The same holds for bacterial leaf blight, be it that knowledge levels about this disease were previously very low, and stayed low after training. The rice seed bug is a pest known by all but one farmer even before training, because of its obvious appearance, smell, and plant symptoms.

In a field school session organised as a follow-up activity in Senengsari one year after training, the participants were given a test in recognising samples of several rice pests, symptoms and natural enemies, and asked to write down the name and the function of the specimen. Although naming insects and symptoms, especially the less common ones, was not so easy, the field school graduates were, in almost all cases, able to identify the function of the specimen shown, distinguishing harmful from beneficial ones. This is considered an important achievement, since knowledge about the function of organisms in the rice field is more important than about their names and appearance. Another achievement observed is that field school graduates, when talking about pest insects, considered population densities more than they did before when a single insects was seen as a reason to take a control measure. Field school graduates advanced much compared to non-trained farmers who considered either all insects in the rice field harmful, or did not ascribe any function at all to harmless insects.

Field management decision making
Farmers possess a certain perception about the rice crop and other ecosystem elements which influence their decision making. Of old, farmers consider environmental factors such as seasonality, climatic conditions, and water supply in their crop management decisions (section 5.1), showing that many factors of IPM knowledge are, actually, farmers' own indigenous knowledge. Young farmers, on the contrary, were brought up in farming with irrigation time schedules and preset recommendations, but many young farmers were met who still lay their parents' and grandparents' words to heart. In the IPM farmer field schools, the older participants in the groups played an important role in reviving the indigenous rice cultivation knowledge. But even though they learned to revalue the importance of, for instance, seasonality for crop health and pest occurrence, farmers in the study area still depend on water supply from the irrigation schemes. Perceptions of the ecosystem are influenced by the prevalent conditions, not only environmental but also socioeconomic ones. Changes in perceptions and their reflection on pest control decision making as a result of IPM training are, therefore, hard to assess. IPM farmers, as all farmers, consider soil, water and weather conditions of major importance to their rice crop. An extra value was given by IPM field school graduates to the natural enemies in the fields, and also to (insect) pests as food for the natural enemies. Especially this last point is a radical change in farmers' perception, since they previously could not stand a single pest insect in the field (section 5.3.1).

According to the competency objectives of the programme as outlined above, field school graduates are expected to consider population densities in relation to economically damaging levels. In the field school sessions, however, the relation between pest and natural enemy populations was emphasised as the main factor determining a pest control decision. In a confusing session about the economic threshold level (see Chapter 6), the farmers were acquainted with this concept, but no usable lessons were learned. The use of the economic threshold concept for pest control decision making is very complex indeed (Mumford and Norton, 1984). In practice, field school graduates do not appear to use this concept in their decision making. Except for the ones still applying pesticides prophylactically, most IPM farmers consider the relation between pests and natural enemies important, in addition to their previous experiences and habitual considerations regarding crop status, water and climatic conditions, availability of money, and a variety of minor factors. This approach actually corresponds better with the ecological principles of the new IPM approach.

Rat management
Early season action against rats when previous season damage was high is considered feasible in the study area for the dry season rice crop only which immediately follows on the wet season crop and, therefore, inherits the full rat population.

The wet season crop follows a secondary food crop or a fallow season so that rat population are drastically diminished through starvation and/or migration. Previous season damage as a measure for early season action does not seem efficient for the wet season, which is definitely the reason why this type of action is never practiced by farmers. At the moment of soil preparation, rice farmers get a good idea of the prevalent rat population, and often take action accordingly by digging out rat burrows individually or, in the better cases, by organising rat control community activities (Van de Fliert et al., 1993). More appropriate than the competency objective on early season action is the opportunity the IPM field school model provides to organise collective action to control rats.

Senengsari is the village most severely attacked by rats in recent history, although rats are an ever occurring pest in the whole study area. Since 1980, the year when the previous village head in Senengsari was elected, it had become harder and harder to organise any collective agricultural activity. Polarising relations in the village due to the policies of the previous and the current village head made people reluctant to collaborate (Box 4.2; Winasis, 1992). The need for collective activities to control rats, and the difficulties in the village to organise these activities were a favourite topic of discussion in the IPM field school in Senengsari. Nevertheless, at the end of the training season, the field school participants managed to organise some community rat drives which were fairly successful. An even better success was attained in the following season after a reorganisation of the farmer groups in the village in which each group was chaired by a few field school graduates. Main activity of the new groups was to organise community rat control activities, which resulted in much lower rat populations in the following seasons. As soon as the success was attained, these activities diminished and, finally, stopped. Rat populations did not build up again before the end of the study, which would have allowed to examine whether the organisation of rat control activities was sustainable.

Injury caused by rat attack as perceived by farmers in Senengsari and Kuduagung is depicted in Figure 7.1, and compared with the two main control measures, rat drives and poison baiting. It should be noted that perception of pest damage is influenced by farmers' previous experiences (Norton and Mumford, 1983). In this case the same injury index can imply different damage levels when the two villages are compared, considering their different rat histories. What matters are the changes over the seasons within one village. The injury index diagrams in Figure 7.1 show that in both villages IPM farmers, after having followed the field school, assess the level of rat damage somewhat lower than non-IPM farmers do. There are two possible explanations for this difference: (1) actually lower damage, maybe due to better control practices, or (2) a change in perception of injury by the trained farmers. The frequency of rodenticide applications seems to correspond with the level of rat occurrence, although high variations in number of applications among villages, and among individuals within a village were observed. Some farmers are known for their habit of placing rat baits as a routine measure. In Senengsari, the IPM farmer field school apparently had a considerable impact on the participants' rat baiting practice. The high number of applications given by these farmers

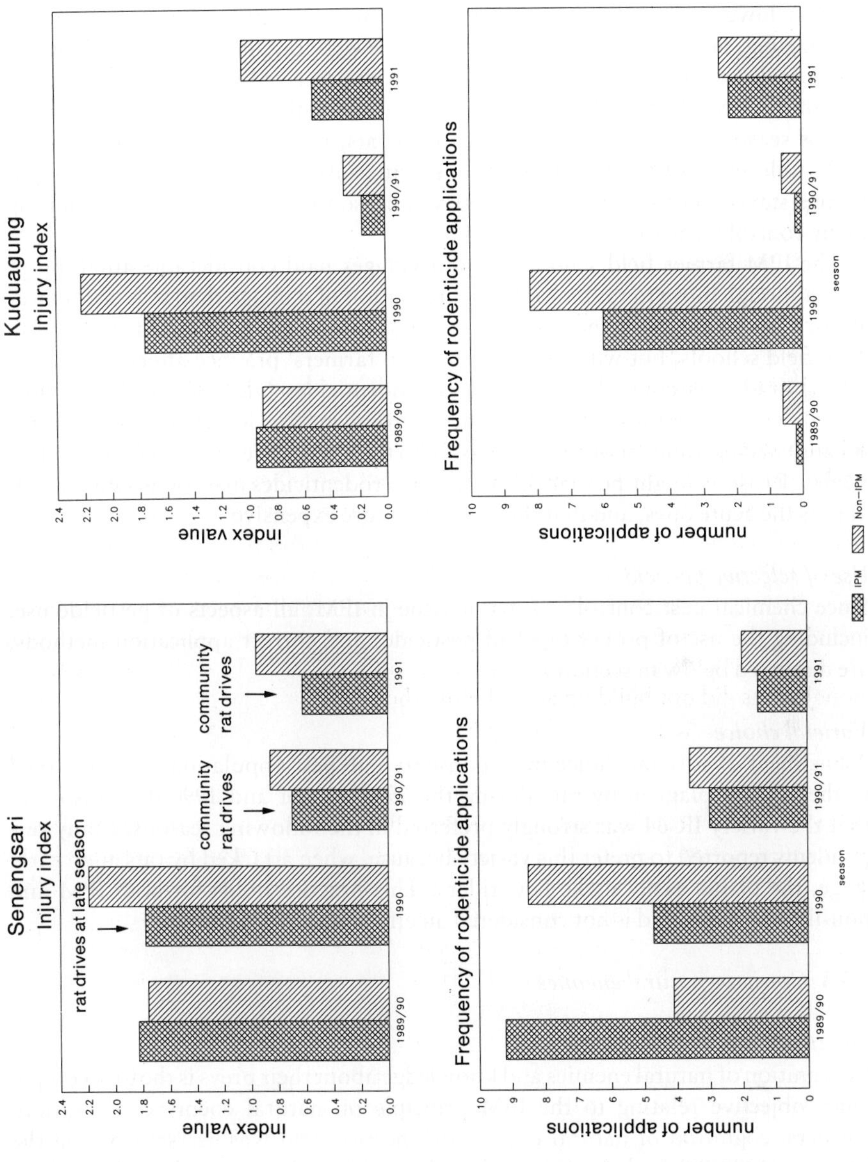

Figure 7.1: Index for injury caused by rat attack as perceived by IPM (left bars) and non-IPM (right bars) farmers, and rat control measures applied over four seasons in Senengsari and Kuduagung. An injury index value 1 indicates an average assessment by farmers as light damage, a value 2 as moderate damage, and a value 3 as severe damage.

before training decreased sharply in the field school season and thereafter, when they started to organise community rat drives. Although seasonal variation should not be effaced either, the rat drives contributed to lower rat attack, and, accordingly, to a lower rodenticide application frequency, also by non-IPM farmers. In Kuduagung, no collective rat control activities were organised, and frequency of rodenticide application seems to follow levels of perceived injury, with the exception of the 1989/90 season when rats appeared suddenly after a few quiet seasons. In this season, farmers were late to control rats, resulting in a high population in the following (1990) dry season. After the intermediate season of 1990, rat populations stayed very low in the 1990/91 wet season crop, therefore not requiring many control measures.

The IPM farmer field schools in both villages paid considerable attention to rats, although more to rat biology and prevention of damage than to control measures. Consequences of the use of acute rat poisons were discussed in detail in both field schools, but without any change in farmers' practice afterwards. Field school graduates continued to use acute rodenticides (aldicarb and zinc phosphide), which are more dangerous to the environment, and less effective for control because of development of bait shyness, than the alternative anti-coagulant rodenticides. Reasons might be that the alternative rodenticides are not as easily available as the acute ones, and that they are (far) more expensive.

Use of selective pesticides

Since chemical pest control is a major issue in IPM, all aspects of pesticide use, including the use of proper types of pesticides and correct application methods, are discussed below in section 7.4.

Varietal choice

Adjustment of varietal choice in response to high pest populations was observed in the villages plagued by rats during the 1989/90 wet and 1990 dry season, in that the variety IR 64 was strongly preferred in the following seasons. Many respondents reported to prefer this variety because, when attacked by rats at an early stage, the plants easily make new tillers. This practice holds for both IPM and non-IPM farmers, and is not considered an effect of the IPM training.

7.3.3 *Conserve natural enemies*

Identification

Recognition of natural enemies and knowledge about their preys is the first competency objective relating to the IPM principle on natural enemy conservation. Farmers' cognition of natural enemies in the rice field was assessed during the baseline (1989/90) and the first post-training (1990/91) season by asking them in individual interviews whether they knew animals in the field that eat rice pests. As with knowledge about pests, IPM farmers had, even before training, a better knowledge about natural enemies than non-IPM farmers: 81% versus 51% already knew about the existence of natural enemies in the rice field (Table 7.5). After

Table 7.5: Knowledge about natural enemies, before (1989/90) and after (1990/91) IPM training, as % farmers reporting.

	IPM		Non-IPM	
	before	after	before	after
Mention: snakes	51%	48%	31%	35%
birds	26%	20%	13%	28%
frogs	31%	66%	15%	36%
spiders	21%	76%	2%	19%
beetles	0%	26%	2%	0%
dragonflies	14%	46%	1%	6%
other [1]	23%	29%	19%	18%
Do not know	19%	0%	49%	33%
N	*94*	*93*	*259*	*233*

[1] 'Other before' includes cat, dog, duck, salamander, chicken, mongoose, fish, crab and cricket; 'Other after' additionally includes waterstrider.
Totals exceed 100% due to multiple response.

training, all IPM farmers could mention at least one natural enemy of rice pests. Additionally, a change in the quality of their knowledge had occurred. Whereas before training only the big, easily observable animals such as snakes, birds and frogs, were known as pest-eaters, a year later spiders, predatory beetles and dragonflies were increasingly mentioned. Knowledge on spiders particularly increased, from 21% to 76%, especially in Sumberagung (96%) (Table V.4). Knowing natural enemies is always the point that farmers mention first as something gained from the IPM field school. Especially the existence of predatory arthropods is a discovery most of them are really excited about. Knowledge level of non-IPM farmers on natural enemies also increased between 1989/90 and 1990/91, but not as much as that of IPM farmers. The increase mainly refers to the big, easily observable natural enemies. This indicates that the knowledge gain of IPM farmers on arthropod natural enemies is a result of the farmer field school.

Remarkable in the Tables 7.5 and V.4 is the decrease of the number of IPM farmers who mention snakes and birds as predators in the post-training interviews. These figures do not indicate that farmers' knowledge about the function of these animals decreased, but that they discovered in the field school that arthropod predators have a more important role in the ecosystem, especially since snakes and birds of prey are no longer abundant due to poaching.

Except for naming many natural enemies, field school graduates could also describe better what (stages of what) pests are eaten by what natural enemies, e.g. ladybird beetles eat brown planthoppers and eggs, dragonflies eat brown planthoppers, caterpillars, moths, eggs, and so on.

Effects of pesticides on natural enemies

Regarding the second objective, the ability to explain the effects of pesticides on natural enemies, it is observed that all field school graduates without exception obtained this ability. As soon as they discovered the beneficial function of natural enemies in the rice field during one of the first field school sessions, they realised that using pesticides would harm these beneficials. In agroecosystem presentations, this argument was often used as a reason not to spray, although it is not always applied in farmers' decision making for their own crops.

The special topic on 'Carbofuran, monocrotophos and natural enemies', presented in the fourth field school week, seemed to bear a negative effect in some villages. The conclusions drawn from this experiment, in which natural enemies were tested with both the contact organophosphate monocrotophos and the systemic carbamate carbofuran, was that carbofuran was less harmful for natural enemies since it took more time to kill these animals than with monocrotophos. In Sugihsari, Kuduagung and Sumberagung, a higher use of these carbamate granules was observed during the IPM training season and the seasons afterwards (Table V.5). An incorrect interpretation of this special topic, possibly due to the way the topic was evaluated by the trainer (the village extension worker), might be one of the reasons of this unintended increase. The second, and maybe more important, reason is promotion of these chemicals by the KUD and extension workers during these seasons. In other field schools visited throughout the country, the same problems in processing this special topic were observed. Ecological studies have shown that carbofuran, one of the pesticides still allowed by INPRES 3/86, strongly reduces natural enemy populations (pers. comm. W. Settle, IPM National Programme). Adjustment of the special topic on carbofuran and natural enemies is, therefore, required.

Management of habitats

Field school graduates were not observed to pay deliberate attention to the promotion of survivorship of natural enemies through managing habitats. Growing short grass on the bunds is usually not possible in the study villages because of the heavy clayish soil resulting in bunds that are either too soft and muddy, or too hard and cracked. Straw is, traditionally, burnt on the field a few days after the harvest, which is a practice good for reducing inoculum of various diseases (Thurston, 1992). During the last study season it was noticed that traders of cattle fodder were actively collecting the straw from newly harvested fields, even in the remotest villages. Both practices, burning or removing the straw, might be favourable to decrease the survivorship of stemborer pupae in the rice straw.

Rather than by managing habitats, survivorship of natural enemies is considered to be more benefitted by no or at least reduced and selective pesticide use, a point repeatedly mentioned in the IPM field schools but not as such included in the programme's competency objectives. This important issue is separately discussed in section 7.4.

7.3.4 Farmers become IPM experts and trainers

IPM experts

When assessing whether field school graduates have become IPM experts, there is no sense in making a distinction between experts and non-experts considering the many competency objectives. In addition, the intensity with which the topics dealing with these objectives were presented in the field school varied greatly from place to place, partly depending upon locally occurring field problems. A more sensible assessment can be obtained from the qualifications of field school graduates in the descriptions of IPM practices and knowledge given in the previous sections, that will not be repeated here. In summary, some competency objectives are achieved, some partly, and some not at all. Relations could often be detected between achievement and aspects of the training, of farmers' previous perceptions and habitual practices, and of external forces like the village extension worker.

My own assessment (be it subjective) rates all twenty-five field school graduates in Senengsari as IPM experts at a level that is reasonable considering the training process experienced, although none of them fulfils all competency objectives. They all understand the IPM principles, but committed implementation is limited to about fifteen people. During the last study season, only one farmer was observed to use pesticides prophylactically, and two others panicked at the slightest symptom of pest attack. Three of the field school graduates often leave the village during the growing season to work in the town. In Kuduagung, only ten of the twenty-five field school graduates are found to understand and implement IPM satisfactorily. Another nine persons apparently understand the IPM concept well, but are not fully convinced that they can do without pesticides, so they use them easily at the slightest pest symptoms. Four persons are qualified as indifferent to IPM since they have too many other duties, and do not seem to consider their rice crop important. Another two persons do not cultivate rice any more. The influence of the village extension worker who continued to promote pesticides is obviously an inhibiting factor in this village. On the whole, I obtained the impression, based on observations and repeated discussions with field school graduates over a period of two years, that something has established in the minds of the field school graduates in both villages that is likely to develop by itself. This 'something' includes a different perception by farmers of the rice crop and of themselves being the managers of it. Most of the field school farmers have learned to learn by themselves, and to perceive new experiences as new opportunities to learn more.

More important than my qualifications are those of the IPM farmers themselves. In an exercise during group interview sessions almost two years after training, field school graduates in Senengsari and Kuduagung evaluated each other according to objectives that they had listed. In Senengsari, somewhat more than half of the group interview participants was qualified as a good IPM farmer, whereas the remainder was called moderate. Good IPM farmers were valued for routine field monitoring, extending IPM to non-IPM farmers, and active involvement in collective rat management. Explanations for moderate IPM farmers included insufficient field monitoring, many other duties, and an incorrect field monitoring technique. In Kuduagung, the field school graduates valued each other in the same proportions of good and moderate IPM farmers as in Senengsari, in addition to one fair IPM implementor. This last person was a village official who hardly ever visited his field, but was still qualified for the fact that he talks about IPM to other people in the village. Good IPM farmers included those who regularly observed their fields, and understood the IPM principles. Explanations for the qualification moderate included preventive use of pesticides, and insufficient field monitoring.

The groups in Sugihsari and Sumberagung were not observed closely and, therefore, no sound estimates can be made on the level of IPM implementation and understanding of the individual field school graduates. Looking at their practices as reported in the surveys only, the achievement among IPM farmers in Sugihsari is considered comparable with Kuduagung, in that a relatively good understanding about IPM is obtained, but farmers are not yet completely loose from the old government recommendations. The pest observer who trained these farmers perceived the field school graduates in Sugihsari as not fully committed to IPM. He ascribed this to the influence of the village head (being manager of the KUD) still promoting high-external-input technology to the farmers. Effects of IPM training in Sumberagung are valued at a lower level, partly a result of the strong influence of the village extension worker there. A second reason for less successful effects might be the lower achievement of the IPM training due to the incompatibility of the (relatively) scientific approach applied by the trainers and the (relatively) low intellectual level of the participants.

IPM trainers

In the IPM groups studied in Grobogan, no field school graduates were observed to give training in IPM to fellow farmers in their own or neighbouring villages, but it was not encouraged by the trainers either. Although several field school graduates complained in the last survey interviews that many non-IPM farmers were still not knowledgeable about IPM, no one took the initiative to organise something to extend IPM more deliberately than through individual communication. Considering the fact that follow-up activities petered out within two seasons after the training, it is not likely that initiatives will be taken in the future.

Farmer-to-farmer training was observed to take place in several other areas in Indonesia where the IPM field school was conducted (Anonymous, 1991). The training model with its curriculum (the field problems) and training materials (the rice ecosystem) available in every rice-growing community, allows farmers to replicate what they have experienced themselves. In many cases, the initiatives from field school graduates to train more farmers in their village or in neighbouring villages are supported by local governments through donations. The IPM programme takes advantage of these initiatives, and supports them with special trainings for farmer trainers to become good facilitators.

7.4 Chemical pest control

Change in chemical control practice by field school graduates is an important evaluation point, also applied in other evaluation studies of the IPM programme (Pincus, 1991; Winarto, 1992; Pusat Penelitian Sosial Ekonomi Pertanian, 1991).

In this section, frequency of pesticide use, reason for chemical control, type of pesticides, and method of application are considered. An initial remark colouring the setting is that pesticide use in the study area was previously not very high, but IPM farmers' attitude towards pesticide use before IPM training was, similarly to non-IPM farmers, one of dependence. In the baseline survey, 78% of IPM farmers expressed that lower yields would be obtained if no pesticides were used on the rice crop, of whom 21% believed that it would imply great or even total yield loss.

Frequency of pesticide use

The frequency of pesticide application by IPM farmers before training (1989/90) was not different from that of their non-IPM colleagues, except for Sugihsari where significantly more ($\alpha = 0.5\%$) applications were given by IPM farmers (Figure 7.2 and table V.5a). Among the four villages studied, the same variation in pesticide use is visible for IPM as for non-IPM farmers. Zadoks (1985) called local variation in pesticide application a result of social control, noticing that farmers in a certain area influence each other to be either risk-avoiding or risk-accepting. In Senengsari, farmers were generally low pesticide users (on average, 1.1 application per season for IPM, and 0.9 for non-IPM farmers), whereas in the other three villages they were (relatively) high pesticide users (ranging from 1.4 to 2.5 applications per season). This variation becomes more obvious when the proportion of farmers not using pesticides at all is considered (Figure 7.3 and Table V.5b). In the baseline season, this proportion among IPM farmers was high in Senengsari (48%), low in Sugihsari and Kuduagung (8% respectively 4%), and medium in Sumberagung (20%). A medium percentage of farmers using no pesticides and a high average application frequency in Sumberagung indicates the high variation among the farmers there in the use of pesticides. Many farmers in this village gave more than the average number of applications: 60% gave three applications or more, compared to 48% in Kuduagung, 28% in Sugihsari, and 13% in Senengsari.

In the IPM training season (1990), the average frequency of total pesticide application given by IPM farmers decreased only significantly in Senengsari (to 0.4 application per season, $\alpha = 2.5\%$). In the other three villages, pesticide use by IPM farmer stayed at the same level as before training. A difference, however, is visible when spray and granular applications are considered separately (Figure 7.2 and Table V.5a). Then, a significant ($\alpha = 5\%$ and lower) decrease in pesticide sprays in all villages can be seen, and an increase in the use of pesticide granules (predominantly carbofuran), which is significant ($\alpha = 0.5\%$) in the case of Sugihsari and Kuduagung. The reasons for increased granule use were explained above, i.e. more active promotion of granular pesticides in this season, and the possible impact of the special topic on 'Carbofuran, monocrotophos and natural enemies' in the field school.

The trend of a reduced frequency of pesticide sprays continued during the following wet (1990/91) and dry (1991) season in Senengsari and Sugihsari. The opposite is true for Kuduagung and Sumberagung where an increase of the spray frequency was visible during the 1990/91 wet season. The trend of increased granu-

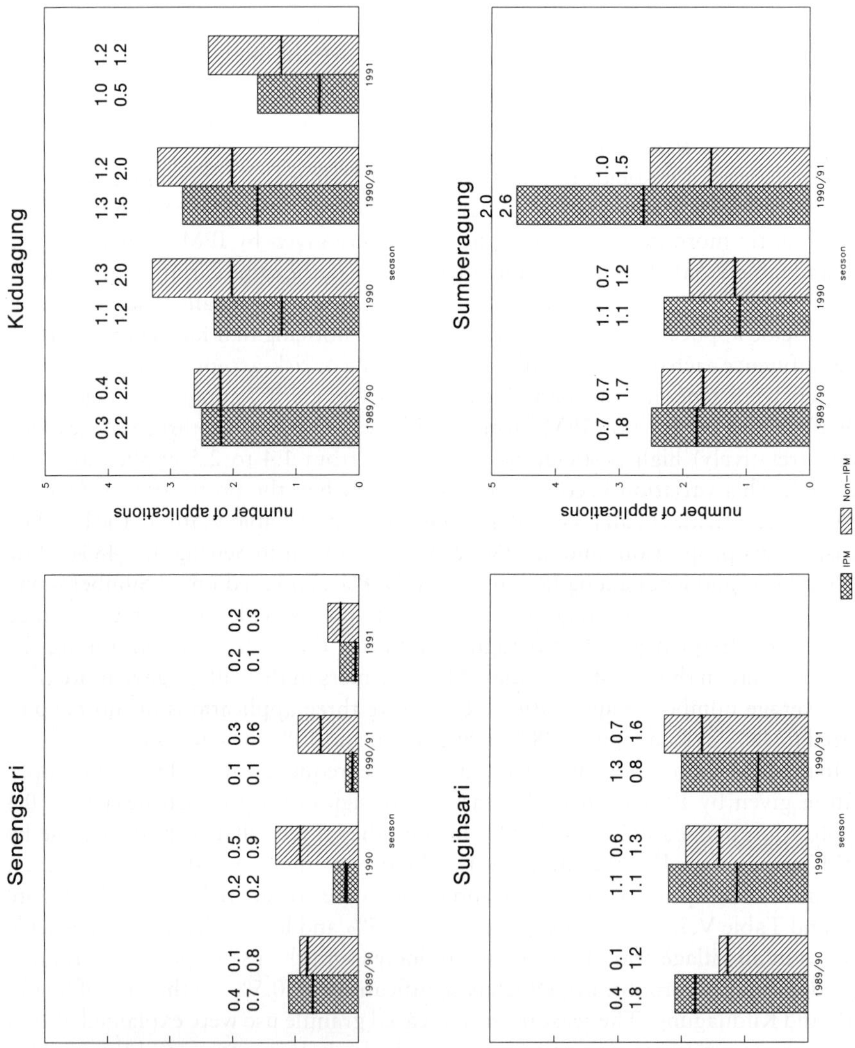

Figure 7.2: Frequency of pesticide application by IPM (left bars) and non-IPM (right bars) farmers over four seasons in the four IPM villages. The part below the line in each bar represents pesticide sprays, and the part above the line granular pesticide applications. 1989/90 wet season = baseline, 1990 dry season = IPM training, 1990/91 wet and 1991 dry season = post-training.

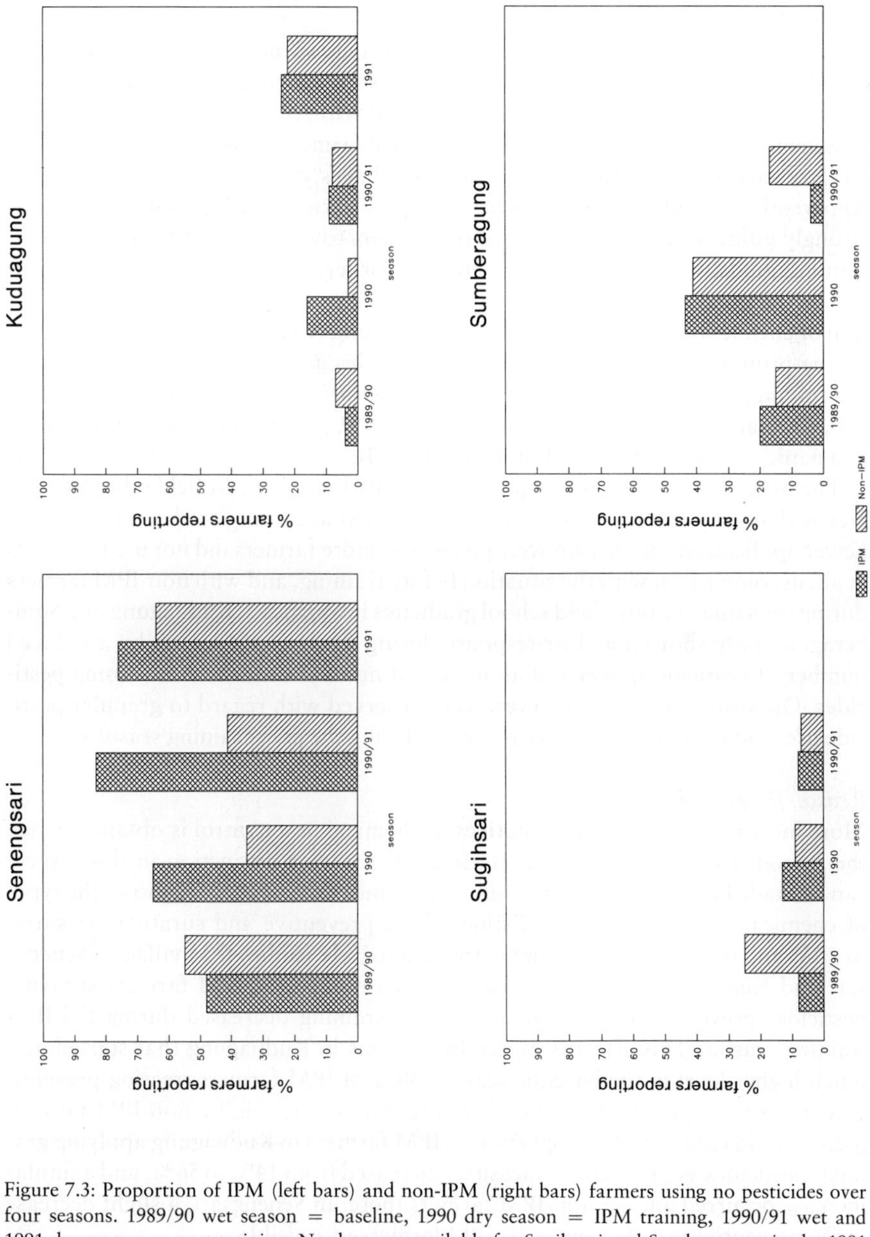

Figure 7.3: Proportion of IPM (left bars) and non-IPM (right bars) farmers using no pesticides over four seasons. 1989/90 wet season = baseline, 1990 dry season = IPM training, 1990/91 wet and 1991 dry season = post-training. No data are available for Sugihsari and Sumberagung in the 1991 dry season.

lar pesticide applications also continued in the post-training seasons, especially in the villages Sugihsari and Kuduagung. To a large extent, farmers' chemical pest control practice was influenced by pest occurrence during particular seasons, i.e. heavy stemborer attack during the 1990/91 wet season, and hardly any pests during the 1991 dry season.

Different reactions to pest occurrence among field school graduates in the different villages were visible. For the 1990/91 season, Senengsari and Sumberagung formed two contrasting examples in this regard with, on average, 0.2 applications by Senengsari farmers and 4.6 by Sumberagung farmers, even though comparable levels of stemborer damage occurred in these villages, if not lower in Sumberagung. Apparently, confidence in the IPM principles varied considerably, supposedly strongly influenced by the attitude of the trainers towards pesticide use, and post-training guidance by the village extension worker. In Sumberagung and Kuduagung, the extension worker had rather promoted pesticides than supported reduction of chemical control in the season after training (Chapter 6).

The proportion of farmers not using pesticides at all, which increased during the training season, further increased in Senengsari only, but decreased in Sugihsari, Kuduagung and Sumberagung in the 1990/91 post-training season, probably as a panic reaction to the stemborers (Figure 7.3).

The overall conclusion on frequency of pesticide use by IPM field school graduates is that only the Senengsari group performed according to the expectations. Fewer applications per season were given, and more farmers did not use pesticides at all, in comparison with the situation before training, and with non-IPM farmers during the same seasons. Field school graduates in Sugihsari, Kuduagung and Sumberagung only showed a clear response during the training season by a reduced number of pesticide sprays and an increased number of farmers not using pesticides. Opposite reactions, however, were observed with regard to granular pesticide use, and to chemical control practices during the post-training seasons.

Reason for pesticide application

More information on farmers' motives in chemical pest control is obtained when the reasons for pesticide use are considered. A major distinction in this respect can be made between preventive and curative measures. Table 7.6 shows the types of chemical control measures, distinguishing preventive and curative measures, by IPM and non-IPM farmers in the two intensively studies IPM villages (Senengsari and Kuduagung) over four seasons. The number of IPM farmers spraying pesticides preventively in the season before training decreased during the IPM training season. This effect is especially obvious in Kuduagung that started at a much higher level in the baseline season (38% of IPM farmers treating preventively) than Senengsari (9%), and where preventive spraying by non-IPM farmers increased. In contrast, the proportion of IPM farmers in Kuduagung applying granular pesticides as a preventive measures increased from 14% to 56%, and a similar increase occurred among non-IPM farmers there. In Senengsari, a slight decrease in the preventive use of granules by IPM farmers was visible.

In the two post-training seasons (1990/91 and 1991), preventive pesticide use

Table 7.6: Type of chemical pest control measure applied, as % farmers reporting.

	Senengsari							
	IPM				Non-IPM			
	1989/90	1990	1990/91	1991	1989/90	1990	1990/91	1991
Preventive measures:								
pesticide sprays	9%	4%	9%	0%	23%	18%	41%	2%
pesticide granules	30%	22%	13%	19%	9%	26%	19%	15%
Curative measures:								
pesticide sprays	17%	9%	0%	5%	27%	32%	12%	13%
pesticide granules	17%	0%	0%	0%	0%	11%	2%	4%
No control measures	48%	65%	82%	76%	55%	35%	41%	64%
N	*23*	*23*	*23*	*21*	*66*	*62*	*59*	*55*
	Kuduagung							
	IPM				Non-IPM			
	1989/90	1990	1990/91	1991	1989/90	1990	1990/91	1991
Preventive measures:								
pesticide sprays	38%	12%	61%	33%	19%	42%	74%	0%
pesticide granules	14%	56%	61%	57%	9%	58%	56%	62%
Curative measures:								
pesticide sprays	62%	56%	26%	10%	79%	70%	28%	23%
pesticide granules	14%	24%	9%	10%	7%	25%	5%	2%
No control measures	4%	16%	9%	24%	7%	3%	8%	22%
N	*25*	*24*	*23*	*21*	*68*	*67*	*61*	*60*

Totals exceed 100% due to multiple response.

by Senengsari IPM farmers stayed at relatively low levels. In Kuduagung, on the contrary, preventive pesticide sprays increased again, and preventive granule application remained at a high level. Remarkably, during the 1991 dry season with low pest pressure, still one-third of IPM farmers there sprayed pesticides preventively, in contrast to none of the non-IPM farmers.

Curative chemical control measures, or pesticide applications with a defined target, decreased among IPM farmers in Senengsari to (almost) zero in the post-training seasons. This result indicates that these field school graduates were confident with their own pest observation and decision making abilities, and did not panic at the first sight of insects or symptoms in the field. The pest occurrence and damage symptoms they observed were not considered worth a pesticide application, although a few of them were still not loose from prophylactic treatments. The message of the need for observation before taking action was apparently better received by farmers than that of the ineffectiveness and inefficiency of preventive pesticide applications. Although there is a habit of low pesticide use in Senengsari, which is shown by the figures for the non-IPM farmers there, a change in (most) IPM farmers' practices and perception relating to chemical pest control as a result of the field school is obvious in this village.

Such a statement cannot be made in the same deliberate way for Kuduagung. Favourable is that curative measures applied by field school graduates decreased sharply over the post-training seasons. The same trend, but less obvious, was visible among non-IPM farmers in this village, and, therefore, might as well be partly ascribed to the seasonal variation in pest pressure. The difference between IPM and non-IPM farmers in the 1991 dry season with a low pest pressure, however, indicates a change in the practice of at least some of the IPM farmers in this group. The field school graduates were observed to know more about ecological principles in the rice field, but apparently the feeling of dependence on pesticides imposed upon them by the village extension worker was for many farmers still a strong aspect in the decision making process.

Results of the control IPM villages Sugihsari and Sumberagung with regard to the type of chemical pest control measures are presented in Table V.6. Since these villages were covered in only two study seasons (baseline and first post-training season), less information is available about changes in farmers' practice. On the whole, the results are not very favourable, possibly due in part to the unlucky pest conditions in the evaluation season (1990/91 with stemborer outbreak). In Sugihsari, the number of IPM farmers applying preventive pesticide sprays decreased with 18% after training, but those applying preventive granules increased with 63%. In Sumberagung, both measures were more frequently applied after training. Curative measures stayed at the same level in Sugihsari, but increased in Sumberagung. As far as the observations reached in these two villages, the disappointing effects are supposed to be a result of a rigid attitude of IPM farmers in Sugihsari towards agricultural development, imposed by their frequent involvement in intensification programmes, and in Sumberagung of the strong, adverse influence of the village extension worker.

Coming back briefly to the increased use of granular pesticides, it seems very unlikely that the preventive applications of these pesticides were mainly an effect of the IPM training. Although misinterpretations of the special topic on 'Carbofuran, monocrotophos and natural enemies' definitely occurred, the increase is assumed to be mainly a result of extension efforts (if not enforcement in the case of Kuduagung and Sumberagung) by the KUD c.q. extension worker to farmers taking the complete KUD input package. This statement is based on the fact that increase of granules occurred only in the three villages with strong KUD influences (Sugihsari, Kuduagung and Sumberagung), where many (IPM) farmers took KUD input packages on credit. Several farmers applying carbofuran granules reported that they bought the chemicals because they were a compulsory part of the package, and they felt sorry not to use them.

Among the pesticide using field school graduates, no distinction can be made between the ones applying chemicals preventively or curatively. Some apply one measure, some both. This indicates that behavioural change relating to pesticide use induced by the IPM training is not a matter of difference in perception of the training contents by the individual trainees. More likely, it is a difference in suitability of the different field school messages to farmers' former practices and attitudes, definitely influenced by factors such as, for instance, habits, a pesticide promoting extension worker, or the economic situation of the farm family. The diversity among the four groups studied which can be traced back to differences among the villages in pest management history and in driving forces in agricultural development, support this statement. Reduction of pesticides seems suitable in the situation of Senengsari farmers who previously did not have the habit to use many pesticides, and who are often facing shortage of money. In contrast, Kuduagung and Sugihsari farmers have a rational attitude towards farming and towards the avoidance of risk, an attitude imposed upon them for years. In the latter case, a decision for no pesticide use at low pest levels seems more logical (more suitable) because it is based on rational considerations and skills gained in training. Opposed to this, deciding not to apply habitual preventive measures cannot be based on any skills, but is a more matter of attitude towards pest control. The IPM farmer field school curriculum, as implemented in the villages observed, apparently has not adequately anticipated to counteract various opposing factors in the farming community.

Target for pesticide application

In spite of the fact that more pesticide applications (mainly granules) were given preventively in the first post-training (1990/91) season compared to the baseline and the training season, more curative applications by field school graduates were given at a high injury level, and against better defined targets (Table 7.7). Whereas before training no or low injury levels were sufficient to induce 87% of the pesticide sprays applied, after training farmers better dared to observe pest population development first and wait for a pesticide application until a higher level injury occurred. Similarly, in the baseline season pesticides were applied for a whole range of pests including those of no economic importance, whilst a year later this range was narrower, and mainly included potentially damaging pests. The observation indicates that (a part of the) field school graduates take better informed decisions for curative application of pesticides than before training. The higher percentages for applications at no injury and no target are mainly caused by Sumberagung farmers, which is another indication for a low effectiveness of IPM training in this village.

Method of pesticide application

Many of the pesticides banned for rice cultivation by INPRES 3/86 are still available in the local kiosks for use in home gardens and secondary food crops. Since no control whatsoever on pesticide usage is exercised in rural areas, these banned pesticides are still widely used on rice crops. Farmers often do not know what types are actually allowed for rice, and pesticide salesmen just sell whatever the farmers ask. In many cases, farmers do not know the brand names either, since these are usually difficult names that do not make sense to the farmers. One brand

Table 7.7: Injury level and target pest at time of pesticide application by IPM farmers before (1989/90) and after (1990/91) IPM training, as % of total number of pesticide applications.

	pesticide sprays		pesticide granules	
	before	after	before	after
Injury level:				
no injury	41%	55%	57%	96%
low	46%	24%	18%	3%
medium	10%	8%	13%	1%
high	3%	13%	13%	0%
Target pest:				
no target	37%	55%	0%	95%
rice seed bug	22%	4%	0%	0%
stemborer	20%	17%	23%	0%
caterpillars	12%	18%	13%	3%
planthoppers	11%	1%	0%	0%
tungro virus	5%	0%	5%	0%
grasshoppers	4%	1%	0%	0%
molecricket	3%	0%	20%	0%
rats	3%	0%	0%	0%
crabs	1%	0%	8%	0%
gall midge	1%	1%	0%	1%
leaf diseases	0%	3%	0%	1%
other	2%	0%	3%	0%
N	*158*	*95*	*40*	*101*

Totals exceed 100% due to multiple response.

often recalled by farmers was 'Brantasan', one of the few with an Indonesian name which literally means 'exterminator'. Although this BPMC/diazinon mixture has not been available in the market for the last couple of years, farmers still ask for it in the kiosks, probably because the name appeals to the imagination. The pesticide salesmen interviewed in Grobogan reported unanimously that whenever a farmers wants to purchase unavailable brands, they inquire after the crop to be treated. When it is rice, they sell one of the pesticides allowed by INPRES 3/86.

In the IPM farmer field schools observed, brief attention was given to the brands of pesticides allowed for rice cultivation. Slight changes in pesticide selection behaviour by field school graduates can be observed, in that proportionally more allowed pesticides are used after training (Table 7.8). Although the proportion of applications with banned pesticide decreased with only 3% after training, the absolute number of applications using these chemicals went considerably down (from 76 to 43 applications). This result is particularly encouraging because many farmers, especially in Senengsari and Sugihsari, select the type of pesticide themselves rather than being advised by someone else. In Kuduagung and Sumberagung, the influence of the extension worker on farmers' pesticide choice is greater. On the whole, IPM farmers were found to better know the types of pesticide they

Table 7.8: Types of pesticides (active ingredients) used by IPM farmers before (1989/90) and after (1990/91) IPM training, as % of total number of applications.

	allowed/banned by INPRES 3/86	before	after
Pesticide sprays:			
BPMC	allowed	54%	68%
carbofuran [1)]	allowed	3%	0%
buprofezin	allowed	0%	1%
subtotal allowed		57%	69%
phentoate	banned	17%	20%
BPMC/diazinon	banned	6%	1%
2,4-D	banned	6%	0%
diazinon	banned	5%	1%
chlorpyriphos	banned	4%	0%
fenthion	banned	4%	0%
monocrotophos	banned	3%	1%
BPMC/chlorpyriphos	banned	2%	4%
carbaryl	banned	1%	4%
cyphalothrin	banned	0%	8%
other	banned	0%	5%
fenitrithion	banned	0%	1%
subtotal banned		48%	45%
N		*158*	*95*
Pesticide granules:			
carbofuran	allowed	98%	100%
other/unidentified	banned	5%	0%
N		*40*	*101*

1) Carbofuran granules (3% a.i.) were soaked, diluted in water, and sprayed.
Totals exceed 100% due to multiple response.

used than non-IPM farmers.

Pesticide application with the proper dosage and delivery is important in IPM to prevent development of pesticide resistance by insect pests and pest resurgence. In the IPM field schools observed, no attention was paid to correct pesticide application. Although it is difficult, if not impossible, to collect reliable information from farmers through retrospective interviews on quantities of pesticides applied, (estimated) records were made with the IPM farmers on the amount of pesticide concentrates used per tank and per field, and tanks sprayed per field, in order to detect trends in their methods of pesticide application. With this information, three values could be calculated: (1) the concentration of the pesticide solution in the tank in ml or g/liter water, (2) the volume of water as a measure for delivery in liter water/ha, and (3) the dosage per unit area in liter or kg pesticide concentrate/ha, or kg granules/ha. These measures were compared with the recommended ranges of concentrations, dosages and volumes given on the labels of the pesticides concerned.

Table 7.9 displays how the methods of application by IPM farmers relates to the recommendations. Only a small proportion of pesticide solutions (15%) were

Table 7.9: Method of pesticide application by IPM farmers related to recommendations before (1989/90) and after (1990/91) IPM training, as % of total number of applications. Concentration in tank in ml or g pesticide concentrate per liter water; volume in l water/ha; dosage in l or kg pesticide concentrate/ha or kg granules/ha.

	pesticide sprays						pesticide granules	
	concentration		volume		dosage		dosage	
	before	after	before	after	before	after	before	after
no response	15%	4%	1%	8%	10%	8%	0%	8%
lower	25%	53%	97%	83%	68%	83%	95%	76%
within range	15%	28%	1%	3%	13%	11%	3%	11%
higher	46%	15%	1%	5%	10%	6%	3%	5%
N	*175*	*131*	*158*	*95*	*168*	*123*	*40*	*101*

within the recommended range of concentration during the baseline season. Twenty-five percent of the solutions are too diluted, and 46% too concentrated. More remarkable is that almost all applications (97%) were sprayed with a far too small volume of water as a result of too few tanks per area unit. Spraying a small volume is a common practice, also by non-IPM farmers, in order to reduce labour cost. The overall result is that the majority of spray applications (68%) are underdosed, and only 13% are given with the correct dosage. Pesticide granules were applied in too low dosages in 95% of the cases, mainly due to the high price of these chemicals. Granule application in too low dosages is completely ineffective, and therefore a great waste.

Some shifts in the post-training figures are visible in the table, not showing better practices by the field school graduates, though. Fewer highly concentrated pesticide solutions are prepared, resulting in a higher proportion of underdosed applications, which is dangerous since too low dosages may induce pesticide resistance. It is not unlikely that the use of smaller quantities of pesticides per application was based on the willingness of some field school graduates to reduce, or save on chemicals, but they actually did not yet dare to take the risk of not using pesticides at all. Considering these practices, and the possible consequences, it is suggested that attention in IPM training be paid to the consequences of incorrect methods of pesticide application.

7.5 Reward and risk of IPM

After three seasons of independent IPM implementation by the field school graduates (1990/91, 1991 and 1991/92), the groups of Senengsari and Kuduagung were called together again for a group interview discussing reward and risk of IPM. The 1990/91 wet season had been a very bad season with disastrous yellow stemborer attack that suddenly occurred at heading stage of the crop, whereas the 1991 dry and 1991/92 wet seasons had been exceptionally good with favourable weather conditions and hardly any pest damage. In the 1991 season survey, 69% of the IPM farmers ascribed the good result of that season to low levels of pest attack, 36% to timely irrigation water supply, and 24% to favourable climatic conditions. In the group interviews, farmers compared the two wet seasons (1990/91 and 1991/92), and tried to identify to what extent IPM or other factors had contributed to the contrasting results of the seasons. Boxes 7.1 and 7.2 describe how the field school graduates perceived the two contrasting seasons, and what factors they attached to the similarly contrasting results.

Reward of IPM

The contribution of IPM to the good yield and good profit in the 1991/92 season was perceived by field school graduates as follows:

- higher profits were obtained because farmers did not spend (much) money on chemical control measures, since better decisions could be made based on observations on pests and natural enemies;
- money saved on pesticides was used to buy more fertiliser which increased the yields (expressed by farmers in Senengsari where fertiliser doses already exceeded the recommendations, either confirming the inefficient use of fertiliser as stated in Chapter 5, or making the validity of their statement doubtful);

Box 7.1
A good season

All field school graduates participating in the group discussions qualified the 1991/92 wet season as the best season they ever had. Factors identified by the participants that had favoured the good result, both yield and profit, included:

- the preceding very long dry season resulting in a more fertile soil;
- optimal weather conditions: a lot of sunshine, no hard winds at the time of pollination, and not much rain;
- sufficient and timely water supply, allowing timely planting;
- no pests, and a lot of natural enemies;
- very low use of pesticides;
- IPM as guideline for rice cultivation (Senengsari);
- good price for the rice (Senengsari).

Although most subgroups in this exercise mentioned that there were no major inhibiting factors during the past season, a few minor points came up, including high prices for rent of land, shortage of labour for harvesting or planting, and high wages for harvest labour because it happened to be the fasting month. Some farmers believed that low stemborer damage (whiteheads) had reduced their yields slightly. Counts of whiteheads in 20 observation plots on 20 plants per plot showed injury levels between 6% and 16% whiteheads of all panicles, whereas yields of 5.8 to 9.0 tons wet, unhusked rice per hectare were achieved, with an average of 7.7 tons/ha. No relation existed between percentage of whiteheads and yields indicating that these levels of stemborer damage do not affect the yield, probably due to high compensation by the rice plant under the favourable weather conditions. The lower yields were obtained on fields where water management had been difficult.

Box 7.2
A bad season

In contrast to the 1991/92 season, the wet season one year earlier (1990/91) yielded one of the worst results since the disastrous brown planthopper outbreaks in 1985/86. Very low yields were obtained causing negative incomes, loans could not be paid off, and new loans had to be contracted to buy inputs for the next crop. Factors causing the bad result, according to the participants, included severe and sudden stemborer attack at the early reproductive stage of the crop, and bad weather (rain and wind) at flowering stage resulting in even more empty seeds. In Senengsari, gall midge attack and low rice prices were mentioned as additional, unfavourable factors. The stemborer occurrence was explained by the fact that irrigation water was late due to major repairs on irrigation channels, resulting in late planting (late December until mid January, see also Box 5.1). The field school graduates had not noticed high stemborer populations in the form of moths and egg masses to be able to foresee, and prevent, damage.

- application of balanced fertilisation (although farmers' practices show an incorrect perception of the concept of balanced fertilisation, namely the application of different fertiliser types in the same quantities);
- routine field monitoring through which better attention was paid to crop health and development;
- better knowledge about pests, natural enemies, rice ecosystem, and chemical pest control;
- higher motivation of farmers to tend their crops.

It is extremely interesting to notice that the factor knowledge always shows up in discussions with field school farmers on the use, profit, and reward of IPM training and implementation, which is described in some more detail in Box 7.3. The same result was found in an IPM programme with wheat farmers in the Netherlands (Zadoks, 1989b; Rossing et al., 1985). Knowledge is not a hard economic factor, which makes the importance of this aspect for farmers more interesting, since it contrasts with most adoption theories.

The economic profit of IPM is a second important reward mentioned by field school graduates. This second position might be due to the relatively low expenditures on chemical pest management even before IPM training: only 6% of the total expenditures for rice cultivation. After IPM training, field school graduates in Kuduagung and Sumberagung even started to use more granular pesticides and foliar fertilisers, in which case economic profit is definitely absent. This behaviour is observed to be under strong influence of the counteracting village extension worker, as one of the IPM farmers in Kuduagung expressed; 'IPM is good for the poor farmers, but the well-to-do's still follow the extension worker'. Although, generally, better field observations were made, and better informed decisions were taken by field school graduates, apparently, pest management practices were still not adequate to prevent yield loss by pests completely, as was painfully apparent for farmers during the 1990/91 wet (stemborer) season. As mentioned earlier, farmers consider pest damage as the main factor determining their yields. When

Box 7.3
Knowledge increase: a reward of IPM training

In both group and individual interviews with IPM field school farmers, knowledge, or knowledge increase, is mostly the first aspect mentioned as a reward of IPM training. The farmers are usually not able to explain in what way the increased knowledge physically contributes to their farm profitability, but they find it worth mentioning. Admitting that they do not always put their knowledge into practice, perhaps influenced by other factors and conditions, the IPM farmers feel that they possess something valuable that can benefit them, sooner or later. By possessing agroecological knowledge, farmers feel they master their farms as independent managers, as was nicely expressed by one of the field school graduates:

> 'Since I followed the IPM farmer field school, I have peace of mind. Because I know now how to investigate, I do not panic any more when I discover some pest symptoms in the field'.

Knowledge increase reduced farmers' feelings of uncertainty. In return, they obtained feelings of expertise, responsibility, independence and self-confidence, which definitely contributed to the higher motivation for farming as was experienced by the field school graduates.

yield loss by pests could be adequately prevented, economic profit would be much more tangible.

Risk of IPM

Interesting discussions among field school graduates in Senengsari and Kuduagung took place when the contribution of IPM during the disastrous stemborer season (1990/91) was reviewed in the group interviews. Various controversial statements were made, such as:

- there is no influence or contribution (easiest answer to avoid conflict);
- farmers have already implemented IPM, but the pest 'was stronger', could not be controlled ('at least, we did our best');
- farmers are aware of the meaning of IPM (implementation and success are something else);
- non-IPM farmers consider IPM not being socialised yet (others are to blame);
- many farmers were still doubtful about the IPM concept and, therefore, still applied pesticides in order to feel at ease. But finally they realised that their doubt (and action accordingly) only resulted in a greater (economic) loss since the pest could not be controlled anyway (learning-by-doing);
- the field school did not cover control methods for stemborer whiteheads, but only identification of the pest ('we should have known if only they taught us');
- it is not enough to rely on field monitoring only;
- if field monitoring is done correctly, but pests, especially those that are difficult to observe such as stemborer, occur suddenly, the only way out is to use pesticides;
- because pests were more abundant than natural enemies, IPM could not be implemented;

- although pesticide use can be reduced, some farmers are not content, or do not feel at ease if they do not use pesticides at all;
- everything depends on gifts from God ('*anugerah*'); farmers can implement IPM well, but if God does not wish a good yield, farmers are, no matter what they do, the ones who suffer.

The way farmers react to the pest outbreak gives extra information about the extent to which field school graduates master the IPM principles, and weigh the risk and reward of IPM. Several reactions can be analysed from the statements above which are supported by field observations and individual discussions with farmers. The first reaction observed, and best from IPM point of view, is that farmers remained convinced that they just did what they thought was good (routine observations, no panic spraying), then, realised that it was too late to take a control measure after the whiteheads had occurred, and drew a lesson from the bad experience. The second reaction shows an almost compulsive action in such a panic situation, making them fall back on previous practices (applying pesticides), if not only to find peace of mind by doing the thing that was taught them for years as the adequate means against pests. The third reaction comprises submission to fate.

One remarkable point that became clear from these group interviews is that several field school graduates perceive IPM as a field monitoring technique only. They value the knowledge and skills gained, but do not feel competent (yet) to react adequately to pest occurrence. Some farmers mentioned that they missed topics on control measures in the field training (also Winasis, 1991). The IPM field schools in Senengsari and Kuduagung, indeed, did not pay much attention to control aspects. In the case of the stemborer outbreak, observation skills of IPM farmers were, apparently, not sufficient since no stemborer egg masses, which are easily detectable on the rice leaves, were noticed that should have alarmed the farmers in time. This is due to the restriction of an experiential learning and field-based approach of the IPM farmer field school. The training curriculum contained mainly field problems locally occurring during the training season, which are not necessarily all the problems that farmers will face in the next seasons. When field school graduates emphasise field monitoring instead of agroecosystem analysis, informed decision making and adequate pest management measures, the sustainability of IPM practice becomes doubtful. Therefore, attention should be paid to this issue in further monitoring and evaluation activities of the programme.

7.6 Horizontal communication

Since not all 18 million farm families in Indonesia can be trained directly in IPM by the programme nor by the extension service, horizontal communication to the farming community is of major importance. Dissemination of the training content to other rice farmers might take place through various channels, and at various times or occasions that can be divided into three categories: (1) the training (something exceptional happens in the village which attracts the attention), (2) the implementation and communication by the field school graduates who give the example, and (3) the successes obtained by implementors (demonstration effect). The second situation is supposed to be most important since it involves the field school graduates actively, and takes place over longer periods, in contrast to the first point where contact is limited to the ten field school sessions. The third occasion is not assumed to have a great impact since the result of IPM implementation is not easily observable (Rogers, 1983). Obviously, the quality of the dissemination depends on the way the IPM message is communicated. Since IPM applies principles and field practices, and seeks to enhance such farmer behaviour as monitoring and decision making, dissemination should best take place in the field, through the same principles of facilitation of the learning process as were used in the IPM training.

Channels of dissemination
Several channels through which the surrounding community might learn about IPM can be expected. The field school graduates are the most likely and hoped-for channel, since they are the medium which can disseminate IPM most completely by repeating what they have learned themselves. Table 7.10, however, shows that a disappointingly low number of farmers was informed about IPM by the field school graduates themselves. In Senengsari, somewhat more than one-third of the farmers aware about IPM mention the field school graduates as their main source of information, with no considerable changes over the three evaluation seasons. These figures suggest a situation in which IPM trainees inform their nearest friends and neighbours about their experience in the field school, but they do not take deliberate action to extend it to a wider audience. This situation is not surprising considering the low coherence among villagers in Senengsari (Box 4.2). In Kuduagung, a different situation is encountered. The proportion of farmers aware of

Table 7.10: IPM dissemination channels and sources of information, as % of non-IPM farmers in intensively studied villages aware of IPM.

	IPM villages				Non-IPM villages			
	Senengsari		Kuduagung		Jayasari		Mulyoagung	
	1990	1991	1990	1991	1990	1991	1990	1991
IPM field school graduates	36%	38%	11%	45%	0%	0%	0%	4%
extension worker	18%	5%	33%	32%	36%	35%	21%	27%
village assistant/investigator	7%	29%	11%	9%	14%	25%	57%	46%
village officials	14%	24%	0%	5%	14%	0%	29%	15%
radio/TV/newspaper	7%	5%	0%	5%	32%	25%	14%	8%
neighbours/other farmers	7%	0%	33%	5%	14%	15%	0%	0%
IPM field school	14%	0%	11%	0%	0%	0%	0%	0%
N	*28*	*21*	*9*	*22*	*22*	*20*	*14*	*26*

Totals exceed 100% due to multiple response.

IPM through the field school farmers increased over the seasons, although starting at a much lower level of dissemination: from 11% at the end of the training season, via 17%, to 40% in the second post-training season. In Sugihsari and Sumberagung, comparable low levels of contribution by field school graduates to IPM dissemination were observed (Table V.7).

The implementation of the IPM farmer field schools in Senengsari and Kuduagung drew some attention from bypassing farmers, but not as much as was expected. Only in a few cases a farmer stopped and had a closer look at the training process that was going on. At the end of the training season, only 14% of the non-IPM farmers in Senengsari knowing about IPM, and 11% in Kuduagung, obtained this awareness from the field school implementation. A different situation occurred in some other field schools observed outside the study area where the originally selected group was consistently supplemented by several farmers (in one case not less than fifteen persons) who attended one or more training sessions on their own initiative. A relatively low attention from the surrounding community in Senengsari and Kuduagung is supposed to be a result of the non-representative composition of the field school groups, causing no direct affinity of the 'average' farmer to the training process. In Senengsari, feelings of envy among the majority of small farmers towards the large (IPM) farmers may have contributed to the gap, and in Kuduagung the participation of the many village officials. This observation shows again that the selection of field school participants is of major importance, also for the spread of IPM in the farming community.

Whereas mass media were hardly ever mentioned by farmers in the baseline survey as an important information source for rice cultivation issues, these media did contribute to the dissemination of IPM. In the 1990/91 season, 10% of all non-IPM farmers aware of IPM obtained this awareness through radio, television or newspapers. Especially in Jayasari, where farmers are generally well informed, mass media were an important source of information on IPM in the 1990 and 1991 seasons. Although the IPM programme did neither broadcast radio or television programmes by itself nor published articles in the newspapers, the IPM activities often attracted attention from these media.

Other important channels mentioned by non-IPM farmers include the extension workers and village officials, not to forget the strong influence the study itself apparently had on raising awareness about IPM. This influence can be a result of the presence of a village study assistant, a young farmer from the village itself doing weekly field monitoring, occasionally accompanied by the author, and assigned for the study period in the intensively studied villages, the frequent presence of the investigator herself, and the activities of the study such as individual and group interviews. It is obvious that this source of information is more frequently mentioned in the intensively studied villages than in the control villages where only two survey rounds were conducted.

Communication by field school graduates

In group interviews in the four IPM villages one year after training, field school graduates were asked to make a comic describing a situation that they still remembered in which IPM topics were discussed with non-IPM farmers. The majority of stories made by the field school graduates in Senengsari dealt with rat control, which shows again the importance of this topic for this village. In most of these stories fellow farmers were invited to join in the rat drives. Three farmers reported that they encouraged their fellow farmers to observe their crops before taking a pest control measure, and not to use pesticides. One story gave a perfect description of some good IPM methods:

> (Explaining to a neighbour) 'Eggs on rice leaves have to be examined well. After taking them home, they have to be observed and put in a plastic bag. If the eggs hatch and it are caterpillars that appear, these have to be destroyed, but if it are spiders that appear, they have to be released in the field.'

Another story shows the unconcern that field school graduates encounter from the environment, which they consider a constraint in IPM implementation and dissemination:

> (Reaction from a fellow farmer) 'Well, the crops of both IPM and non-IPM farmers are similarly attacked by pests.'

Places where field school graduates talked about IPM included mainly their homes, parties, and sometimes the rice field. Unlike in Senengsari, religious meetings seem to be an important place in Sugihsari to talk about IPM to other farmers. Topics mostly discussed include natural enemies, synchronized and uniform planting, row planting, and other good cultural practices to prevent pests. A consistency between what is told to other farmers and what they report to do themselves is visible. In Kuduagung, field school graduates generally talk about the existence of, and relation between, pests and natural enemies in the rice field: not all insects in the crop are pests. They warn their friends that they should not spray indiscriminately, mainly for economic reasons, and recommend they should monitor their fields first. Three people mentioned rat control in particular. In Sumberagung, rat problems and control are a more favourite topic, and, indeed, the greatest problem in rice cultivation. Other topics are the use of suitable varieties, and crop health. It is striking that none of the field school graduates there mentioned the role of natural enemies in the agroecosystem, their relation to pests, nor the effects of pesticide application to natural enemies, which was a popular topic in the other villages. Also, the economic profit of IPM was not mentioned. A concrete message of the field school recalled by a couple of farmers is the threshold level of five rice seed bugs per square meter below which they should not spray. The stories of a few farmers in Sumberagung gave the impression that the IPM message was not received so well:

- 'Have you already applied pesticides to your crop?' 'Yes of course I did, I did not forget what you ordered me to do.'

– I told a friend about my last fertiliser application in which I mixed pesticide granules with the fertiliser, and applied this to the crop at the age of about 25 days. His answer: 'In that case, I'll follow your IPM field school experience for my next crop.'

Although the training process in this field school group was not observed, and, therefore, nothing can be said about reasons of an incorrect perception of the IPM concepts, misinterpretation by some people might help explain the inconsistent practices of Sumberagung field school graduates during the post-training seasons.

The contents of the stories told about IPM by field school graduates to fellow farmers show more clearly the advantage or usefulness of IPM as experienced by farmers: rat control in Senengsari and Sumberagung, the role of natural enemies and no indiscriminate spraying in Kuduagung, and good cultural practices in Sugihsari. The recognition of pests and natural enemies is a topic often discussed in all villages. Where follow-up activities were still organised by the time of the group interviews, dissemination of IPM appeared to be incorporated in regular agricultural and religious meetings. In the other villages, dissemination depended more on incidental discussions in the field, in the houses of farmers, or on parties. The comics showed that IPM is usually communicated in other places than the rice field, which supposedly weakens the (field-oriented) message to a considerable extent.

Awareness by non-IPM farmers

In the standard work on diffusion of innovations (Rogers, 1983), five stages in the innovation-decision process are suggested:

1. knowledge (or awareness) about the innovation;
2. persuasion;
3. decision (either adopt or reject);
4. implementation;
5. confirmation.

It is debatable whether the traditional diffusion theories can be applied to IPM which we do not want to call an 'innovation', since it relies on experiential learning and enhances changes in farmers' decision making behaviour, rather than adopting recommendation of a package technology. Nevertheless, an attempt is made to recognise some of the stages mentioned above considering the spread of IPM in the study villages. Attention is paid mainly to the first stage, 'awareness' about IPM among non-IPM farmers, because we focus on activities generated by the field school graduates. Since inventories of farmers' practices were made in the seasonal surveys as a point of reference for evaluation, these data can also give some information about changes in non-IPM farmers' practices which belong to the implementation stage. To say something about the stages of persuasion, decision, and confirmation, more in-depth and long-term studies among non-trained

Table 7.11: Awareness about IPM by non-IPM farmers over three seasons, as % farmers reporting.

	IPM villages				non-IPM villages				Total	*Total N*
	Seneng-sari	Sugih-sari	Kudu-agung	Sumber-agung	Jaya-sari	Plosok-sari	Mulyo-agung	Lempung-agung		
1990	45%		12%		26%		16%		24%	*299*
1990/91	39%	38%	39%	45%	35%	14%	28%	10%	30%	*549*
1991	42%		42%		36%		34%		38%	*263*

farmers need to be done, which was beyond the purpose of this study.

Awareness about IPM among non-IPM farmers in both IPM and non-IPM villages was estimated in the seasonal surveys. At the end of the training season (1990 dry season), 24% of the non-trained respondents had heard about IPM, which is considered quite high (Table 7.11). The proportion was relatively high in Senengsari (45%), and low (even lower than in the non-IPM village) in Kuduagung (12%). Over the two post-training seasons surveyed (1990/91 wet and 1991 dry season), the number of farmers aware about IPM gradually increased, via 30% to 38%. This increase was especially visible in Kuduagung, confirming that dissemination mainly took place through communication (afterwards), rather than through attention to the field school implementation, probably due to the factor of non-representativeness of participants mentioned above. In Senengsari, the number of farmers knowing about IPM slightly decreased in the post-training seasons, suggesting a situation in which communication throughout the seasons is restricted to the same group of people. This is likely considering the high level of inequality in this village. The same levels of awareness are obtained in both Senengsari and Kuduagung at the end of the study period, but through remarkably different processes. Striking is that none of the women respondents in the surveys ever heard about IPM (Box 7.4).

A remarkably high awareness was found among farmers in the non-IPM villages, in particular Jayasari and Mulyoagung which were intensively studied, in contrast to Plosoksari and Lempungagung where only two surveys were conducted. Since there was no difference among the intensively studied and control villages regarding exposure to other sources informing about IPM, such as mass media and extension workers, it can be concluded that the study itself strongly influenced the outcome of this evaluation point. Presumably, the presence of a village study assistant in the intensively studied villages was the main factor causing higher awareness among farmers. The village study assistants, when out in the field for their weekly observations, were often questioned by farmers what they were doing and why. In the end, they were even asked for advice on pest control topics. A second factor might be the implementation of certain additional activities in the intensively studied village, such as more frequent surveying, group interviews, and field observations of and interviews with farmers. Figure 7.4 shows the great difference in awareness between intensively studied and control non-IPM villages at the end of the first post-training season: 32% versus 12%. This observation puts the use

Box 7.4
Women farmers and IPM communication processes

All farmer respondents reporting in the surveys that they had ever heard about IPM, were men. What can be the reasons that women farmers are not easily reached through the existing communication channels in the village, even though the issues communicated are relevant to them?

First, several categories of women farmers need to be distinguished: (1) women who operate the farm together with their husbands, (2) single women who bear the responsibility for farming alone, and (3) women whose husbands leave the village after planting to work in the towns, and come back before harvesting, leaving the responsibility for crop care to their wives. Most of the women respondents in the surveys could be categorised under the second and third categories. It is not surprising that these women, on average, belong to the lower socioeconomic layers in the community. They operate smaller farms, usually not owned, and earn a very small income. Those women who operate the farm (partly) together with their husbands, usually discuss cultivation problems together, and make decisions together. Since many women manage the household (and farm) money, deliberation between husband and wife is necessary.

An obvious observation is that women in the study villages have no contact with the extension worker, and are never invited to attend agricultural extension meeting. But women farmers are supposed to meet men farmers in the rice field when they are working. Apparently, these two categories do not communicate about agricultural issues here, as confirmed by both men and women. Non-communication with IPM field school farmers, in particular, can then be explained by the fact that all field school participants were men. In addition, these men were primarily the more affluent villagers. Considering local social structures, communication is not likely to occur between these IPM (male) farmers from higher socioeconomic layers and women from the lower layers.

A special category of women interviewed about communication on IPM were the wives of the IPM farmers. Remarkably, most of them did not know (much) about IPM. The reason for this should again be sought in the status of IPM farmers in the village. Since they are the better-off residents, there is, in most cases, no need for the wives to be involved in farming, and, therefore, no need to be informed about agriculture. Relatively many of the IPM farmers' wives even had off-farm employment. No communication from IPM farmers via their wives to the women farmers in the villages can be expected. In order to reach women farmers on agricultural issues, deliberate action will have to be undertaken.

of 'comparison' villages in programme evaluation studies to relate situations with and without intervention, in a dubious light. In general, it is extremely difficult to quantify the extent to which the study itself influences the outcomes. The use of non-intensively studied control villages in this evaluation study gave some indications, although the few activities in these villages might have had a certain influence, too.

Perception of non-IPM farmers

Although the awareness levels among non-IPM farmers over the study seasons seem high, this result becomes somewhat disappointing when the perception about IPM is considered. Twenty-five percent in Senengsari, and even 44% in Kuduagung, of the non-IPM farmers who reported to have heard about IPM in the 1990 season survey, had no idea what it actually implied (Table 7.12). This percentage stayed the same in Senengsari in the period of one year, but increased in Kuduagung to 75%, due to a larger number of people who heard about IPM, but nothing

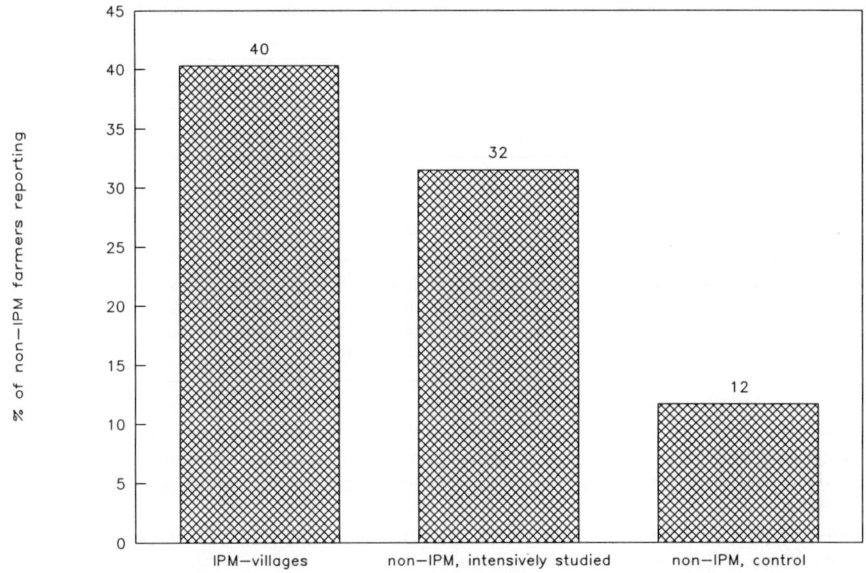

Figure 7.4: Awareness about IPM (as % farmers reporting) among non-IPM farmers in IPM and non-IPM villages at the end of the first post-training (1990/91) season.

more than that. Some more farmers expressed incorrect views, such as collective spraying and government pesticide aid. Non-IPM farmers in Senengsari complained that the field school graduates are stingy in informing others about what they have learned, wanting to keep the knowledge for themselves (Winasis, 1992). Similarly disappointing responses were given in Sugihsari and Sumberagung (Table V.8). Correct perceptions by non-IPM farmers mainly included collective rat control (especially in Senengsari), aspects of need-based and correct pesticide use, field monitoring, and pest and natural enemy identification. These topics correspond fairly well with the subjects communicated by the field school graduates.

Considering the type of messages conveyed by IPM farmers to non-IPM farmers, and particularly the way IPM knowledge was disseminated mostly without any field demonstration, no substantial changes in non-IPM farmers' practices can be expected. Relating to the main objective of field monitoring and informed decision making, no changes were observed, indeed. A practice often communicated and easily measurable is pesticide use. Figure 5.4 (Chapter 5) showed that the frequency of pesticide application as an average of all non-IPM farmers of the eight villages over the four study seasons could, to a certain extent, be related to pest occurrence. An obvious decrease in number of applications is visible during the 1991 dry season when pest populations were of minor importance. When looking at the averages per village (Figure 5.5), a distinct variation can be detected that covers more information than a relation with pest occurrence only, not unlikely an additional effect of dissemination of IPM knowledge. The diagrams in this

Table 7.12: Perception about IPM, as % of non-IPM farmers in intensively studied villages aware of IPM.

	IPM villages				Non-IPM villages			
	Senengsari		Kuduagung		Jayasari		Mulyoagung	
	1990	1991	1990	1991	1990	1991	1990	1991
pest control – correct [1)]	68%	83%	44%	17%	64%	50%	57%	35%
pest control – incorrect [2)]	7%	9%	12%	4%	36%	29%	7%	4%
monitoring/identification [3)]	18%	22%	22%	8%	9%	7%	7%	4%
agronomy [4)]	7%	4%	0%	0%	9%	7%	7%	4%
not clear/do not know	25%	26%	44%	75%	9%	7%	36%	65%
N	*28*	*23*	*9*	*25*	*22*	*26*	*14*	*26*

Totals exceed 100% due to multiple response

1) Includes collective rat control, aspects of need-based and correct pesticide use, integrated measures.

2) Includes collective control measures (spraying), preventive measures, government pesticide aid.

3) Includes field observations, identification and function of pests/natural enemies, (collective) decision making.

4) Includes good cultivation, synchronised and timely planting, balanced fertilisation, resistant varieties.

figure show a possible effect of IPM training on the chemical control practices of non-IPM farmers in Senengsari, Jayasari and Mulyoagung. Frequency of pesticide sprays decreased over the four seasons in these villages. It is disappointing that this effect was seen in only one of the four villages that received IPM training. Inhibitory factors in the other villages, especially in the subdistrict Sumberagung, were already mentioned several times.

The decreased frequency of pesticide applications, particularly sprays, by farmers in Jayasari and Mulyoagung over the 1990/91 wet and 1991 dry season, is assumed to be merely a result of the study activities in these intensively studied, non-IPM villages. Most likely, as was the case in awareness raising about IPM, the village study assistants were the main source of information causing change in farmers' spraying behaviour. Inquiring why the village study assistant is monitoring the field, farmers learned that there is no use in applying pesticides indiscriminately. This message was apparently well received by the rice farmers in these two villages, as it suited their crop management. However, nothing changed in their field monitoring behaviour and agroecological knowledge, the main objective of the IPM Programme. The fast reaction to the message that pesticide use can be reduced, is striking, though. A change in chemical pest control practices was not visible in the control non-IPM villages Plosoksari and Lempungagung, which indicates that the influence of the study activities on the outcomes in these control villages is negligible.

7.7 Conclusions

IPM knowledge and implementation

In this chapter, several behavioural aspects of IPM field school graduates in the four villages observed in Grobogan were evaluated by relating them to the programme's (competency) objectives and to farmers' own standards for a good IPM farmer. Competency objectives, however, do not appear to be fully satisfactory as a standard for evaluation of the non-formal training approach, for two reasons. First, the training curriculum is not fixed which is inherent to the experiential learning approach, therefore, not all objectives are necessarily covered. Second, the objectives contain various practices that are traditionally being applied by Javanese farmers. The unfortunate situation occurred that the main study season to measure training effects in farmers' behaviours (1990/91 wet season) was not representative due to severe stemborer outbreak. The results described should be perceived within this context.

IPM knowledge of field school graduates reached a satisfactory level, especially with regard to pest and natural enemy identification, ecological functions in the rice field, and negative effects of chemical pest control to the rice ecosystem. A different perception of the rice ecosystem elements was observed, especially in that pest organisms are no longer seen as harmful agents only, but also as food for the natural enemies, and necessary for the ecological balance in the field. The existence of the many predatory arthropods in a rice field was a discovery most IPM farmers were excited about. An interesting observation is that the factor knowledge, or knowledge increase, was usually mentioned as the main reward of IPM training and implementation, before economic profit. Knowledge increase was highly valued by the field school graduates because it provided them with more expertise, responsibility and independence in their farm management.

Effects of IPM training on farmers' practices left a few things to be desired, in particular with respect to control measures. Field monitoring was practised well by the larger part of the field school graduates. Better informed decisions than before training were noticed with regard to timing and target of control measures. A decrease in curative application of pesticides sprays was observed, but this result contrasted with an increase in (preventive) granular applications. The message of the need for observation before taking a measure was apparently better received by farmers than the message about the ineffectiveness and inefficiency of preventive pesticide applications. An attitude of pesticide dependency was still common among part of the field school graduates, although total pesticide use decreased. No adequate measures were taken when a rice stemborer outbreak occurred, for which the training approach is partly to blame. A weakness of the experiential and field-based learning process is that the training does not cover all problems that farmers' will encounter later.

Suitability of the IPM messages seemed to vary from place to place, depending on current perceptions and prevalent conditions. Particularly, the role of the village extension worker and the KUD in input distribution were found to be factors inhibiting the full adoption of IPM principles by the trained farmers. As in most extension programmes, a mix of measures will be needed, supplementary to farmer training, to create favourable conditions for the implementation of what is learned in training (Röling, 1986).

Despite the fact that practices were not always completely consistent with the

IPM principles, my own observation is that something has established in the minds and habits of most of the farmers who attended the farmer field school. Because of the different learning process they went through, it is not unlikely that the changed perceptions and knowledge can further develop, and at some stage be put into practice. Several observations were made in this regard during the post-training seasons, when farmers showed to continuously analyse and learn from experiences.

Collective action

The IPM farmer field school model is designed to enhance collective action in farmer groups and farming communities. In Senengsari, the IPM field school farmers became the organisers and motivators for collective rat control, resulting in a considerable decrease of rat damage to the rice crop for several seasons. An attempt was made by the IPM group leader in Kuduagung to mobilise the field school graduates to coordinate collective irrigation maintenance work. This attempt failed due to the authoritarian way of designing and presenting the plan. Although desired by the group members, no collective action ever materialised in the IPM groups observed in Sugihsari and Sumberagung.

Horizontal communication

Despite a fairly high level of awareness about IPM attained among non-IPM farmers, the perception of what it implied was quite limited. A high awareness, but low understanding, also found in studies in other countries (Hussein, 1987), indicates that the message is of interest for farmers, but that extension requires adapted channels. Although the IPM message was extended to the surrounding community through various channels, which supposedly contributes to a higher knowledge level (Lawson, 1982), IPM was not disseminated through the way field school graduates learned themselves, in the field. This is perceived as a weak chain in the communication process since field practice is considered crucial for IPM implementation consisting of field-oriented, knowledge-intensive practices.

A relation was observed between the content of IPM communication and the favourite topics at the time of training. Considering the fact that perception, which influences the decision making process (Norton, 1982; Mumford, 1987), about IPM by non-IPM farmers was often incorrect or incomplete, and that communication was of a low quality, it is not surprising that few changes in practices relating to IPM were observed among non-IPM farmers as a result of dissemination of IPM. Written materials distributed to field school farmers to take home would probably have contributed to a better quality of dissemination.

The role of field school graduates in the dissemination process was disappointing. Not unlikely, this is a result of the non-representative composition of the training groups. IPM farmers represent mainly one layer in the farming community, that of more affluent and better informed farmers. Intensive communication with other layers is not likely if no explicit efforts are made to that purpose. Studies on the relationship between diffusion and equity showed that extension directed to the higher social categories (which is usually the case in agricultural extension programmes) does not reach the lower ones through diffusion (Röling et al., 1976). The group of farmers most disadvantaged in this respect are the women farmers, who did not have any representative among the trainees, and who generally belong to the lower socioeconomic layers in the village communities. Other theories suggest that larger farmers and formal or informal leaders play an important role in the dissemination process at the village level (Gibbons et al., 1980). In some of the study villages, and some of the IPM groups visited in other areas, village officials, indeed, seemed to have a major function in the spread of IPM knowledge. The level and quality of communication through this channel, however, depends upon existing social relationships in a community. Kuduagung is an obvious example where many village officials participated in IPM training, but hardly contributed to the spread of IPM among non-trained farmers.

Variation among villages

High variation in behavioural change occurred among the four IPM villages observed. Senengsari farmers, in particular, performed well. Community rat control activities were organised by the IPM farmers, and consequently rodenticide use decreased. A sustainable reduction of pesticide use over the whole study period was visible in this village, whereas in the other villages a reduction occurred only during the training season. Little effect of the IPM training on farmers' practice was measured in Sumberagung. Factors relating to both training process and external conditions were identified to have caused this low result.

A small contribution to variation in training effects among villages can be ascribed to the trainer in the field school (the pest observer), as was most obvious in the case of Sumberagung. Several trainees in this village apparently misinterpreted some of the IPM principles, probably due to a language barrier. The trainer used the national language (Indonesian), and quite a few scientific terms, whereas most farmers in this group felt only comfortable with the local language (Javanese). The same trainer led the field school group in Kuduagung, where this language problem was not so apparent, as a result of the different (higher educated) composition in this group.

A stronger influence of IPM training effects was found through the village extension worker who was supposed to guide the field school graduates afterwards. The attitudes of the extension workers in the two subdistrict studied contrasted in this regard (Chapter 6). An aspect partly related to the extension worker's influence was the role of the KUD, which was strong in Sugihsari, Kuduagung and Sumberagung, and of no importance in Senengsari. It is repeated here that the majority of the farmers selected for IPM training belonged to the group of more affluent, well informed farmers who were used to rely on the extension worker's advice. Apparently, the IPM field school, especially in Kuduagung and Sumberagung, had not been convincing enough to change farmers' practices and combat contradicting external forces, as those exerted by the extension worker and KUD. In Sumberagung, rather adverse pest management measures and adverse IPM communication messages were even observed in the post-training seasons. In this regard, the role of the extension worker as an IPM trainer becomes dubious. On the other hand, the extension worker can have a very stimulating influence on farmer group activities, as was obvious in Senengsari. An agricultural extension programme has to work with the field level extension workers, and it is impossible to distinguish officers with different interests. Change in farmers' practice can only be achieved if farmers are aware and convinced of the profit of a certain change, and if they are not dependent upon external factors such as extension officers, although these might be supportive.

Another important difference among the villages with regard to sustainable implementation of IPM and horizontal communication is found in the prevailing crop cultivation practices, and the suitability of the IPM approach to those practices. A field school trainer should be able to detect aspects of suitability, and adjust the training process accordingly.

Some change in pest management practice was observed in the intensively studied, non-IPM villages, particularly with regard to the frequency of pesticide sprays. The message that pesticide use can be reduced was received mainly as a result of the study activities in these villages, leading only to changed practices, not to a gain in agroecological knowledge or skills, and to considered decision making. The influence of the study on its outcomes is an important observation considering the use of control villages in evaluation studies.

8 The promise of IPM: profitable and safe

8.1 This chapter

A promotion brochure presenting the results of the Indonesian pilot (crash) IPM training programme in 1986-87 reports a decrease in pesticide use from 4.5 to 0.4 applications per season, and an increase in yield from 6.1 to 7.4 tons per hectare on fields managed by IPM trained farmers (Food and Agriculture Organisation, 1988). Although these results represent a sample only, they indicate that changes in pesticide use and yields, standing for improvement of farm economics and safety, can be achieved as a result of IPM training. However, as described in Chapter 2, the crash training did not result in major changes of trained farmers' pest management independency and expertise, due to the top-down training approach. Sustainability of the achievements is, therefore, questionable. The different training approach as applied by the National IPM Programme from 1989 onwards, which makes farmers the central decision makers and independent IPM implementors, is expected to yield the same improvements on farm economics and safety, but at an increased level of sustainability.

Whereas Chapter 7 gave a description and analysis of the effects of IPM training through the farmer field school model at the level of the individual farmer as observed in Grobogan, this chapter looks into the effects at the farm level as a result of changed farmer behaviour. Three important farm level effects can be distinguished: farm economics, farm ecology, and human health. This chapter will mainly elaborate on economic farm level effects measured. The design of the study, longitudinal but only covering four seasons, restricts the validity of the conclusions since effects at the levels of economy should actually be measured over much longer periods. To better understand the general economic picture that will be given in section 8.3, the rice production practices and economics of one IPM farmer throughout one season is presented first as a case in section 8.2.

Non-validity of information due to a short study period is all the more true for possible effects on farm ecology and human health. Therefore, no substantial measurements were done with regard to ecological and health aspects. This chapter describes only a few observation done on these aspects in the study area.

Chapter 8 uses the same terminology as Chapter 7. It will consider the two intensively studied IPM villages, Senengsari and Kuduagung, only. Unlike the two control IPM villages, data collection was complete over four seasons in the intensively studied villages, and intensive observations there could support interpretation of the data. All tables belonging to this chapter are presented in the text (indicated with Table 8.*). Significance of differences was statistically tested by means of Student's t-test.

8.2 A case of IPM practice and economics

During the 1990/91 wet season, Mr. Agus, a young farmer in Kuduagung who graduated from the IPM farmer field school in the 1990 dry season, kept daily records of all practices and expenditures in his rice crop. He was one of the field school graduates qualified by his colleagues as a good IPM farmer. Agus operates two fields, which he owns himself, on a walking distance of five minutes. One field measures 0.29 ha and the other 0.53 ha. In addition to rice farming, he has a small home garden and raises one cow. He is engaged in a small non-farming job, arranging wedding ceremonies in the local mosque.

The two rice fields that Agus operates have comparable soil and water supply conditions. In the 1990/91 season, he planted both fields with the brown planthopper resistant rice variety IR 64. Week by week, throughout the season, the following actions were taken:

Week 1: preparation of seedbed;
Week 2: sowing 40 kg seed, hoeing, seedbed observation (2 times);
Week 3: hoeing, seedbed fertilisation at 9 days after sowing (9 kg urea/ha field, 4 kg TSP/ha field), field preparation (ploughing by tractor);
Week 4: field preparation (ploughing by tractor), hoeing, levelling, pulling out seedlings, transplanting field 1;
Week 5: hoeing, levelling, pulling out seedlings, transplanting field 2;
Week 6: monitoring (2 times);
Week 7: 1st fertilisation field 1 at 15 DAT (68 kg urea/ha, 51 kg TSP/ha, 34 kg KCl/ha), weeding field 1 at 16-18 DAT, 1st fertilisation field 2 at 15 DAT (94 kg urea/ha, 94 kg TSP/ha);
Week 8: weeding field 2 at 17-18 DAT, 2nd fertilisation field 1 at 24 DAT (51 kg urea/ha, 34 kg TSP/ha, 34 kg KCl/ha);

Week 9: monitoring (twice, found little rat damage), 2nd fertilisation field 2 (38 kg urea/ha, 75 kg TSP/ha);
Week 10: monitoring (once, little rat damage);
Week 11: monitoring (once, little rat damage);
Week 12: monitoring (once, 2-3 tillers damaged by rats), placing poisoned rat bait in both fields;
Week 13: monitoring (once, all rice leaves have red tips);
Week 14: monitoring field 1 (once, few brown spots on leaves), monitoring field 2 (looked alright);
Week 15: monitoring field 2 (3-5 whiteheads/m^2);
Week 16: --
Week 17: monitoring (once), harvesting field 1;
Week 18: drying harvest field 1, harvesting field 2, drying harvest field 2;
Week 19: drying harvest field 2.

The seed Agus used was for 50% his own product of the previous season and for 50% newly bought, certified seed. He made one seedbed for both fields, but planted the two fields with one week time interval in order to spread labour demand. On his small field (called field 1 above) he wanted to do an experiment with the recommended fertiliser dosages, and to compare these with his usual practices on the larger field 2. The transplants from the certified seed were planted on the experiment plot. Fertilisation on the two fields differed with regard to KCl application and TSP dosage, especially at the second application. His total fertiliser dosages were lower than the recommendations due to wrong calculations. According to the size given on the land certificates, he thought that his fields were smaller than they actually are, which he discovered when we measured them for the records. The yields obtained were 5.66 tons/ha for field 1 and 5.16 tons/ha for field 2. Both fields were slightly damaged by rats during the vegetative stage, but loss because of this damage was negligible. Field 2 was attacked by stemborer at the reproductive stage, which might have caused some yield loss. Because of the many factors varying between field 1 and 2 (seed, planting date, fertiliser types and dosages, and pest occurrence) he was not sure what was the determining factor having caused the higher yield in field 1.

Throughout the season, Agus monitored his fields once or twice a week, except for week 16 when both crops were ripening. During monitoring, on which he mostly spent one hour for a visit to either one or both fields, he always observed a sample of rice plants closely, and repaired the bunds if necessary. Based on his observations, he did not find it necessary to apply a pest control measure except one baiting round against rats. The decision not to apply pesticides when pest populations were low was an achievement from the IPM field school for Agus. Previously, he definitely gave one or two pesticide applications to his rice crop. Table 8.1 shows what Agus spent on labour and purchased inputs throughout the season. He did most of the work himself, weeded together with his wife, and only hired outside labour for ploughing, transplanting and harvesting. Harvest labour was paid in kind by one-eighth (field 1) or one-ninth (field 2) of the harvest.

Table 8.1: Labour and purchased inputs of rice production, 1990/91 wet season.

	Labour			Purchased inputs	Total paid-out costs
	Family		Hired		
	(hours/ha)	(Rp/ha)	(Rp/ha)	(Rp/ha)	(Rp/ha)
Field preparation	38	9,500	28,900	0	28,900
Seedbed preparation	12	3,000	0	28,200	28,200
Transplanting	34	8,500	34,700	0	34,700
Monitoring	30	7,600	0	0	0
Fertilisation	16	3,900	0	72,200	72,200
Weeding	49	12,100	0	0	0
Pest management	1	300	0	1,200	1,200
Harvest	78	19,400	40,100	0	40,100
Post-harvest	50	12,500	0	0	0
Total	308	76,800	103,700	101,600	205,300

Costs involved for harvesting were spent on food for the labourers. Purchased inputs included seed, fertiliser, and rat poison and bait.

Looking at the output of Agus' rice farm (Table 8.2), he received a gross return of more than one million rupiahs per hectare, which is about one third higher than the average of owner-operators during the same season (see Table III.27). This high return is a result of the relatively good yield he obtained. His field was not severely attacked by stemborer, unlike those of most farmers in Kuduagung,

Table 8.2: Return of rice production, 1990/91 wet season.

Output		
yield (tons/ha)		5.33
share for harvest labour (tons/ha)		0.64
gross return (Rp/ha)		1,036,500
Costs (Rp/ha)		
land:	opportunity cost	378,000
	tax	17,000
	subtotal	395,000
labour:	family labour	76,800
	hired labour	103,700
	subtotal	180,500
purchased inputs (Rp/ha)		101,600
Paid-out costs (Rp/ha)		205,300
Farm (rice) business income (Rp/ha),		831,200
return to:	land	378,000
	family labour	76,800
	capital and management	376,400

because he planted quite early. Additionally, he sold the harvest in parts, and could therefore wait for good prices in the market. Due to his high gross return and relatively low paid-out costs (since he did not use much fertiliser and no pesticides at all), his farm business income of rice production reached a level of more than twice the average for owner-operators in the 1990/91 season. The return to the opportunity cost for family labour of his farm business income is very low. When related to a wage for hired labour, Agus spent only about forty-two working days on his farm over a growing season of four months.

Agus can be characterised as a farmer who is rational in his farming: he continuously experiments to optimise his practices and returns. He does not spend much time on farming, but does it regularly and accurately.

8.3 Effects on farm economics

Scope of analysis

An economic analysis of rice production inputs and outputs was made with the respondents in Grobogan for the 1989/90 wet (baseline), 1990 dry and 1990/91 wet seasons. The last survey season (1991) focused on qualitative aspects of cultivation practices, and detailed economic data were collected on farm output and pest management only. Many factors appeared to influence the levels of farm inputs and outputs, which is a general problem in conducting on-farm research (Shaner et al., 1982). Most difficult to cope with in drawing conclusions about the data collected over four seasons was the seasonal variation with respect to climatic conditions, market conditions and pest occurrence (see Chapter 5). Whereas climatic conditions and pest pressure during the 1989/90 wet and the 1990 dry seasons were fairly comparable, the 1990/91 wet season was extremely bad due to stemborer outbreak, and the 1991 dry season was extremely good. Prices fluctuated considerably over the four seasons. Rents for land and fertiliser prices increased. Rice prices went up and down mainly following the market principle of demand and supply, but also influenced by factors such as the quality of the product, the distance of the local market to the central storage facilities, competitive supply in the area, and the way in which the product is sold (standing crop, wet grain or dry grain).

When assessing economic effects of IPM training, the complexity of influencing factors tends to disturb the picture. The most important expected economic effect of IPM implementation includes an increased return on rice farming as a result of decreased pest management costs. An increased gross return as a result of higher yields is likely, according to experiences in many IPM programmes. At least, yields are not expected to decrease. Farmers in the study area perceive pest damage as the main factor determining yield loss. More effective pest management is therefore supposed to contribute to higher yields. Yield variability, as a measure for risk in rice cultivation, is expected to become smaller after IPM implementation. In order to show the economic effects of IPM implementation as clearly as possible, the analysis below focuses on yields, yield variability and pest management costs.

It seems justifiable not to consider the complete economic picture of IPM versus non-IPM farmers, since no major differences in cultivation practices, except pest management, were found between these two categories of farmers (Chapter 7). Regarding fertilisation, for instance, the same trend of increasing fertiliser dosages over the four seasons was visible among both IPM and non-IPM farmers, thus increasing fertilisation costs for both groups. Before training, IPM farmers tended to use less family and more hired labour than the average non-IPM farmers, which is conform the land tenure classification (Chapter 5), since IPM farmers mostly belong to the larger, mixed status farmers. During the post-training seasons, a shift in labour use was visible showing an increased use of hired labour and a decreased use of family labour by all farmers. This shift was stronger among non-IPM farmers, resulting in an almost similar labour use during the last study season. Changes in labour and fertilisation patterns are considered beyond the scope of the IPM decision making process and are, therefore, not of interest for the analysis below.

Since complete data sets over all seasons were obtained from the IPM villages Senengsari and Kuduagung, the discussion below only considers IPM and non-IPM farmers in these two intensively studied villages. The conduct of the IPM field school during the 1990 dry season is likely to have influenced the outcome for this season. Effects are supposedly stronger at the time farmers are still engaged in training activities.

Yield

The average yield obtained by IPM farmers in Senengsari and Kuduagung from the baseline season crop was slightly higher (5.3%) than that of non-IPM farmers in the same villages, but this difference is not significant (Table 8.3). Cultivation practices applied by IPM farmers before training were not considerably different from those by non-IPM farmers. The reason for the somewhat higher average yields can possibly be explained by the fact that the farmers selected for IPM training primarily and inevitably belong to the better informed farmers (Chapter 6). A second aspect might be an often occurring phenomenon in Java that the elite

Table 8.3: Average yields (in tons/ha) of IPM and non-IPM farmers over four seasons. T-test significance level is given for significant differences between IPM and non-IPM farmers.

	before training 1989/90	training season 1990	after training 1990/91	after training 1991
IPM	5.38 ± 1.80	5.77 ± 1.29	3.70 ± 1.88	6.38 ± 1.02
N	*47*	*48*	*45*	*40*
non-IPM	5.11 ± 1.69	4.64 ± 1.65	3.11 ± 1.62	5.68 ± 1.44
N	*127*	*129*	*119*	*109*
% difference	5.3%	24.4%	16.9%	12.3%
significance level	–	0.05%	5%	0.5%

of a village, to which a part of the IPM farmers (e.g. the village officials) belonged, usually operate the best pieces of land with regard to soil and water conditions (Hüsken, 1989). The yield difference between IPM and non-IPM farmers before training was stronger (and significant) in the two control IPM villages (Sugihsari and Sumberagung), which underlines the statement that those selected for IPM training represent the better farmers in the villages.

A possible effect of IPM implementation on yields can best be estimated by comparing the yields obtained by IPM versus non-IPM farmers in the (post-)training seasons, in relation to the yield difference between these two groups during the baseline season. A comparison of yields across seasons would be useful too, if there had not been such a great seasonal variation with regard to pest pressure and climatic conditions over the study period.

The training (1990) season crop gave slightly higher average yields for IPM farmers, but lower yields for non-IPM farmers compared to the baseline season. This difference caused IPM farmers to produce 24.4% more rice than the non-IPM farmers, a highly significant difference ($\alpha = 0.05\%$). Non-IPM farmers blamed rat damage as the main factor for the lower yields, whereas IPM farmers ascribed their good result to the farmer field school. As a result of participation in the field school, they were more aware of problems in their own fields and reacted more adequately. The rat control activities in Senengsari are an obvious example.

The trend of higher yields by IPM farmers compared to non-IPM farmers continued during the two post-training seasons, although somewhat less obviously. Yield differences of 16.9% in the 1990/91 season (be it at a low averages with high standard deviations due to the stemborer outbreak), and 12.3% in the 1991 season between IPM and non-IPM farmers were achieved, which were both significant. These results suggest that an effect on yields occurred as a result of IPM implementation.

Yield variability

A second indication for possible effects of IPM on the yields of trained farmers can be found when looking at the yield variability, which gives some information on risk. Yield variability is depicted by a frequency distribution diagram of obtained yields. A situation in which risk reducing practices are applied is displayed by a narrower and steeper curve of the diagram, compared to broader and lower curve for a situation without those practices.

In Figure 8.1, yield distributions of IPM farmers over the four study seasons are presented against those of the non-IPM farmers surveyed in the IPM villages. The diagram for the baseline (1989/90) season shows that the yield distribution for IPM farmers follows almost the same curve as the one for non-IPM farmers. The distributions are quite broad indicating high variability in this season, probably a result of varying rat attack, and similar for IPM and non-IPM farmers. The situation changes in the training (1990) and post-training (1990/91 and 1991) seasons. The curves for IPM farmers become narrower and steeper than those of non-IPM farmers, and somewhat skewed to the right. The latter implies that more IPM than non-IPM farmers harvest the 'attainable yield' (Zadoks and Schein,

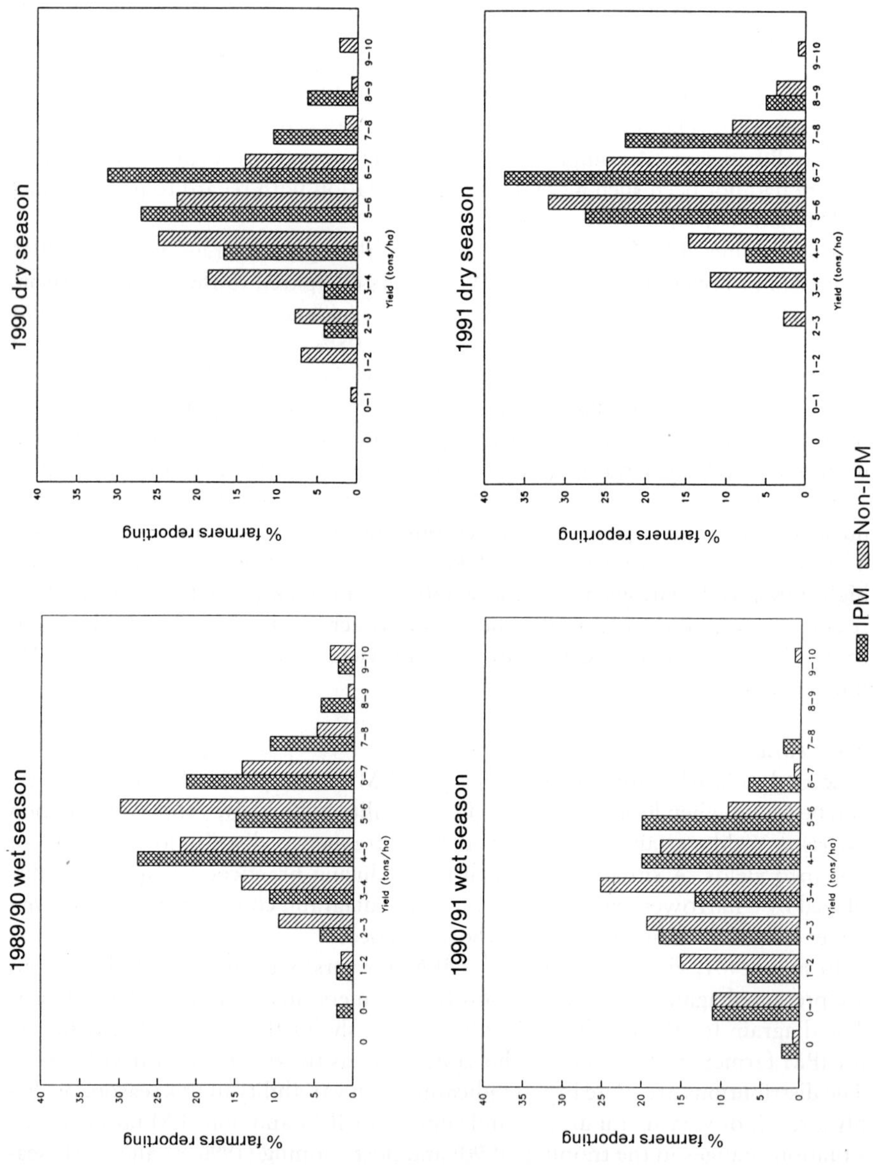

Figure 8.1: Yield distribution of IPM (left bars, N = 48) and non-IPM (right bars, N = 132) farmers over four seasons.
1989/90 = baseline, 1990 = IPM training, 1990/91 and 1991 = post training.

1979), and fewer suffer from avoidable hazards.

Seasonal influences, however, should not be effaced. Broad curves for both IPM and non-IPM farmers in the 1990/91 wet season, starting at very low yield levels and with peaks also at relatively low levels, clearly represent the hazardous situation of the stemborer outbreak. In contrast, the excellent dry season of 1991, with good weather conditions and low pest pressure, is reflected by narrow and high distributions. Nevertheless, the yield distribution diagrams for IPM farmers stand for a less hazardous and therefore more valuable management system.

Remarkably, Figure 8.1 shows that mostly non-IPM farmers obtained the highest yields over all seasons. Although the bars for the highest yields (9-10 tons/ha) in the diagrams represent only a few farmers, they represent more non-IPM than IPM farmers. These few farmers supposedly cultivate their rice fields under very good conditions and management. No consistent characteristics of these farmers could be identified, but probably they can be categorised under the good, diligent farmers, referred to in section 5.6 (Chapter 5). These high yields indicate that their is still a yield gap and thus room for improvement.

To understand what factors caused the increased average yields and more favourable yield distributions of IPM farmers, cultivation practices likely to determine yields by IPM and non-IPM farmers need to be compared (Chapter 7). As mentioned above, average dosages of fertiliser applied by IPM farmers were not significantly different from non-IPM farmers over the four study seasons, rather somewhat lower except for TSP quantities in Senengsari. IPM farmers reported similar levels of pest attack occurring in their crops as non-IPM farmers. The frequency of pesticide application, especially pesticide sprays, by IPM farmers decreased, but a higher use of granular pesticides by both categories of farmers was observed. Based on farmers' and my own comparative observations in fields with and without granular treatment, the increase of granular applications is not supposed to have contributed to the yield increase. One aspect of farmers' practices considered contributive to yield increase is the more intensive use of high quality seed by IPM than non-IPM farmers.

Summarising these aspects, the main reasons for the achievement of higher yields by IPM farmers should probably be sought in practices that are more subtle and more difficult to measure, such as adequacy and timeliness, as was also stated in Chapter 5. The regression analysis shown there showed that no correlation was visible between yields and farmers' practices relating to fertilisation, pesticide application or weeding.

Expenditures on pest management

Pest management expenditures are divided into costs on insect pest control and on rat control, since the two categories of pests require completely different control methods. Money on insect pest control is mainly spent on purchased inputs (chemicals, sprayer), whilst rat control involves much (family) labour for which an opportunity cost is counted. Saving on insect pest control is considered more crucial than on rat control from the IPM point of view, since the former implies mainly pesticides. Table 8.4 shows the average expenditures on insect pest and rat control

Table 8.4: Average expenditures (in Rp/ha) on insect pest and rat control over four seasons.

	1989/90		1990		1990/91		1991	
	Seneng-sari	Kudu-agung	Seneng-sari	Kudu-agung	Seneng-sari	Kudu-agung	Seneng-sari	Kudu-agung
IPM								
insect control	9,800	53,600	8,500	27,300	3,000	33,800	2,100	16,200
rat control	7,800	800	25,500	14,400	18,300	900	9,500	8,900
total pest control	17,600	54,400	34,000	41,700	21,300	34,700	11,600	25,100
N	*23*	*22*	*23*	*25*	*23*	*23*	*21*	*21*
non-IPM								
insect control	9,600	34,400	20,500	41,200	9,400	33,000	4,800	32,800
rat control	8,800	4,800	26,500	33,200	12,500	3,900	10,300	9,000
total pest control	18,400	39,200	47,000	74,400	21,900	36,900	15,100	41,800
N	*65*	*67*	*62*	*67*	*59*	*61*	*55*	*60*
% difference	4.5%	38.8%	38.2%	78.4%	2.8%	6.3%	30.2%	66.5%

by IPM and non-IPM over the four study seasons. A classification is made distinguishing Senengsari and Kuduagung farmers, which is considered most relevant due to the different habits in pest management in these two villages, as discussed in Chapter 7. Chapter 5 already stated that there is no difference in pest management expenditures among the various tenure status categories, which justifies a location-specific classification.

In the (1989/90) baseline season, no significant differences were visible between pest management expenditures between IPM and non-IPM farmers in Senengsari. Expenditures on insect pest control (mainly insecticides) accounted for only 2% of the total paid-out costs of rice production for IPM farmers. In Kuduagung, IPM farmers spent 38.8% more than non-IPM farmers on pest management, particularly on chemical insect control. This difference is not significant due to the very large standard deviations, indicating that the presence of some high-pesticide users existed, especially among the IPM farmers. Expenditures on insect pest control accounted for 14% of the total paid-out costs on rice production for IPM farmers in Kuduagung, which is considerably higher than in Senengsari. Although pest management costs did not seem to be a big item in the total rice production budget, the Rp 17,600/ha spent by Senengsari farmers in the baseline season can be compared with the costs of a primary school fee for seven months.

During the IPM training season, IPM farmers in Senengsari spent about the same (relatively small) amount of money on insect control as before training, but increased their inputs on rat control. Their non-IPM colleagues' expenditures on insecticide treatments in that season were significantly ($\alpha = 2.5\%$) higher, probably as a prevention against higher pest levels usually occurring during the dry season. A sharper contrast was visible in Kuduagung in the 1990 season. IPM farmers reduced their costs on insect control by almost 50% compared to the previous season, whereas an increase occurred among non-IPM farmers, resulting in a significant (α = 2.5%) difference in expenditures between the two groups. Both spent more on rat control, the main pest problem in Kuduagung during that season.

The next (1990/91 wet) season brought along low insect control costs for Senengsari farmers, the IPM farmers in particular spending significantly (α = 2.5%) less than non-IPM farmers, and somewhat higher rat control costs compared to the previous wet season. Neither IPM nor non-IPM farmers reacted strongly with pesticides to the stemborer outbreak, since they knew it was too late. The IPM farmers in Kuduagung, however, spent more than ten times as much on chemical insect control than those in Senengsari, the same amount as their non-IPM colleagues. These expenditures were mainly a result of promotion and obligatory purchase of granular pesticide by farmers during this season, as described in detail in Chapter 7. Costs on insect pest control accounted for 0.5% of the total paid-out costs for rice production of IPM farmers in Senengsari, and for 7% in Kuduagung, a considerable decrease in both villages.

In the last study season (1991), Senengsari farmers continued spending negligible amounts of money on insect control, whereas Kuduagung IPM farmers reduced their insecticide use by half. Especially for the latter group, this season proved, after all, that IPM graduates did learn from the field school, as well as from the bad experience of the stemborer outbreak and the useless indiscriminate granular pesticide application in the 1990/91 season. Under the low pest pressure during the 1991 season, allowing field school graduates to better rely on the IPM principles, IPM farmers spent significantly less on insect control than non-IPM farmers (α = 5% in Senengsari and 0.5% in Kuduagung). Rat control costs were the same.

All in all, IPM farmers in Senengsari consistently spent significantly less on insect pest control than non-IPM farmers in the training and post-training season, whereas in Kuduagung this was also true except for the 1990/91 season. Regarding rat control, no great differences were observed among IPM and non-IPM farmers in Senengsari over the study seasons, which may relate to the collective activities organised in this village. In Kuduagung, IPM farmers tended to spend less on rat control than their non-IPM colleagues during all except the 1991 dry season.

A more detailed pest management cost analysis was recorded with the respondents during the last study season (Table 8.5). The table shows that most money is spent on granular pesticides, which are relatively expensive indeed. Labour is a small item in insect control, and in the case of granular application usually not counted at all since the granules are mixed with fertiliser. In rat control, labour counts for more than half of the costs, mainly in the form of rat drives. Table 8.5 again shows the difference between the two villages, Senengsari and Kuduagung. Senengsari farmers, both IPM and non-IPM, spent most on rat control, whereas Kuduagung farmers emphasised insect control. Senengsari farmers can be categorised as risk-accepting, Kuduagung farmers as risk-avoiding, although the IPM farmers in these villages gradually accepted risk after having followed the IPM field school.

Table 8.5: Average expenditures (in Rp/ha) on insect and rat control activities, 1991 dry season.

	IPM		Non-IPM	
	Senengsari	Kuduagung	Senengsari	Kuduagung
Spray applications:				
Chemicals	300	2,800	2,300	9,800
Labour	100	1,000	600	2,500
Total [1]	400	4,100	3,100	13,200
Granule applications:				
Chemicals	1,600	12,000	1,800	19,600
Labour [2]	0	100	0	0
Total	1,600	12,100	1,800	19,600
Rat control:				
Poison baiting [3]	3,500	3,300	4,500	4,900
Rat drives [4]	5,400	4,500	5,800	3,800
Total [5]	9,500	8,900	10,300	9,000
N	*21*	*21*	*55*	*60*

1) Total includes chemicals, labour, rent for sprayer.
2) Labour is not counted when granules are mixed with fertiliser.
3) Includes poison, bait and labour.
4) Includes labour and implements.
5) Total includes baiting, rat drives, fumigation, plastic fencing.

8.4 Effects on farm ecology

Brown planthopper outbreaks in 1985/86 occurred mainly in intensive rice growing areas in Java and Sumatra. Under the rice intensification programmes, whole rice tracts in villages were uniformly treated with pesticides, killing most pests and natural enemies in the whole area, which in turn allowed brown planthopper populations to build up very fast. These outbreak areas served as infestation sources for surrounding areas, where the brown planthopper outbreaks also triggered excessive pesticide applications, making problems only worse. In fact, as often expressed by the coordinator of the FAO Inter-country IPM Programme, 'it were the pesticide applications that triggered pest outbreaks', rather than the other way round. Farmers in Grobogan reported that large predators in the rice fields, such as snakes and birds of prey, almost disappeared, partly because of pesticide poisoning, and partly because of increased poaching activity. A result is that rat populations are much higher than before, often causing major problems to the rice crop.

When pesticide load is reduced through IPM implementation, positive effects on the rice ecosystem and surrounding natural environment can be expected. Contamination of the environment is caused directly by pesticide application, but also indirectly by the cleaning of knapsack sprayers in irrigation ditches, and the disposal of wrappings or bottles in the field. Spray applications of persistent contact pesticides were always thought to be most destructive to the natural enemy population in the rice ecosystem (Hassan, 1984), but recent studies carried out within the National IPM Programme showed that the systemic granular insecticides have tremendous negative effects on the development of beneficial organisms in the field (pers. comm. W. Settle). Farmers themselves are well aware of the killing capacity of these chemicals on earthworms, and subsequent unfavourable effects on soil texture.

As described in Chapter 7, field school graduates reduced their frequency of pesticide spray application, although an increase of granular pesticide use was visible in the 1990 and 1990/91 seasons. After training, pesticide applications were better targeted, and lower pesticide dosages were used by the IPM farmers. Pesticide reduction and better targeting contribute to a lower pesticide load in, and less severe effects on the environment, although much remains to be improved. Collective preservation of the environment through reduced pesticide use by all farmers in an area is supposed to bear a greater impact on the improvement of the natural balance of the whole rice ecosystem than the effect of only a handful of trained farmers. Except for slight changes regarding pesticide use among non-IPM farmers in Senengsari, no influence was observed yet among the surrounding farming community in the other IPM trained villages within the (limited) study period. No initiative for collective action was undertaken by the field school graduates, except for community rat control in Senengsari.

Records of my own field observations throughout the 1990, 1990/91, 1991 and 1991/92 seasons in the four intensively studied villages did not show any differences in pest and natural enemy populations between IPM and non-IPM managed fields (for methodology see Chapter 3). No relation whatsoever could be detected in the observation records between the level of pesticide application and the levels of natural enemy or pest populations. This result is supposed to be strongly influenced by the surveillance method used. Sample fields were fairly small, on average less than 0.25 ha, and dispersed over the total rice areas (ranging from 107 to 337 hectares) of the villages. Most pest and beneficial animals in the rice field are highly mobile. Therefore, migration of all kinds of animals to and from neighbouring fields with unknown pest management regimes strongly determined population changes in the observation plots. In addition, pesticide load in the sample fields was fairly low. Only a few fields received up to three pesticide applications in the first and second observation season, but the majority of fields was not treated at all, or only once. After pesticide application, a temporary reduction of pest and natural enemy populations was often visible in the records, but both populations never fell down to zero, and recovered quickly. Chemical pest control measures applied by farmers on the observed fields had no visible pest control effect. Considering the low volumes and dosages generally used by farmers, it is not unlikely that this low control effect is a result of poor application techniques. Monitoring in farmers' fields under the conditions mentioned is not considered a suitable technique to assess ecological effects of IPM implementation. The finding that migration of arthropods among the small fields of farmers strongly determines populations (and thus strongly influences the effectiveness of pest control measures) confirms the need for the collective implementation of IPM by all farmers in an area.

An interesting observation resulted from the records of the sample fields in the four study villages. Pest populations up to certain levels, which are usually alarming for farmers, did not effect the yields thanks to the abundance of suppressing natural enemies in most rice fields, and to the compensatory capacity of rice plants. This observation is consistent with studies done on yield loss assessment (e.g. Rubia et al., 1989; Arida and Heong, 1988).

8.5 Effects on human health

Many studies reported on the hazardous effects of pesticides on human health (e.g. Loevinsohn, 1987; Pimentel, 1991). Examining health issues relating to pest control was beyond the purpose of this evaluation study. An extensive study on the impact of pesticide use to human health is being implemented in a high-pesticide-use vegetable-growing area in Central Java by a special study team of the National IPM Programme. Human health can be affected by pesticides through either occupational, accidental or intentional exposure (Hogstedt et al., 1987). High incidental pesticide exposure (or attempts to suicide), although frequently occurring, will not be considered here. An interesting case of incidental exposure is described in Box 8.1. A few observation with respect to effects of pesticides on health made in Grobogan will be mentioned here.

Accidental exposure is usually a result of careless storage of pesticides. Not uncommonly, poorly closed pesticide packages or bottles were found in the houses

Box 8.1
Medicine or poison?

In June 1992, a special case of pesticide poisoning was reported to the IPM Programme's secretariat in Jakarta. During their field work, enumerators of the programme's IPM field laboratory in an intensive rice production area in West Java were confronted with a group of forty-three people, including children of ten years and older, screaming for help. The group of people had just transplanted a rice field, and had treated the parts of the body that were itching (arms and legs) with a 'medicine' that was provided by the owner of the field, as usual. Unusual was the reaction to the 'medicine', this time. Many of them started shaking, sweating, vomiting, and had head and stomach aches. Eight persons were in a very bad condition, and had to be treated with an injection and some medicines by the regional medical officer. After half a day rest, and drinking milk with tamarind and sugar that was sent by the owner of the field, everybody felt better, and used the anti-itch 'medicine' again on the following day.

This 'medicine', supposed to remove itch on the skin due to rice plants, was actually the very poisonous insecticide dichlorvos which has an oral LD_{50} value for rats of 20 mg/kg. For several years, farmers and labourers in the area had been using the insecticide for this purpose, until then without any bad effects. According to the people interviewed, this time the chemical had a different smell and colour. It could not be discovered why the dichlorvos looked and reacted differently on that occasion. Amazing is that the labourers who had just experienced how poisonous their 'medicine' actually was, used it again without hesitation as soon as they worked in the rice fields again.
(Source: Laboratorium Lapang PHT, 1992)

occasion during an interview, the farmer noticed a row of ants marching through the room. He took a plastic bottle of some powder pesticide, sprinkled the contents on the row of ants, and left it there while several children were playing, barefooted, in the same room. Another example was found with a farmer who had prepared poisoned rat bait by mixing cooked rice with the highly poisonous nematicide aldicarb. He put the bowl with the poison bait on a cupboard to be placed in the field the next morning. When his wife came home, she found the rice, not knowing that it was mixed with the cream-coloured poison, and started eating this. Fortunately, she could be brought to a hospital and treated in time, which is often not the case for people living in remote villages. Low awareness about the poisonousness of pesticides is the cause of careless handling. Pesticides are usually called 'medicine' in the local languages, which clearly describes the wrong perception people have of these chemicals.

Occupational exposure can occur during manufacturing, processing, use, and as a result of residues in the harvested product (be it negligible for rice). Pesticide poisoning during or after application was discussed with both IPM and non-IPM farmers in group interviews. Generally, farmers know the symptoms of poisoning well, and most can tell a story of pesticide poisoning that (usually) happened to somebody else. Poor spraying techniques, such as spraying against the wind, or omission of protective clothing are often mentioned by farmers as the cause for symptoms of poisoning. Wrong application was found to cause poisoning in a case where the farmer had soaked carbofuran granules (that actually have to be broadcast), diluted them in water mixed with an emulsifiable pesticide concentrate, and sprayed this mixture on the rice crop. Not uncommonly, farmers were observed to use their arm, instead of a stick, to mix the pesticide solution, but no one ever reported that this practice had caused problems. Granular pesticides are usually mixed with fertiliser by bare hand, if not broadcast directly without using gloves or some tool. Protective clothing while spraying is found uncomfortable, and supposed to be superfluous. Some people think, and are proud, that they are already resistant to the chemicals. Others use a handkerchief or cloth to close off the mouth and nose, not a very effective protection as soon as the cloth becomes wet of the spray dust, or they wear a T-shirt with long sleeves.

In the study villages, pesticide spraying is not uncommonly executed by a contracted labourer. Several of these spray labourers explained that it is mostly the operator of the rice field who determines the moment of spraying, the type of pesticide, and the dosage. Some of the spray labourers prepare a more concentrated solution when they see a lot of pest insects in the field. Volumes (number of tanks per area unit) sprayed by these labourers are in general far too low because they want to save labour, and they are not aware, and probably do not care, about the consequences. On the whole, they know the major rice pests fairly well, but have little knowledge about the existence of natural enemies in the field. Although some of them also work for IPM farmers, none of the spray labourers interviewed had any idea about IPM. According to their own opinion, they are not informed because they are only labourers and do not operate rice land themselves. Some of them experienced pesticide poisoning symptoms, but got easily cured by drinking coconut water.

Before training, 52% of the IPM farmers (against 43% of non-IPM farmers) were aware of hazardous effects of pesticides on human health. Although not checked quantitatively, this proportion increased considerably after training. Many of the field school graduates did no longer talk about 'medicine' for pesticides but about 'poison'. Health issues were sometimes mentioned in post-training group interviews by field school graduates, but mostly tangentially and not as a major point pro IPM. Reduction of pesticide sprays by IPM field school graduates is supposed to contribute to better health of these farmers and their labourers. Since the IPM field schools did not pay attention to less dangerous ways of pesticide application, no detailed observations were made on spraying techniques or pesticide storage habits by field school graduates. A higher awareness about the hazardous effects of pesticides among the IPM farmers is expected to have its impact on the way they handle these chemicals.

8.6 Conclusions

Effects of IPM at the farm level are difficult to measure in a relatively short evaluation study in which seasonal variation strongly influenced the outcome. Especially conditions during the first post-training (1990/91) season, when heavy stemborer outbreak occurred, obscured possible economic effects. Despite these constraints, it can be concluded that IPM fulfilled the profitability part of its promise to the trained farmers in the area studied. Consistently over three seasons during and after training, IPM farmers obtained significantly higher yields with more favourable yield distributions than non-IPM farmers.

Lower expenditures on (insect) pest control by IPM farmers contributed to a positive effect on the economics of trained rice farmers, especially in Kuduagung where expenditures used to be high before training. Although pest management costs accounted for only a small percentage of the total paid-out costs, the saving of money as a result of reduced expenditure was tangible in most farm households.

As in other IPM programmes (Blokker, 1984; Zadoks, 1989b), the economic effects measured are not overwhelming but they are encouraging. IPM training helped farmers to become better farmers who make more effective use of their investments through more adequate and timely practices. Considering pest management in particular, timeliness and adequacy of control measures based on more regular and thorough field observations resulted in a more effective pest management. Relatively minor changes in farmers' behaviour, as a result of knowledge increase, could bring about visible economic effects, while no investment or extra resources were required from farmers. This aspect is supposed to be favourable to the dissemination of the IPM practice. The rather low observability of the effects of IPM implementation at the farm level, however, might have consequences for the diffusion of the message (Rogers, 1983), and should therefore be anticipated by IPM training programmes. More detailed observations on farmers' pest management decision making and control practices than were possible within the scope of this study are needed to verify the findings above, obtained from only two villages during four seasons.

No direct improvements of environment and health, the second part of the IPM promise, could be discovered within the scope of the study. A somewhat lower and more proper pesticide use by IPM farmers, and a higher awareness about the hazardous effects of pesticides are expected to contribute to a higher level safety for the environment and health. However, for more rigorous results with regard to safety, IPM implementation of a better standard and at a larger scale is required.

The IPM farmer field school, the medium through which farmers learned and experienced IPM, seems to be an effective training model to procure change in farmers' perceptions and practices. Despite counteracting forces, such as a pest outbreak and persistent pesticide promoting efforts by extension workers, KUD and local governments, effects of the IPM training on farmer behaviour, and subsequently on farm output, were tangible in the study, although maybe not as much as one should wish. In my own estimation, sustainability and improvement of IPM implementation at the village level can be expected in the future as a result of the changed perceptions of most IPM field school graduates relating to farm management, and increased self-confidence. These farmers learned to learn from each experience, which was already visible during the last study seasons.

Collective implementation within the farming community, which was not observed sofar, would greatly reinforce IPM. Trained farmers will be encouraged when IPM becomes broadly accepted, and the ecological balance in the rice ecosystem will be strengthened. Much will depend on the development of agricultural (extension) programmes, including policies of local governments, in the future.

9 Experiences leading to learning: conclusions and discussion

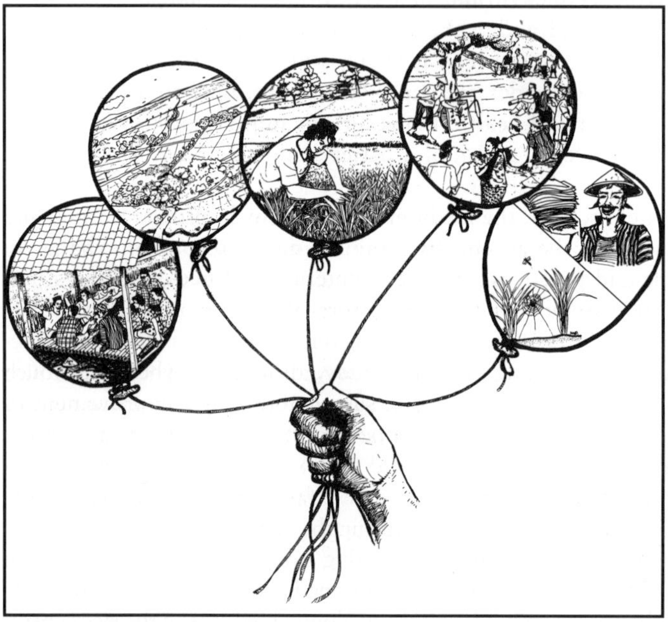

9.1 This chapter

This book described the findings of a case study in Central Java evaluating the implementation and effects of the Indonesian National IPM Programme. In contrast to IPM training conducted sofar in Indonesia, the National IPM Programme developed an IPM approach using ecological principles instead of preset recommendations, and a training approach based on nonformal education principles. It was implemented as a government programme through the existing extension system in a favourable political climate created by specific policy measures. Investigation of the introduction of the IPM Programme at the village level was considered an interesting subject. An evaluation study was designed to describe and analyse the processes and effects taking place at the village level as a result of IPM training and dissemination, and to provide useful management information to the programme for further implementation. The study posed the question what processes and effects would result at the village level when sustainable practices, which contrast in many respects with the prevailing high-external-input technology, are introduced through nonformal farmer training in conditions created by policy measures.

The study focused on eight villages in one district in Central Java. Therefore, it does not allow generalisation across the National IPM Programme's field implementation, in view of the high diversity in cultures, social structures, and crop cultivation behaviours among the provinces involved in the programme. The consistency of the aspects which can be considered to define a training model served as a leading supposition in conducting the evaluation study. These aspects include the policy context, the role of the farmer in extension, characteristics of the training contents, desired farmer practice, the nature of the intervention, and the institutional framework (Röling, 1992c).

The previous chapters described in detail the setting, emphasising diversity among villages (Chapter 4), the current practice of (Green Revolution) rice farmers (Chapter 5), the IPM farmer field school implementation (Chapter 6), and training effects at farmer (Chapter 7) and farm level (Chapter 8). In this last chapter, the main findings of the study are discussed referring to the research questions posed in Chapter 3. Final conclusions are placed in a larger perspective, leading to suggestions for further investigation. The last section of this chapter evaluates the evaluation study and discusses some methodological issues that emerged during the conduct of the study.

Since the study examined a case only, it is not representative for IPM training and implementation throughout Indonesia. The findings of the study sometimes led to suggestions for programme improvement, which are given in the sections below. Considering the non-representativeness of the study, these suggestions are not valid in themselves, but should be considered within the context of the whole programme.

9.2 Major findings

9.2.1 The context

Rice farmers with a Green Revolution history
For more than twenty years, rice farmers in Grobogan have been exposed to the 'Green Revolution' approach of food crops intensification, which primarily aimed at production growth, and consisted of a package of high-external-input technology. Extension approaches to promote the package were top-down. A good farmer was perceived to be an adopter of the technology who applies the recommendations.

Although in most villages studied coercion to adopt the high-external input technology was initially exerted, farmers soon discovered the benefits of the new technology. Within a few years' time, high-yielding varieties, nitrogen and phosphorus fertilisers, and pesticides were accepted by the majority of farmers, and are used until now. Within fifteen years, yields increased from an average of 2 to 5 tons/ha. Adoption of the technology to date, however, seems to be restricted to the adoption of the types of inputs. Farmers' practices differ considerably from the recommendations when other aspects of technology implementation, such as quantities of inputs applied and time of application, are considered. The more recently introduced inputs, for instance potassium fertiliser, are still not widely used. Most farmers base their practices on their own or fellow farmers' previous experiences, rather than on extension recommendations. Some even conduct simple experiments to find out what practice is most suitable in the conditions of their farm.

Recommendations from the extension service, received by only a small proportion of farmers, are not highly valued, since the village extension worker (who is not a farmer and seldom visits the farmers' fields) is not considered a serious discussion partner. In addition, these recommendations are mostly not adapted to farmers' current practices, conditions and unit measures, and often incomprehensible. An example of a commonly misunderstood concept is 'balanced fertilisation', perceived by farmers as the application of all types of fertiliser in the same quantities. This misunderstanding often leads to inefficient use of especially phosphorus fertiliser.

Despite the low appreciation farmers express of the extension service, to a certain extent they seem to be influenced by the top-down system, not unlikely as a result of the cultural sensitivity to hierarchal structures. The (selective) group of farmers who are invited for extension meetings, do attend these, even though they do not expect to gain much. Generally, they just listen quietly and pick up whatever they find applicable. Recommendations that look valuable are sometimes tried out first on a small plot, or adapted according to the farmer's own opinion (also Winarto, 1992). When farmers are commanded by the village government to participate in an intensification programme (which mostly implies obligatory purchase of inputs), hardly anybody dares to refuse participation. But actual implementation in the field is often adapted to what the farmer thinks to be best. Farmers show an obedient attitude, but wherever they can, they go their own way. The fact that farmers do experiment and adapt recommendations to their own conditions is an indication that they are not submissive and docile. But being sensitive to power structures (which in the context of the current extension system implies that they are the actors at the lowest level), farmers do not feel self-confident enough to openly reject somebody else's opinion (such as the extension worker's) and defend their own practice.

After twenty-year long experience of hearing about and experimenting with high-external-input technology, farmers have become 'addicted' to it, whether implemented in accordance with the recommendations or not. Many of them cannot imagine farming without pesticides, or with lower fertiliser dosages. Farmers feel that they are much better off now than they were before, and want to maintain (or still increase) the production level they have achieved, 'at all costs'. The present economic picture of rice farming does, however, no longer show a favourable return to investment, especially not for those who operate under rent or sharecropping agreements. High levels of rice production 'at all costs' is no longer profitable, and thus economically not sustainable. Farmers do not seem to be aware of this situation yet, which is shown by their practice of still increasing fertiliser dosages, whereas no correlation is visible between yield and quantity of fertiliser applied within the range practised by the farmers surveyed.

Conflicts at the introduction of sustainable practices

When more sustainable practices, such as IPM, are introduced into a rice production system that emphasises high-external-input technology, conflicts at various levels can be expected. Practices aiming at the sustainability of production and returns of the individual farmer (Sriskandarajah et al., 1989) might interfere with the goals of the Government pushing for maximum production (only) to maintain food security. As observed in Grobogan, IPM field school graduates obtained higher yields than non-IPM farmers, yield variability decreased implying lower risk, and expenditures were lower, all in all leading to a higher return of rice farming for IPM trained farmers. Both goals, sustainability and maximum production, are then achieved, without any conflict. Longer-term investigation, however, is needed to see whether the achievements as a result of IPM field school implementation with respect to production and returns will hold, and will also apply to the categories of farmers not involved in field schools sofar.

A clear conflict at the introduction of IPM in the study area is visible at the level of the village extension worker. In the current system, extension workers are at the interface between the government's food crop intensification policies and the farmers as final implementors. They are intermediary in input distribution between the KUD or salesmen and the farmers. They receive a commission for these services, often an indispensable extra source of income. Some extension workers take advantage of their position and strongly promote inputs, or even sell them from their houses. One extension worker was observed to (incorrectly) recommend pesticides in the IPM field school when the pest observer was not there, a phenomenon also observed elsewhere (Winarto, 1992). Afraid of losing his extra income, this person more aggressively promoted pesticides when IPM was introduced in his jurisdiction, resulting in an increase of granular pesticide use by both trained and untrained farmers. In addition, the effects of the IPM field school on the pesticide application practices of farmers were relatively poor in the villages where the extension worker was not committed to the IPM concept.

Not only the action of the village extension worker, but the role of the KUD, in general, is found to be inhibitory to the implementation of IPM. Farmers who want to obtain inputs on credit via the KUD are still confronted with a preset technology package. In some places, farmers are obliged to take the complete package including unwanted inputs such as pesticides and foliar fertilisers, despite central government regulations stating that inputs from the KUD can be achieved on a need-basis. The position of village heads or other influential people in the villages as KUD managers is observed to strongly influence local government policies in agricultural development. This, in turn, influences farmers' attitude towards farming, favouring the high-external-input technology, hence hampering IPM.

Diversity in farming communities

Among the eight study villages, located in a seemingly homogeneous and limited area of only two subdistricts, high diversity was observed with respect to direction and level of (agricultural) development. Various aspects were considered to explain this diversity, including policies and activities of the village head (in interaction with the general attitude of the villagers towards development on one side, and traditional ceremonies on the other side), coherence of the farming community, availability of inputs, timeliness and reliability of irrigation water supply, frequency and direction of activities of the village extension worker and input distributors, and land and income distribution. Especially the first point mentioned, the either favourable or unfavourable influence of the village head, seemed to be of major importance.

In addition to village policies, farmers themselves strongly influenced each other in their rice cultivation practices through direct social control. They considered the advice of fellow farmers more important than that of village extension workers, and often followed practices of certain 'model' farmers. Considering pest management, a division could be made between villages with high and low pesticide use, or risk-avoiding and risk-accepting farmers (Zadoks, 1985).

Diversity at the individual farmer level was particularly linked to farm size and land tenure status, as found in many other studies (e.g. Hüsken, 1988; Lipton and Longhurst, 1989). The most obvious observations showed that small fields were more intensively cultivated than larger fields, and sharecropped farms were operated less intensively in terms of input use, probably because the return is only a share of the harvest.

Dealing with diversity is a difficult but important task of facilitators in a training programme, especially when the training material is based on farmers' experiences. Trainers have to be thoroughly prepared to recognise diversity, identify factors favourable to their programme, and adapt their programme to these factors. As the team leader of the National IPM Programme expressed it, they should see the sparks in a community, and then fan these to cause a fire. For example, this may imply the active involvement or support of certain people in the community, or the identification of a collective problem.

9.2.2 *IPM farmer field school*

Selection and composition of field school farmers

Villages in Grobogan selected for the IPM farmer field school were all located on or close to the main road, and close to the subdistrict town. The non-representativeness of IPM villages was a direct result of the programme's selection procedure which used criteria of accessibility and activity of farmer groups. The two criteria are to some extent related, since easily accessible villages generally receive more attention from the agricultural extension service than remote villages, which in turn encourages farmer group activities. People in the more easily accessible villages have better access to inputs and information, and more opportunities for off-farm employment than those in remote villages. The field school facilitators (pest observers and village extension workers), who were in charge of the selection, chose those villages which they knew best and which were likely to be successful in the field school.

Farmers selected for participation in IPM field schools were, on average, the more affluent and better informed people in the villages. They operated relatively larger, mostly owned, fields, had received more education, and many were engaged in off-farm enterprises. No women were involved in the field schools observed, though they play an important role in rice farming. Many village officials participated as IPM trainees. The non-representative composition of field school groups was not typical for Grobogan, but occurred in most groups trained in Central Java during the first implementation cycle. Mostly, farmer group leaders, in consultation with extension workers and/or village officials, were in charge of the participant selection. Criteria used included the ability to read and write, willingness and ability to attend the field school throughout the season, and ability to disseminate what was learned to the surrounding community. In practice, selection resulted in people usually involved in village activities, mostly those with a certain status in the village, again in order to guarantee success of the training. This procedure, strongly based on customary social patterns in Javanese society, excluded several potential groups from participation, such as women farmers, illiterates and resource-poor farmers.

The non-representativeness of IPM villages and trainees during the first implementation cycle implies a non-conclusive test of the IPM farmer field school model. This does not necessarily mean that the field school model would work differently for other groups of farmer not involved sofar. It is not unlikely that a training based on experiential learning is received even better by the groups of farmers under-represented sofar, such as farmers in remote places, small farmers, sharecroppers, women farmers, or in general, those who are fully occupied with rice cultivation, and whose subsistence is totally dependent upon a good crop. Field school facilitators should be trained to deal with different situations and groups of farmers other than their usual audience consisting of the better informed (male) farmers. Stricter guidelines for the selection procedure are needed to encourage participation of groups of farmers neglected sofar.

Field school implementation

IPM farmer field schools observed in Grobogan were conducted successfully. Attendance of trainees was high. Although initially appointed to participate, and not knowing what to expect, field school farmers were highly motivated to continue attendance and participate in the various activities, because they felt rewarded by the knowledge they gained. The farmers appreciated to be trained in the field instead of in the class-room, and through facilitation instead of instruction. Indigenous knowledge about non-chemical pest control measures and on seasonality was revived by older participants in the groups. Farmers found the training material relevant to their daily farm practice, a key determinant for knowledge utilisation (Glaser et al., 1983). This suggests that the field school model might be highly adequate for the groups of farmers mentioned above who were not reached sofar. The facilitation approach had a visible effect on farmers' self-confidence in taking decisions and in expressing their views openly in the group. This effect, however, was inhibited where village officials or other trainees with higher education or status dominated the group.

Pest observers performed well in the role of facilitator. They seldom lectured, and were committed to guide the learning process of farmers, but sometimes needed to be stricter in group leadership with regard to managing time schedules, leading discussion, and curtailing the dominance of some participants. Pest observers seemed excited about what they had learned themselves, and were enthusiastic to repeat it with the farmers. On the whole, they were well acquainted with the ecological IPM approach, but when uncertain (as was, for instance, the case when presenting the topic on economic threshold level) they fell back on the old, preset concepts. The pest observers were well accepted by the farmers for their facilitation skills and practical knowledge.

The performance of village extension workers as co-trainers varied depending on the personal interaction with the pest observer, and on the extension worker's interest in IPM. An extension worker actively involved in (and earning an extra salary from) pesticide sales showed a non-supportive, indifferent behaviour during the training and acted adversely in follow-up activities. Initial insecurity was observed among extension workers since the facilitation approach changed their status as trainer. Most of them, having adapted to the approach, felt rewarded through the appreciation shown by the farmers.

Group dynamics exercises were an attractive element of the IPM field school. They were appreciated by the participants and had a visible effect on group coherence and collaboration. These exercises tended to vanish in self-supporting field schools, because of time and money constraints, but sometimes also because the pest observers felt insecure and awkward to do the 'games' with the (older) farmers. The programme should stimulate continuation of these exercises in follow-up trainings.

Language used by the trainers was important to achieve full understanding of the training contents by the field school farmers. A better understanding by all participants was achieved where the local language was used instead of the national language, and where scientific terms were avoided.

Consistency of the field school implementation

The assumption that, in effective training, the nature of the training process is consistent with the nature of the training contents, can be confirmed with regard to the National IPM Programme (Röling and Van de Fliert, 1993). In the IPM farmer field school model, IPM uses principles which are promoted through training methods based on experiential learning, with the expectation that farmers take better (informed) decisions. The actual implementation of the model at the village level was observed to be, for the greater part, consistent in this respect. The trainers presented IPM principles and only occasionally gave preset recommendations. Training was done mainly through facilitation, and trainees performed as independent decision makers in the field school. To a large extent, the training process allowed its contents to be effectively applied. Only in exceptional cases did trainers give recommendations, teach in the form of lecturing, and did farmers routinely apply standard recommendations without informed decision making.

9.2.3 *Effects at farmer level*

Decision making and practices

IPM knowledge of field school farmers increased considerably, especially with regard to pest and natural enemy identification, ecological functions in the rice field, and negative effects of chemical pest control for the rice ecosystem. Rice ecosystem elements were perceived differently after field school graduation. Many organisms in the rice field were known to be beneficial, and pest insects were no longer seen as harmful agents only, but also as food for the natural enemies. The existence of the many predatory arthropods in a rice field was an exciting discovery for most IPM farmers.

Regular field monitoring was practised well by the larger proportion of the field school graduates. Equipped with the new knowledge and observation skills, field school graduates took better-informed decisions than before. Not only pest occurrence, but pest densities and the balance between pests and natural enemy populations were considered in pest management decision making. This resulted in a decrease of curative applications of pesticide sprays, and a better timing and targeting of chemical control measures. Similar effects were measured in a nation- wide impact study of the IPM Programme (Pincus, 1991), and in the IPM programme in the Philippines (Kenmore, et al., 1987). Lower pesticide dosages were applied which, on the one hand, reduced pesticide load in the environment and showed farmers' willingness to reduce pesticides, but on the other hand, might have a negative effect by permitting development of pesticide resistance in pest organisms. It is suggested that the field school curriculum pay attention to the dangers of incorrect pesticide application methods to the ecosystem. Sofar, topics dealing with pesticide application were avoided in the field school curriculum, because they might be perceived by farmers as an implicit approval of pesticide use. Stressing the consequences of wrong application methods, however, I consider a higher priority.

The decreased use of pesticide sprays contrasted with a (temporary) increase of granular pesticide (carbofuran) applications in some villages. This increase was partly a result of promotion and obligatory purchase of these chemicals in KUD input packages during the first post-training season. IPM principles were apparently not sufficiently convincing for the risk-avoiders among the IPM farmers to resist these counteracting forces of pesticide promotion. Additionally, higher carbofuran use might have been influenced by an inadequate evaluation of the special topic ' Monocrotophos, carbofuran and natural enemies'. in the field school. This topic made some farmers believe that carbofuran granules were not as harmful as pesticide sprays, resulting in a replacement of sprays by granules. The increased use of carbofuran had no effect on crop health, on the contrary. During the second season with high carbofuran use (1990/91), a stemborer outbreak caused severe crop loss. Through this experience farmers learned that they had wasted their money, and carbofuran use considerably decreased in the next seasons.

No adequate action was taken by field school graduates when the rice stemborer outbreak occurred during the wet season after they had been trained. Farmers (and the extension workers who should have assisted them) lacked the observation skills to forecast the outbreak, and did not know what measures to take. This situation shows a weakness of the training approach using experiential and field-based learning, since not all problems that farmers will encounter later can be covered during the training season.

In the field school, no attention was paid and subsequently no differences were observed with regard to other cultivation practices of field school graduates. Fertilisation would have been a topic relevant for the field school in view of ecological and economic aspects of IPM. Additionally, the focus on learning through experimentation seems highly applicable to fertilisation topics. Farmers' current fertilisation behaviour often seems to be inefficient and wasteful, especially with respect to application methods (e.g. urea in standing water) and timing (before weeding), and with respect to TSP dosages.

Collective action

Farmer group reinforcement through IPM field schools is expected to have a positive impact on collective action relating to rice growing. Collective implementation is important for the effectiveness of IPM, especially with regard to the maintenance of a balanced ecosystem. This implies collective conservation of natural enemies by not using pesticides and by prevention of poaching. Other typical examples are rat control, for which individual measures are not sufficient but communal action is beset with problems (Van de Fliert et al., 1993), irrigation management, and synchronised planting and harvesting.

In one of the villages observed in Grobogan, community rat control was initiated through the farmer field school group. The (non-active) farmer groups in the villages were reorganised on a hamlet basis, and headed by the field school graduates. These farmer group heads were made responsible for organising the villagers to participate in collective rat management, which was implemented successfully. Initiatives in another village to reinforce irrigation maintenance through the farmer field school group failed due to the authoritarian attitude of the farmer group leader. Before the end of the study period, follow-up activities of the IPM field school groups petered out.

None of the IPM groups studied made deliberate efforts to extend the IPM principles to the non-trained farmer community, which would have encouraged a broader-based, thus more effective IPM implementation.

Horizontal communication

A fairly high level of awareness about IPM among non-IPM farmers was attained in the seasons after training, but the perception of what it implied was limited. IPM was mainly communicated verbally, not by field practice, and to a selective audience in the villages. Consequently, this resulted in a low effect of the horizontal communication on practices of untrained farmers.

The non-representative composition of the training groups seemed to have hampered dissemination, in that farmers representing the lower socioeconomic layers and women farmers were not reached, a phenomenon often occurring in extension programmes (Röling et al., 1976). In IPM villages in West Java, communication about IPM was also observed to be restricted to those persons who regularly interact through work or social networks (Winarto, 1992). Conversely, due to the status difference between trained and untrained farmers, the latter group did not show affinity with the field school. Probably for this reason, not much attention was given by the community to the IPM field day which was organised at the end of the field school season.

A composition of field school groups that better represents all layers of farmers in the community can be expected to lead to better and faster spread of IPM principles and practices. In addition, written materials distributed among field school farmers to take home are suggested to contribute to a better quality of communication. The programme's efforts in some areas to encourage and guide folk theatre as a means of communicating IPM to the village communities appeared to be another effective way of reducing communication barriers between social layers (Rogers and Adhikarya, 1979). Expansion of these pilot efforts are assumed to contribute to the spread of IPM.

According to Rogers (1983), possible limiting factors in diffusion processes are the complexity of the message and the low observability of effects, both characteristic for IPM practice, in particular with respect to a large part of the non-IPM farmers in the study area. As described above, the non-IPM farmers are, on average, less informed, have lower educational backgrounds, and operate smaller fields than the IPM farmers. Risks on these small farms are higher, so observability of the effects becomes more important. Even though the contents of the IPM field school do not require a certain educational background, but rather farming experience, they seem complex and do not appeal to daily experience when conveyed through vocal communication only, instead of through field practice. These factors might explain the low understanding and implementation of IPM concepts, despite high awareness, by non-IPM farmers. More deliberate efforts by trained farmers to disseminate IPM through the way they have learned themselves (in the field) is expected to improve the quality of communication processes.

9.2.4 Effects at farm level

Economic effects

Despite the relatively short study period and strong seasonal influence on the outcomes, economic effects of IPM implementation were measurable in Grobogan. Consistently over three seasons during and after training, IPM farmers obtained significantly higher yields than non-IPM farmers. This finding is particularly positive considering the fact that regression analysis of yields with some production factors, such as fertilisation and frequency of pesticide application or weeding, did not show any correlation. On the other hand, it should be recognised that yields obtained by IPM farmers before training were already somewhat higher than those of non-IPM farmers, confirming that the ones selected for training were the better farmers in the villages. Nevertheless, this yield difference became larger after the IPM field school. The reasons for higher yields had to be sought in quality aspects of crop management, such as more timely and adequate implementation of practices.

Yield variability for IPM field school graduates was lower than before training, and lower than that of non-IPM farmers in the same season, which indicates that risk is lower in IPM than in non-IPM fields. In addition, lower expenditures on (insect) pest control by IPM farmers contributed to a positive economic effect. This finding corresponds with results of the nationwide impact study conducted for the National IPM Programme in 1991 (Pincus, 1991). Although pest management costs accounted for only a small percentage of the total paid-out costs for rice production, the savings as a result of reduced expenditure were tangible in most farm households. IPM implementation did not cost the farmers extra time, but they spent their time in the field more appropriately than before, for instance by doing focused field observations.

All in all, IPM implementation resulted in higher returns of rice production for trained farmers. Training helped farmers to become better farm managers who made more effective use of their investments through more adequate and timely practices. Considering pest management in particular (and disregarding the improper and ineffective carbofuran applications during two seasons), timeliness and adequacy of control measures based on more regular and thorough field observations resulted in more effective pest management. Relatively minor changes in farmers' behaviour, as a result of knowledge increase, could bring about visible economic effects, while no investment or extra resources were required from farmers.

Effects on farm ecology and human health
Due to the short study period and maybe to inadequate methods, no direct improvements of environment and health could be identified. The somewhat lower and more appropriate pesticide use by IPM farmers, and a higher awareness about the hazardous effects of pesticides are expected to contribute to a higher level of safety for environment and health. Much longer-term and more focused investigations are needed to evaluate effects of IPM training and implementation at the levels of ecology and health. The National IPM Programme, indeed, assigned special study teams to investigate these issues.

9.3 Do farmer field schools generate sustainable practices?

9.3.1 The policy context
The occurrence of a national crisis and the resulting regulatory policy framework appear to be of major importance as a context for extension of IPM. The National IPM Programme with its emphasis on nonformal education cannot be divorced from the conditions created by the Presidential Decree of 1986 (INPRES 3/86), including measures such as the ban of broad-spectrum pesticides, and the removal of subsidies. In the conventional, transfer-of-technology extension model, the market mechanism is considered a sufficient context for extension, although, in practice, often supported by price and other interventions. The introduction of an extension model in support of sustainability within the conventional context is not likely to break through vested interests and market-based incentive structures without the deployment of policy instruments such as INPRES 3/86.

Although the introduction of the IPM Programme was highly favoured by the policy measures mentioned above, actual implementation at the field level seems sometimes to be hampered by conflicting mechanisms, such as the conventional extension system still having considerable momentum. The SUPRA-INSUS programme is still implemented, forcing farmers to apply a complete, preset package of high-external-input technology. KUD managements do not consider needs of farmers, but force them to purchase unwanted inputs, and (some) village extension workers are too much involved in (and dependent upon) input distribution. It is assumed that these conflicts can only be solved through more central government interventions or other focused external pressure, in order to encourage synergetic functioning of the programme elements over a sustained period (Kaimowitz, 1990). However, before this will happen, the national policy context should be oriented to a sustainable rather than to a production-focused agriculture.

9.3.2 The learning

An elaborate farmer training effort is considered necessary, in addition to the regulatory measures. Although these measures proved to be sufficient to reduce pesticide use and maintain food security by effectively controlling brown planthopper outbreaks in the late 1980s, nothing basically changed in individual rice farmers' decision making. The pesticide industry and the input distribution apparatus can be expected to exert continuous pressure on farmers to use pesticides, replacing banned substances with permitted ones. Necessary counter-pressure can only be provided by those farmers who believe in the IPM principles and in their own ability to manage their farms independently. This corresponds with the 'logic' of sustainable agriculture, which relies on knowledge-intensive agroecosystem management, and requires farmers to be experts in their own fields, capable of observation, experimentation, considered decision making and joint deliberation.

The farmer field school model of the national IPM Programme was found to represent a different paradigm of extension for supporting sustainable agricultural practices. Applying a 'Farmer First' philosophy (Chambers et al., 1989), the training contents emphasised individual, independent application by using principles instead of preset recommendations, and the training process focused on the development of expertise and independence by using experiential learning and empowerment approaches. In every IPM field school visited, farmers can be observed to actively and intelligently discuss their problems, analysing complex agroecosystems, speaking in front of others (including such visitors as the Minister of Agriculture), and making considered decisions about pest control. Basic changes occurred in farmers' perceptions, decision making and practices, in a way highly appreciated by the farmers themselves, which is considered a proof of the effectiveness of the field school model. It is an impressive accomplishment that seems to illustrate the change needed for extension in support of the new wave of sustainable agriculture.

Of course, not all field school farmers become sufficiently convinced of IPM within one season's training. Conviction seems to highly depend on previous and current counteracting forces. But on the whole, after observing a few groups of field school graduates intensively over a period of two years and visiting many more, I obtained the impression that something has established in the minds of (most of) the trained farmers that is likely to develop by itself. This 'something' includes a different perception by farmers of the rice crop and them being the managers of it. Most of the field school farmers have learned to learn by themselves, and to perceive new experiences as new opportunities to learn more, an indication of sustained learning.

9.3.3 The practice

Final goal of the IPM training is that sustained learning will lead to sustained practising of what is learned. Considering the experiential learning process IPM field school farmers went through, it can be assumed that the changed perceptions and the new knowledge may further develop by themselves, and, at some stage, be put into practice. Several observations were done in this regard, during the post-training seasons. Farmers demonstrated an ability to continuously analyse and learn from experience.

Within the limited study period, various levels of internalisation of the IPM practice among field school graduates were observed, as described in section 9.2.3. To a certain extent, achievements varied according to location-specific factors, such as (previous) risk-avoiding or risk-accepting attitudes, village policies in rice production, the role of extension worker or KUD in pesticide promotion, and not to forget the implementation of the IPM field school itself. Variation of achievement at the individual level was found to depend on factors such as the relative importance of rice farming, the economic situation of the farm family, and so on. Depending on the factors mentioned above, farmers' perceptions of the suitability of the IPM practices determined the extent to which they finally implemented IPM. Especially the influence of counteracting factors seemed important in this respect.

Certain messages of IPM were better received by farmers under certain conditions than other messages. For instance, decision making for curative control, based on observation, was better understood by risk-avoiding farmers than the ineffectiveness and inefficiency of prophylactic pesticide application. Although the IPM field school process proved to be effective in changing farmers' behaviour during and after training, more favourable conditions should be created to allow (all) farmers to implement IPM wholeheartedly and to prevent counteracting pressure. As in most extension programmes, a mix of measures in addition to training is needed (Röling, 1986), which would possibly imply more regulatory measures, as those suggested in section 9.3.1.

One remark that needs to be made here is that, as far as observed in Grobogan, IPM training increased the difference between the more affluent and better informed farmers (those selected for training who afterwards obtained better yields and had lower expenditures) and the small farmers, generally with less access to information. Although this gap is a common phenomenon in extension programmes (Röling, 1988), the IPM Programme did not intend to increase the gap. It was an inevitable result of the selection procedure applied. Now, the programme is challenged to close the gap again through deliberate activities to reach those categories of farmers who were excluded sofar, which is presently attempted through emphasising follow-up activities and farmer-to-farmer training. Such large-scale replication needs to be carefully monitored, and adapted. Less well-endowed farmers often have less time, fewer slack resources, and run greater risks than farmers with better access to resources. In addition, a more intensive and deliberate use of mass media and written materials might contribute to reach a larger audience.

Although several aspects of IPM implementation by field school graduates can be improved, I want to repeat my assumption that something basically changed in trained farmers' perception of crop management, that is likely to develop by itself. This assumption should become a hypothesis to be verified through further investigations monitoring field school graduates over a longer period.

9.3.4 The reward

Old studies on adoption of innovations claimed economic variables as the major factors determining the rate of adoption (e.g. Griliches, 1957). Rogers (1983) recognised four characteristics of an innovation affecting the rate of adoption and, in turn, the rate of diffusion: (1) relative advantage, (2) compatibility, (3) complexity, and (4) observability. Many adoption studies showed the importance of these aspects (e.g. Rogers and Havens, 1962; Brandner and Kearl, 1964). Relative advantage, defined as the degree to which an innovation is perceived better than the idea it supersedes (Rogers, 1983), was considered to be determined by economic factors (e.g. profitability, low initial cost, savings in time and effort) and social status aspects. Although it is questionable whether the adoption theories are applicable to IPM training, which emphasises a learning process and not an technological innovation, an attempt is made to analyse the effects of IPM training measured in Grobogan in relation to these theories.

Economic effects of IPM implementation, a result of savings on pest management costs and higher gross returns, were measurable but small. Conform other IPM programmes (Blokker, 1984; Zadoks, 1989b), savings on pest management costs were a negligible proportion of the total paid-out costs on rice production. When translated into daily household expenditures of Javanese farmers, the opportunity costs of savings were tangible but direct observability of economic (and other) effects was low. Yields (and thus gross returns) of trained farmers increased, but the relation between this increase and IPM practice was not directly visible for farmers. Low economic effects (thus small relative advantage), low observability, and the high complexity of the IPM field school contents would, according to Rogers' theory, imply a low rate of adoption of IPM. In the case of the IPM villages observed in Grobogan, adoption of IPM practice by trained farmers was high, but the rate of adoption (diffusion) in the whole community was low.

Against the background of these theories, it is interesting that **increase of knowledge** is mostly mentioned by IPM field school graduates as the first reward of IPM training and implementation, before other rewards such as economic profit. This is in line with experiences in other IPM programmes (Rossing et al., 1985; Zadoks, 1989b). In this context, a recently developed theory on economic growth seems worth mentioning, in which knowledge is considered as a third factor of production, after capital and labour (Romer, 1990; Anonymous, 1992). It is stated that knowledge can raise the return on investment, and, therefore, economies should invest more in knowledge. Although this theory considers a different aggregation level, it gives the extent to which farmers value knowledge relating to farm management a different perspective. Knowledge then becomes an additional aspect of relative advantage. Possession of knowledge also contains aspects of status, thus further adding to relative advantage. Considering knowledge as a determinant of relative advantage, a high adoption among trained farmers but a low diffusion to non-trained farmers seems consistent with Roger's theory.

Another possible explanation for the fact that farmers perceive increased knowledge as an important reward relative to economic profit, was found in the theory of Scott (1989) stating that farmers' management is usually focused more on preventing failure and risk in order to guarantee continuity of their farms, than on trying to obtain the highest profit possible by taking a certain amount of risk. Mumford and Norton (1987) found that farmers consider costs and benefits in relation to a range of objectives, among others maintaining the quality of their lives. In this context, farmers would not be much concerned about the (relatively low) expenditures on pest control, as long as they feel that the efforts contribute to their safety (regardless whether they indeed do or not). This perception was observed among many farmers, especially those in the villages with a risk-avoiding attitude, who expressed to feel more 'at ease' after having applied pesticides. Studies in the Philippines also showed that spraying pesticides provides a feeling of security to farmers, rather than being a necessity for crop health (Medina, 1987). Increased knowledge through IPM training, and especially through the empowerment approach of training, however, helped to reduce feelings of insecurity, thus fear for risk, and increased farmers' self-confidence. This is expressed by trained farmers with the same words of 'feeling at ease' now they know how to monitor their crops and take adequate decisions.

Increased self-confidence is considered a proof of empowerment. As stated above, the farmers studied in Grobogan cannot be called submissive since they mostly 'go their own way', although they appeared to be easily influenced by external forces. An increased self-confidence is expected to have made them stronger with respect to their 'independence' in rice cultivation. The empowerment approach of the IPM Programme can, therefore, be considered highly applicable to the situation of Central-Javanese farmers.

9.3.5 The sustainability

A discussion on 'sustainability' in relation to the Indonesian IPM Programme needs to touch two different themes: (1) sustainability of the IPM farmer field school model and its achievements, as a door to larger issues of farmer training and extension, and (2) sustainability of IPM implementation by trained farmers, as a door to larger issues of sustainable agriculture.

Farmer field school and sustainability
As said before, farmers learned to learn and to perceive new experiences as new opportunities for learning through the IPM field school, which is considered an important achievement that will sustain. A second important achievement likely to support the sustainability of the field school training effect is that farmers obtained expertise, and thereby more self-confidence. These two achievements can never be lost by the person who obtained them, different from isolated bits of knowledge or skills that can be forgotten or become outdated. But sustainability of these two achievements can be encouraged or discouraged through various internal or external mechanisms. Applied to IPM, discouragement will occur when the public opinion in a community still favours high-external-input technology, and, in contrast, encouragement when IPM becomes more widely accepted and implemented. The implicit objective of collective action in the IPM field school model supports encouragement of achievements, and thus reinforces its sustainability. Communication on IPM at the village level was observed to continue over several seasons after training which indicates sustained results of the training model, although it would have been stronger with more deliberate follow-up activities by the field school groups.

The same achievement of increased self-confidence holds for the pest observers, as trainers in the IPM field school. Being thoroughly trained themselves through facilitation of their own learning processes, and having experienced the rewards of this approach, most pest observers became committed IPM field school facilitators. The existence of a large cadre of these committed trainers all over the country, whose job description was changed in favour of IPM training, contributes highly to a continuation of IPM farmer field school implementation. On the long term, however, there seem to be more implications for the institutional framework of IPM training.

The institutional framework required for IPM
One important issue is whether a decentralised system as applied in the National IPM Programme can function in a traditionally hierarchical system as Indonesia's public sector. Will it be possible, for example, to pay salaries commensurate with the high level of training required of decentralised staff in the IPM approach? A decentralised set-up might require local funding. To date, several spontaneous cases of financial support from local sources (district and subdistrict governments) are on record, but these remain incidental. Local IPM trainers need to become adept at 'pulling down' local moneys. A hopeful indication of wider institutionalisation of the approach developed by the IPM Programme is that the proposed revision of the national agricultural extension system is modelled on it.

A second unsolved issue is science linkage. Sofar, the programme has relied heavily on the special input of expatriate specialists and their connections with international research. In future, local experimentation and decentralised expertise must have easy access to sources of scientific knowledge and research capacity.

The multiplication of IPM to the masses of Indonesian rice farmers is still an unresolved issue. Sofar, the farmers reached have been the usual, relatively better-off farmers. The question is how to rapidly expand the IPM practices to the relatively less endowed. The programme has looked for a solution in two directions. In the first place, it is experimenting with ways of more intensively involving the regular extension workers in IPM training (Van de Fliert et al., in prep.). These officers cannot be trained as thoroughly as the pest observers. Yet they have much to unlearn. At best, one can expect the IPM principles to be watered down during implementation. In the second place, the programme has experimented with farmer-to-farmer communication. In some cases, the members of farmer field schools became so enthusiastic that they spontaneously started field schools for other farmers. Farmer-to-farmer training seems a very promising route to large scale multiplication. The programme actively assists such groups by providing the required inputs.

It seems difficult to look at IPM except as a dynamic, emergent movement, fostered by the energy of activism and the excitement of learning, and characterised by different stages, with specific sets of institutions involved at each stage. As the IPM Programme matured, it developed significantly in terms of identifying approaches and developing a cadre of trained trainers, as well as in terms of establishing a model for policy which is gaining acceptance. The next challenge is to institutionalise IPM in the regular structures and procedures of the government so as to introduce it on a large scale. Undoubtedly it will lose some of its activist energy but hopefully it will not be diluted by bureaucratic and transfer thinking to a point where it loses its consistency and character as a post-Green Revolution 'new wave'. From a movement, IPM must necessarily become more of a system.

IPM and sustainable agriculture

IPM implementation by the trained farmers observed in Grobogan resulted in effects supposedly tangible enough to encourage farmers to continue these practices. A continuation of the practices implies a sustained implementation and provides continuing opportunities for dissemination. The IPM principles as applied in the Indonesian programme seem to offer more opportunity to improve rice farmers' management than was realised by the farmer field schools. The first principle of 'Grow a healthy crop' raises the issue of fertilisation which is, indeed, briefly touched upon in most field schools. Considering farmers' uneconomical and inefficient fertilisation practices, more attention to this topic is desired, which fits well with the economic objectives of IPM too. In fact, the ecological approach to IPM requires a broader perspective of total crop management, or sustainable agriculture as a whole.

Sustainable agriculture is often perceived (especially by commercial agriculturists) as a low-input form of food production, therefore considered low-production. A different perspective on sustainable agriculture is obtained when it is considered a form of agriculture that combines modern scientific knowledge with ecologically sound practice to ensure a more affordable and sustainable production (Altieri, 1987; National Research Council, 1989). Or when it is defined as a way of life for rural people, where sustainability means the maintenance of stable, self-reliant, rural communities (Sriskandarajah et al., 1989), which holds to a great extent for the small Javanese rice farmers. Such views on agriculture recognise people and their relationship with the environment as central objects of agriculture, and not production. IPM as applied in the Indonesian programme seems to provide an excellent start of a more sustainable approach to food production. The easily observable ecological processes of pest management (such as predator-prey interactions) are concrete features to start changing farmers' production-oriented thinking towards a more ecological and, at the same time, profit-oriented thinking. Additionally, the farmer field school is assumed to be highly applicable as a method to generate more sustainable practices at the farm level beyond IPM, since the farmer is the central actor and decision maker in this extension model.

This leads us to the suggestion that, in general, the 'logic' of sustainable agriculture is consistent with a 'facilitation model' of extension focusing on human resource development, and not with the conventional transfer-of-technology model of extension focusing on production (Röling, 1992c). Knowledge-intensive, sustainable practices of IPM do not allow to be transferred through instructions in a technology package. The inadequacy (with regard to internalisation and continuity of IPM practice by trained farmers) of the crash IPM training, which used the conventional extension approach, is significant in this respect. A similar experience is recorded by Agudelo and Kaimowitz (1989) for an effort to reduce inputs in irrigated rice in Colombia. Additionally, the farmer is perceived differently in the two models: a central actor (experimenter and independent decision maker) in sustainable agriculture and in the facilitation model, versus a recipient and adopter of the package in the transfer-of-technology model of high-external-input agriculture. As described above, IPM requires a different institutional framework. Applying the experience of the IPM Programme to sustainable agriculture, assuming it also requires a substantially different model of extension, raises important implications for the extension system in total, e.g. with regard to investment in extension, the design of extension institutions, the conception of research-extension linkages, the deployment of other policy instruments along-side extension, staff training, and, of course, extension strategies and methods.

9.4 Suggestions for further research

The conclusions and discussions above lead to several suggestions for further research:

- Longer term investigations seem needed on how the IPM farmer field school is implemented by the graduated pest observers, the trained extension workers, and by farmer trainers. Is the quality of the training maintained when no longer intensively supported and supervised by the central programme?
- Longer term investigations on how IPM practice is implemented, communicated and expanded at the village level seems interesting, both for programme management and from a scientific point of view. Communication processes are supposed to be different at later stages (Cancian, 1979), especially after being more

socially accepted. IPM is expected to become more effective if implemented by more people. Do farmers continue to apply IPM principles in their farm management, and are the effects measured sofar sustainable? Does IPM become a broadly supported pest management strategy at the village level?

- Information about the suitability of IPM training and practice for groups not reached sofar (women, small farmers, farmers in remote villages, illiterates) needs to be generated since the programme will definitely encounter these groups in its audience in later training cycles. The need to adapt training approach and contents, as well as the preparation of the field school facilitators to different audiences must be considered. Women involvement and participation should be given special attention, considering the important role of women in rice farming.
- The programme obviously relies on activist energy for its initial effect. The question is whether such initial momentum can be transformed into regular government routines without compromising its essence. The transfer of technology model, still applied in the current extension system, has a much better fit to the top-down bureaucratic structure of Indonesia than the nonformal education approach, which is more consistent with the needs of IPM. How can nonformal training be institutionalised without being outcompeted at the bottom levels by the prevailing institutionalised approaches?
- An intrinsic aspect of sustainable agriculture seems to be the need for farmers to form platforms for collective decision and action which are consistent with the level of the agroecosystem to be managed in a sustainable manner. Such farmer decision making platforms are an exciting area of extension research.
- A final question to raise is whether the idea of the need for an alternative extension model for sustainable agriculture in general is valid. This idea was based on the careful micro-analysis of the field impact of IPM farmer training in Indonesia. Investigations within a broader perspective of sustainable agriculture are needed to confirm the statements above.

9.5 Evaluating the evaluation study

Conduct and utilisation

As stated above, the case study evaluating the IPM Programme in Grobogan aimed at describing and analysing the processes and effects taking place at the village level, and at delivering useful management information to the programme with respect to implementation and achievements as a result of IPM training and dissemination. The methodology consisted of a variety of methods, including qualitative and quantitative ones, in order to obtain a full understanding of rice farmers' decision making and cultivation practices. The longitudinal character of the study allowed continuous adaptation of study objectives and methodology, and provided a good opportunity to build up a long-lasting contact with the farmers in the (intensively studied) villages. The duration and intensity of the contact supported a relationship with many respondents based on trust, which favoured the reliability of the study outcome. Reliability was also supported by the variety of methods used, which allowed intensive cross-checking. While during the first season, and in group discussions, some farmers were found to give socially acceptable answers to questions, their own opinions seemed to be openly expressed in later seasons and in individual interviews. Qualitative information on reasons and processes were complementary to quantitative data on practices, inputs and outputs, and vice versa.

Since the study was conducted in only eight villages in one Central-Javanese district, findings are not considered representative for the National IPM Programme which is implemented in eight Indonesian provinces with highly varied cultures and conditions. The high diversity of situations encountered in the limited study area confirms the non-representativeness of the study. In addition, many phenomena observed seemed to be typical for the cultural context of Central Java, which became increasingly obvious after visiting and comparing villages in other areas of the archipelago. The value of this case study for the IPM Programme (and other training programmes) is expected to be in the description and analysis of processes that occurred at the village level when IPM was introduced through nonformal training, as a context for the effects of training measured. This description of phenomena observed is intended to create higher awareness among programme stakeholders, which in turn might assist in recognising similar, or rather different, phenomena in other places.

In order to encourage evaluation utilisation during and after the study, it was attempted to adjust the conduct of the study and the analysis of findings to the (programme and evaluation) objectives of the programme's main stakeholders (i.e. managers and field implementors). Communication with these stakeholders was facilitated by the fact that I was an internal evaluator assigned for a longer period to the programme. Time constraints of the always fully occupied people in this dynamic programme, however, often hampered intensive communication. This was partly remedied by regularly reporting findings of field activities. It is hoped that final utilisation of the evaluation findings has been favoured by this regular reporting.

Diversity in evaluation studies

High diversity in communities has consequences for development programmes and their evaluations. Programme evaluation studies mostly seek to generalise, but generalisations may not always be valid for certain groups of people when high diversity is recognised. The existence of so many factors with high variation among only eight villages, as encountered in Grobogan which is seemingly homogeneous, gives some insight in the complex situation to be dealt with by the programme and its evaluation. The consequence of this diversity, the difficulty of generalisation, is considered an advantage for this evaluation study which aimed at explaining processes. The diversity observed actually provided excellent study material. Therefore, it was attempted to emphasise and exploit variation of phenomena observed, rather than to eliminate it. Existence of diversity contributed to the adjustment of the study objectives.

By emphasising diversity, I experienced the study to be enriched rather than hampered, since it could continuously reckon with the real situation. The consideration of the complexity of processes and mechanisms helped to explain reasons why things happened at the village level in a certain way, in a certain place, at a certain time. These explanations highly contributed to the aim of the study to be interpretative. Analysing results and drawing conclusions, however, was not always easy because of the complexity of the situations encountered. Phenomena observed and effects measured had to be related to their context. In most cases in the Grobogan study, this required a consequent classification according to village.

It is expected that emphasising diversity has a value for the programme by giving insight in the complexity of programme implementation at the village level and by presenting a variety of exemplary processes that can occur. Although the examples described are specific for the villages studied and far from exhaustive, I hope that they will sharpen the observation ability of those further involved in improving the programme.

Comparison villages in evaluation studies

As described above, villages selected for IPM training were those located on or close to the main road, therefore not representative. Conditions with respect to many crucial aspects of farming are supposed to be different in such villages compared to more remote places. Non-representativeness of villages involved in a training programme has consequences for the evaluation. Evaluation findings are not likely to predict what can be expected when the training process is replicated in more remote villages. In addition, the use of comparison villages, usually meant as a reference to measure effects of a certain intervention by comparing situations with and without the intervention, is questionable. Comparison villages are not easily equivalent to the few selected, typical villages involved in the programme.

Recognising diversity among villages brings up another problem in the use of comparison villages. Except distance from the main road, other factors determining development processes in a village, such as the village head's position and policies, make it difficult to compare villages.

Despite the two arguments above, investigating comparison villages in a programme evaluation study is not considered useless. Using comparison villages in Grobogan yielded many additional insights in possible processes occurring during agricultural development and in farmers' reasons to act in a certain way. Interpretations, however, always had to be done within the individual context of a certain village. In no way could conclusions be drawn on the effects of the IPM field school by comparing IPM and non-IPM villages, considering the factor training only.

As a conclusion I want to argue that, in general, the relative value of comparison villages has to be recognised in evaluation studies. Processes observed in villages should first be interpreted within the specific, individual, contexts of those places. Recognition of these individual contexts of villages studied is then assumed to help identify and analyse reasons for success or failure of a programme.

Influence of study activities on its outcome
The four (control) villages with a low contact intensity between investigators and villagers, included in the study to assess the influence of the study's presence on its outcome, served an important function. With respect to some aspects, results obtained in the intensively studied villages differed from those in the control villages which must (partly) be attributed to study's influence. The IPM farmer field school groups that had been subjected to continuous investigations during four seasons performed better than the control IPM groups. Effects on farmers' (IPM) practice and farm economics were more obvious in the former groups. This influence on trained farmers seems logical. Attention from outside encouraged their activities, since their practice and self-confidence is regularly confirmed. Continuous contact appeared particularly supportive to resist counteracting forces, such as pesticide promotion by the village extension worker. The phenomenon that farmers are encouraged after being given special attention was observed in many field school groups that were visited by a delegation from the IPM Programme.

The influence of the study activities on non-IPM farmers was most visible in the non-IPM, intensively studied villages. Higher awareness about IPM was measured there, and a reduction of pesticide use was visible in the post-training seasons, in contrast to the control non-IPM villages. This influence was mainly the result of the presence of village study assistants who did weekly field observations. These assistants were often questioned about what they were doing and why. The main message perceived from their answers was that pesticide use could be reduced, which reflected on farmers' control practices, but not on their monitoring and decision making behaviour. In IPM villages, the influence of the study activities on non-IPM farmers' behaviour seemed to be insignificant compared to the influence of the field school.

The impact of study activities on its outcome is not considered to reduce the value of the evaluation, since it was attempted to describe and analyse processes. Obviously, the processes and mechanisms observed leading to positive effects were a result of IPM training activities in the area, and only to some extent encouraged by the evaluation activities. In general, evaluators need to be aware of the impact of their presence and to attempt to asses the extent to which study results are influenced. The use of control, non-intensively studied, villages in this evaluation study proved to be a valuable method.

Evaluation as a learning experience
Evaluating the experiential learning approach of the National IPM Programme in Grobogan became a learning experience in itself. In the course of time, the study determined its own needs supported by the relative absence of limitations to methods of study, and the interactions with the programme managers and implementors. The longitudinal design allowed to adjust the topics to be investigated and the methods to be used to the experiences obtained throughout the study seasons. The emphasis on diversity taught me to better understand the complexity of factors and processes at village level, which I found most helpful in recognising similar or different mechanisms in different situations. Considering this experi-

ence, the case methodology served a generalising function. It is hoped that something of the learning experience is reflected in this book. This would allow the reader to apply the results of this descriptive case study in her or his own purposes and situation.

Summary

In the late 1960s, the Government of Indonesia introduced new high-external-input technologies in rice cultivation through a series of large-scale intensification programmes. The programmes included subsidised packages with high-yielding rice varieties, fertilisers, and pesticides to be applied on calender-based routine, all supplied on heavily subsidised credit. A strongly top-down extension system applying a transfer-of-technology model was designed to support the utilisation of the input packages at the farm level. The success of the intensification programmes became obvious in 1983 when Indonesia achieved self-sufficiency in rice, a status that appeared difficult to maintain, though. National food security was soon threatened by brown planthopper outbreaks, which appeared to be induced by pesticides creating resistance in pests, and destroying natural enemies.

In 1986, the Indonesian government recognised the hazardous effects of increased pesticide use, and announced a presidential instruction (INPRES 3/86). This decree included the ban of fifty-seven broad-spectrum pesticides, and declared Integrated Pest Management (IPM) as the national pest control strategy. A second important policy measure was the gradual abolition of the high subsidies on pesticides. These measures created a favourable climate for the implementation of IPM training on a large-scale: the National IPM Programme was launched in 1989.

Based on the experiences of IPM training and implementation in several Asian countries over more than a decade, the Indonesian programme developed an IPM approach using an ecological perspective. Instead of a set of recommendations, the approach applies four principles: (1) grow a healthy crop, (2) observe the field weekly, (3) conserve natural enemies, and (4) farmers become IPM experts. With these principles, rice farmers are provided a tool for decision making. This different approach to the training contents required an extension approach different from the conventional top-down farmer training. Farmers could no longer be considered as passive receivers of packages, but had to be seen as active learners and independent decision makers. Therefore, the National IPM Programme designed a strategy that is based on the principles of nonformal education. This strategy is consistent with the requirements of active ecosystem management by farmers implied by IPM. The approach was chosen after implementation of IPM training through conventional approaches had proven to be inadequate. Training is conducted in so-called IPM farmer field schools, during which learning occurs through experience with real field problems and experimentation. The trainer is considered a facilitator of the experiential learning process, and not an instructor. This training approach seeks to empower people to actively identify and solve their own problems by fostering participation, self-confidence, and collective action and decision making.

Since the National IPM Programme, applying unconventional approaches of training contents and process, is implemented as a government programme through the existing structures, the introduction of this programme at the village level seemed an interesting subject for study. An evaluation study was designed which aimed at describing and analysing the processes and effects taking place at the village level, and at delivering useful management information to the programme with respect to implementation and achievements as a result of IPM training and dissemination. To allow in-depth investigations, the study covered only four IPM and four comparison villages in one Central Javanese district. Therefore, it is a case study which does not allow generalisation across the National IPM Programme, implemented throughout the country. During five rice-growing seasons, rice farmers' practices and perceptions were investigated in relation to the context of agricultural development in the individual villages. IPM training processes were observed, achievements were analysed, and effects at farmer and farm level were assessed.

A high diversity was observed among the villages studied with respect to direction and level of (agricultural) development. Especially the (either favourable or unfavourable) influence of the village head's policies seemed to be of major importance. On the whole, rice farmers in the study area had become high-external-input farmers after being exposed for more than twenty years to the food crop intensification programmes. Practices, however, differed considerably from the governmental recommendations with respect to the actual implementation of the technology, such as quantities of inputs applied and time of application. Farmers value their own experiences and experimentation more than extension recommendations.

Implementation of IPM farmer field schools provided a new perspective on farming to most participants. Field school farmers appreciated to be trained in the field, through facilitation rather than instruction, and to be considered the central farm managers. Achievements were obvious, including improved knowledge on pest and natural enemy identification, changed perceptions with respect to pest occurrence, rice ecosystem management and pest control, and improved skills in field monitoring. What is more, farmers obtained higher self-confidence as a result of increased knowledge, which they themselves considered a major reward. In accordance with the IPM Programme's main objective, trained farmers learned to make considered pest management decisions, based on their observations and agroecosystem analysis. This resulted in effects on the pest management behaviour in their own fields, in that fewer curative pesticide sprays were applied than before (and fewer than by non-trained farmers), and that control measures were better targeted and timed. However, in some villages counteracting forces, such as an extension worker promoting pesticides or the KUD enforcing pesticide purchase, did not lead all farmers to implement IPM committedly.

The changed practices of IPM trained farmers resulted in consistently higher yields obtained in the seasons after training, and in a more favourable yield variability, standing for less risk in farm management. In addition to lower expenditures on pest management, IPM farmers gained measurably higher returns. Although proportionally these gains were only a minor part of the total paid-out costs for rice production, in absolute amounts they were tangible in the household budget of Javanese rural families.

Communication on IPM from trained farmers to the surrounding community seemed hampered for two reasons. First, vocal communication dominated, whereas field practice would have been more appropriate to extend IPM. Second, farmers selected for participation were, on average, the more affluent and better informed farmers, therefore not representative for the farming community. Communication between these farmers and especially those from the lower socioeconomic layers did not occur, as could be expected. Women farmers seemed to be the group most disadvantaged since they were not selected for training, and were not reached through horizontal communication processes.

The IPM farmer field school model seems to provide a start for more sustainable practices in intensive food production, and for more sustainable, farmer-oriented extension methods. What is more, the experience of the IPM Programme suggests that a facilitation model of extension is consistent with sustainable agriculture. If this is true, the promotion of more sustainable forms of agriculture (which seem to be of increasing importance) has great implications for extension systems, an exciting area for further research.

Samenvatting

Aan het einde van de jaren zestig introduceerde de Indonesische regering hoge externe input technologieën in de rijstbouw door middel van een serie grootschalige intensiveringsprogramma's. De programma's bevatten gesubsidieerde pakketten bestaande uit hoogopbrengende rijstrassen, kunstmeststoffen en chemische bestrijdingsmiddelen, toe te dienen volgens kalenderschema's. Deze pakketten konden met gesubsidieerd krediet door boeren worden verkregen. Een sterk hiërarchisch voorlichtingsapparaat volgens het 'Transfer-of-Technology' model werd ontwikkeld om de promotie van de input pakketten aan de rijstboeren te ondersteunen. Het succes van de intensiveringsprogramma's werd duidelijk in 1983, toen Indonesië's rijstproduktie het nivo van zelfvoorziening bereikte, een toestand die echter moeilijk te handhaven bleek. De pas bereikte voedselveiligheid werd spoedig ondermijnd door epidemieën van de 'brown planthopper', een plaaginsekt, die geïnduceerd waren door chemische bestrijdingsmiddelen. Een hoog gebruik van deze middelen had resistentie in de plaaginsekten opgewekt en natuurlijke vijanden uitgeroeid.

In 1986 onderkende de Indonesische regering de schadelijke effecten van het toenemend bestrijdingsmiddelengebruik en kondigde een presidentieel besluit af (INPRES 3/86). Dit besluit bestond onder andere uit het verbod op vijfenzeventig breedwerkende insecticiden voor de rijstbouw. Tevens verklaarde het 'geïntegreerde bestrijding' (IPM) tot nationale strategie voor gewasbescherming. Een tweede belangrijke maatregel was het geleidelijk afschaffen van de hoge subsidies op chemische bestrijdingsmiddelen. Door deze maatregelen werd een gunstig klimaat gecreëerd voor de uitvoering van een grootschalig trainingsprogramma voor geïntegreerde bestrijding: het 'Nationale IPM Projekt' werd ontwikkeld.

Gebruik makend van een tienjarige ervaring met het trainen in en de toepassing van IPM in verscheidene Aziatische landen, ontwikkelde het Indonesische projekt een benadering voor IPM vanuit een sterk ecologisch perspektief. In plaats van een aantal aanbevelingen gaat deze benadering uit van vier beginselen: (1) verbouw een gezond gewas, (2) doe wekelijkse veldobservaties, (3) behoud de natuurlijke vijanden, en (4) de boeren zijn IPM deskundigen. Op grond van deze beginselen wordt de boeren een instrument voor besluitvorming aangereikt. Deze andere benadering voor de inhoud van de training vereiste tevens een andere benadering voor voorlichting dan het conventionele, top-down systeem. Boeren konden niet meer worden gezien als passieve ontvangers van pakketten, maar moesten worden benaderd als aktieve studenten en onafhankelijke besluitvormers. Daarom past het Nationale IPM Projekt een strategie toe die gebaseerd is op de beginselen van volwassenenedukatie. Deze strategie is consistent met het vereiste van IPM, dat boeren aktieve managers zijn van het rijstecosysteem. Een dergelijke benadering werd gekozen, nadat IPM training via de conventionele benaderingen inadequaat was gebleken. De IPM training wordt georganiseerd in zogenaamde IPM veldscholen, waar de boeren leren door middel van ervaringen met werkelijke veldproblemen en van zelf uitgevoerde experimenten. De trainer heeft een funktie als facilitator van het leerproces, in plaats van als instrukteur. Het belangrijkste doel van deze trainingsbenadering is om het mensen mogelijk te maken (en aldus te emanciperen) hun eigen problemen aktief te ontdekken en op te lossen door ze aan te moedigen tot participatie, zelfvertrouwen en gezamenlijke aktie en besluitvorming.

Doordat het Nationale IPM Projekt, met zijn onconventionele benadering voor trainingsinhoud en -proces, wordt uitgevoerd als een overheidsprojekt via de bestaande voorlichtingsdienst, leek de introduktie van dit projekt op dorpsnivo een interessant studieobjekt. Hiertoe werd een evaluatiestudie ontworpen welke tot doel had om processen en effekten, die plaatsvinden op dorpsnivo, te beschrijven en te analyseren, en om nuttige management informatie te leveren aan het projekt met betrekking tot de uitvoering van IPM training, de toepassing en de verspreiding. De studie besloeg slechts vier dorpen met en vier zonder IPM training in één distrikt in Midden Java om diepteonderzoek mogelijk te maken. Hiermee is het een case studie hetgeen geen generalisaties toelaat voor het hele Nationale IPM Projekt, dat over het hele land wordt uitgevoerd. Gedurende vijf groeiseizoenen werden de praktijken en percepties van rijstboeren onderzocht in relatie tot de context van landbouwontwikkeling in de afzonderlijke dorpen. Het IPM trainingsproces werd geobserveerd, de resultaten van training geanalyseerd, en de effekten op het nivo van de boer en het boerenbedrijf bepaald.

Tussen de verschillende onderzoeksdorpen werd een hoge mate van diversiteit waargenomen met betrekking tot de richting en het nivo van (landbouw)ontwikkeling. In het bijzonder leek de (gunstige of ongunstige) invloed van de dorpsleider van essentieel belang te zijn. Over het geheel genomen zijn de rijstboeren in het studiegebied, na meer dan twintig jaar bloot te hebben gestaan aan de intensiveringsprogramma's, geworden tot moderne boeren met een hoog extern input gebruik. Hun toepassing van de technologieën verschilt echter nogal met de aanbevelingen van de overheid, vooral wat betreft de hoeveelheden aan inputs die gebruikt worden en het tijdstip van toediening. Voor boeren zijn hun eigen ervaring en die van hun medeboeren belangrijker dan de aanbevelingen van de voorlichtingsdienst.

De uitvoering van de IPM veldschool leverde veel deelnemers een nieuw perspektief op landbouw op. De boeren waardeerden het om in het veld getraind te worden door middel van facilitatie in plaats van instruktie, en om beschouwd te worden als de centrale managers van hun boerenbedrijf. Resultaten van de training waren duidelijk: kennis over de identifikatie van plagen en natuurlijke vijanden was toegenomen, perceptie ten aanzien van plagen, het rijstveld als ecosysteem en de gewasbescherming was veranderd, en vaardigheid in veldwaarnemingen was verbeterd. Belangrijker nog is, dat de boeren meer zelfvertrouwen kregen door middel van de vergaarde kennis en vaardigheden, wat voor henzelf één van de voornaamste voordelen van deelname aan IPM training was. Overeenkomstig de hoofddoelstelling van het IPM Projekt leerden IPM boeren om geïnformeerde beslissingen te nemen, gebaseerd op hun eigen waarnemingen en analyses van het agro-ecosysteem. De genoemde resultaten hadden een aantal gunstige effekten tot gevolg op de gewasbeschermingspraktijken van getrainde boeren in hun eigen veld. Er werden minder curatieve bespuitingen gegeven dan voorheen (en minder dan door niet getrainde boeren), en de bestrijdingsmaatregelen werden beter uitgevoerd met een gerichter doel en op een beter tijdstip. Echter waren er in sommige dorpen tegenwerkende krachten aanwezig, zoals het aktief promoten van bestrijdingsmiddelen door een landbouwvoorlichter, of het verplichten van aankoop van middelen door de KUD (de dorpscoöperatie), waaraan niet alle boeren het hoofd konden bieden, welke aldus een overtuigde toepassing van IPM tegenwerkten.

De veranderde werkwijze van IPM getrainde boeren resulteerde in consistent hogere rijstopbrengsten in de seizoenen na training. Tevens was de variatie in opbrengsten gunstiger dan voor training en dan bij niet-getrainde boeren, hetgeen staat voor een lager risiko in de bedrijsvoering. Eveneens werden lagere uitgaven aan gewasbescherming gedaan, wat tezamen met hogere opbrengsten leidde tot hogere inkomsten voor IPM boeren. Hoewel deze extra inkomsten verhoudingsgewijs slechts een klein deel uitmaken van de totale kosten voor de rijstproduktie, zijn ze, in absolute termen, tastbaar in het huishoudbudget van de Javaanse boerenfamilies.

De communicatie over IPM van getrainde boeren naar de boerengemeenschap leek om twee redenen belemmerd te zijn. Ten eerste werd IPM voornamelijk via mondelinge communicatie doorgegeven, terwijl demonstratie in het veld veel geschikter zou zijn geweest. Ten tweede waren de boeren, die geselekteerd waren voor training, gemiddeld genomen de welvarender en beter geïnformeerde boeren, die niet voldoende representatief waren voor de boerengemeenschap. Communicatie tussen deze boeren en die uit de lagere sociaal-economische lagen van de samenleving vond niet plaats, wat te verwachten was. De vrouwelijke boeren bleken het meest benadeeld te zijn, omdat zij èn niet waren geselekteerd voor deelname in de training, èn niet bereikt werden door horizontale communicatieprocessen.

Het trainingsmodel van de IPM veldschool lijkt een aanzet te zijn voor meer duurzame werkwijzen in de intensieve voedselproduktie, en voor meer duurzame, op de boer gerichte voorlichtingsmethoden. Bovendien duidt de ervaring met het Indonesische Nationale IPM Projekt erop, dat een voorlichtingsbenadering, die gebruik maakt van facilitatie, consistent is met duurzame landbouwmethoden. Indien dit het geval is, dan zal de bevordering van meer duurzame landbouwvormen (wat een toenemende aandacht lijkt te genieten) grote implikaties hebben voor landbouwvoorlichtingsdiensten, een interessant terrein voor verder onderzoek.

References

Adalla, Candida B. and Agnes C. Rola (1986). Participatory Verification and Technology Generation of Location Specific IPM Technology in Calamba, Laguna: A Pilot Project. In: *Filipino Women in Rice Farming Systems*. U.P. at Los Baños, IRRI and the Philippines, Institute for Development Studies, p. 41-56.

Adams, M.E. (1982). *Agricultural Extension in Developing Countries*. Intermediate Tropical Agricultural Series. Burnt Mill (Essex): Longman Group Ltd.

Adhikarya, R. (1989). *Strategic Extension Campaigns: A Case Study of FAO's Experiences*. Rome (Italy): Food and Agriculture Organisation of the United Nations.

Adjid, Dudung Abdul (1983). *Vertical Transfer of Agrotechnology in Indonesia: The Case of the BIMAS Programme*. In: Cagauan, B.G. et al. (op. cit.), p. 48-55.

Agro Economic Survey (1972). *Agricultural Census in 33 Villages Located in the Major Rice Producing Areas of Indonesia*. Research Notes 6. Bogor: Agro Economic Survey.

Agudelo, L.A. and D. Kaimowitz (1989). *Institutional Linkages for Different Types of Agricultural Technologies: Rice in the Eastern Plains of Colombia*. The Hague: ISNAR/RTLL, linkage discussion paper 1.

Alkin, Marvin C. (1985). *A Guide for Evaluation Decision Makers*. Beverly Hills: Sage. 168 pp.

Altieri, Miguel A. (1987). *Agroecology. The Scientific Basis of Alternative Agriculture*. Boulder: Westview Press. 227 pp.

Anonymous (1969). *Leven met Insekten. Het Onderzoek naar een Geïntegreerde Bestrijding van Plagen*. Wageningen: PUDOC. 177 pp.

Anonymous (1991). *Mission Report, Mid-Term Review Mission*. Training and Development of Integrated Pest Management in Rice-based Cropping Systems. Jakarta: IPM Secretariat, Jl. Ki Mangunsarkoro 5, Jakarta 10310, Indonesia.

Anonymus (1992). Economic Growth: Explaining the Mystery. *The Economist*, Jan. 4th: 17-20.

Arida, G.S. and K.L. Heong (1988). Yield Loss Caused by Rice Stem Borers (SB) in Southern Bhutan. *Int. Rice Research Newsletter* 14(6):32.

Austin, R.B., Ed. (1982). *Decision Making in the Practice of Crop Protection*. Monograph No. 25, BCPC Publications. Proc. British Crop Symposium.

Axinn, George H. (1988). International Technical Interventions in Agriculture and Rural Development: Some Basic Trends, Issues and Questions. *Agriculture and Human Values*, Winter/Spring 1988: 6-15.

Barker, R. and D. Chapman (1988). *The Economics of Sustainable Agricultural Systems in Developing Countries*. Cornell University (mimeo.).

Benor, D. and J.Q. Harrison (1977). *Agricultural Extension: The Training and Visit System*. Washington: The World Bank.

Benor, D. and M. Baxter (1984). *Training and Visit Extension*. Washington: The World Bank.

Benor, D., J.Q. Harrison and M. Baxter (1984). *Agricultural Extension: The Training and Visit System*. Washington: The World Bank.

Bhuktan, Jit Pradhan and Glenn L. Denning (1990). *Risk Adjustment Strategies in the Resource Organization of the Small Farmer Household Farming System*. Paper presented in the Int. Workshop on Sustainability through Farmer's involvement in Technology Generation and Diffusion, 9 Febr. 1990, New Dehli, India. 45 pp.

Birch, H.F. (1958). Effect of Soil Drying on Humus Decomposition. *Plant and Soil* 10: 9-31.

Birgegård, Lars-Erik (1980). *Manual for the Analysis of Rural Underdevelopment*. Rural Development Studies No. 7. Uppsala: Swedish University of Agricultural Sciences. 149 pp.

Biro Pusat Statistik (1987). *Survey Pertanian Luas Lahan Menurut Penggunaannya di Jawa 1986*. Jakarta: Biro Pusat Statistik. 37 pp.

Biro Pusat Statistik (1988). *Struktur Ongkos Usaha Tani Padi dan Palawija 1986*. Jakarta: Biro Pusat Statistik. 204 pp.

Birowo, Achmad T. and Joko Budianto (1983). *The Role of Government in the Transfer of Agrotechnology*. In: Cagauan, B.G. et al. (op. cit.), p. 78-80.

Blokker, K.J. (1984). *Computergesteunde Voorlichting. Een Decisiegericht Voorlichtingskundig Onderzoek naar Epipre en andere Geautomatiseerde Informatiesystemen in de Landbouwvoorlichting*. Ph.D. dissertation. Wageningen: Agricultural University. 389 pp.

Booth, Anne (1985). Accomodating a Growing Population in Javanese Agriculture. *Bulletin of Indonesian Economic Studies* 21: 115-145.

Booth, Anne (1988). *Agricultural Development in Indonesia*. Asian Studies Association of Australia, Southeast Asia Publications Series No. 16. Sydney: Allen and Unwin. 295 pp.

Boserup, Ester (1984). *Peranan Wanita dalam Perkembangan Ekonomi*. Jakarta: Yayasan Obor Indonesia.

Bottrel, D.G. (1987). Applications and Problems of Integrated Pest Management in the Tropics. *J. Pl. Prot. Tropics* 4(1): 1-8.

Brandner, Lowell and Bryant Kearl (1964). Evaluation for Congruence as a Factor in Adoption Rate of Innovations. Rural Sociology, 29: 288-303.

Bryk, A.S., Ed. (1983). *Stakeholder-based Evaluation*. San Fransisco: Jossey-Bass.

Bulmer, Martin (1983). *Sampling*. In: Bulmer, Martin and Donald P. Warwick (op. cit.), p. 91-99.

Bulmer, Martin and Donald P. Warwick, Eds. (1983). Social Research in Developing Countries. Surveys and Censuses in the Third World. Chichester: John Wiley & Sons Ltd. 383 pp.

Bureau of Agricultural Extension (1986). *Agricultural Extension Experiences in Indonesia*. Jakarta: Ministry of Agriculture, Govt. of Indonesia. 53 pp.

Byerlee, Derek (1987). *Maintaining the Momentum in Post-Green Revolution Agriculture: A Micro-Level Perspective from Asia*. MSU International Development Paper No. 10. East Lansing: Dept. of Agricultural Economics, Michigan State University. 57 pp.

Cagauan, B.G. Jr., G.Y. Tsuji and L.W. Peterson, Eds. (1983). *Agrotechnology Transfer in Indonesia and the Philippines*. BSP Tech, Rep. 9, HITAHR Res. Series 027. Manoa: Benchmark Soils Project, Dept. of Agronomy and Soil Science, College of Tropical Agriculture and Human Resources, Univ. of Hawaii.

Campbell, D.T. (1971). *Methods for the Experimenting Society*. Paper presented at the meeting of the Eastern Psychological Association, New York, and at the meeting of the American Psychological Association, Washington, D.C.

Cancian, Frank (1979). *The Innovator's Situation. Upper-Middle-Class Conservatism in Agricultural Communities*. Stanford, California: Stanford University Press. 159 pp.

Carson, Rachel (1962). *Silent Spring*. Boston: Houghton Mifflin. 368 pp.

Casley, Dennis J. and Denis A. Lury (1982). *Monitoring and Evaluation of Agriculture and Rurual Development Projects*. Baltimore: The John Hopkins University Press. 144 pp.

Casley, Dennis J. and Krishna Kumar (1988). *The Collection, Analysis, and Use of Monitoring and Evaluation*. Baltimore: The John Hopkins University Press. 174 pp.

Chambers, R. and J. Jiggins (1986). *Agricultural Research for Resource-Poor Farmers: A Parsimonious Paradigm*. Discussion Paper 20. Brighton (Sussex): IDS.

Chambers, Robert, Arnold Pacey and Lori Ann Thrupp (1989). *Farmer First. Farmer Innovation and Agricultural Research*. London: Intermediate Technology Publications. 218 pp.

Chang, T.T. (1976) The Origin, Evaluation, Cultivation, and Diversification of Asian and African Rices. *Euphytica* 25: 425-441.

Colin, R. (1978). *Analyse Comparative et Problématique a Partir de l'Étude de Quelques Dossiers Significatifs*. Paris: Division for the Study of Development Projects, UNESCO.

Collier, William L., Soentoro and Gunawan Wiradi (1973). *Rice Harvesting and Selling Changes in Central Java Which Have Serious Social Implications*. Jakarta: Agro Economic Survey.

Collier, W.L., Soentoro, Gunawan Wiradi, Effendi Pasandaran, Kabul Santoso, Jospeh F. Stepanek (1982) Acceleration of Rural Development of Java. *Bulletin of Indonesian Economic Studies* XVIII(3): 84-101.

Collier, William L., Gunawan Wiradi, Soentoro, Makali and Kabul Santoso (1988). *Employment Trends in Lowland Javanese Villages*. Jakarta: USAID.

Cook, Thomas D. and Donald T. Campbell (1979). *Quasi-experimenting Design and Analysis Issues for Field Settings*. Boston: Houghton Mifflin Co. 405 pp.
Craswell, E.T., S.K. De Datta, M. Hartantyo and W.N. Obcemea (1981). Time and Mode of Nitrogen Fertiliser Application to Tropical Wetland Rice. *Fert. Res.* 2: 247-259.
Cronbach, Lee J. (1982). *Designing Evaluations of Educational and Social Programmes*. San Fransisco: Jossey-Bass. 374 pp.
Dahm, Bernhard, Ed. (1992). *Regions and Regional Developments in the Malay-Indonesian World*. 6. European Colloquium on Indonesian and Malay Studies (ECIMS). Wiesbaden: Otto Harrassowitz. 222 p.
De Datta, Surajit K. (1981). *Principles and Practices of Rice Production*. Singapore: John Wiley. 618 pp.
De Datta, S.K. (1987). *Advances in Soil Fertility Research and Nitrogen Fertilizer Management for Lowland Rice*. In: International Rice Research Institute (op. cit.), p. 27-41.
Denno, Robert F. and T. John Perfect, Eds. (1992). *Ecology and Management of Planthoppers*. London: Chapman and Hall.
Dilts, R. (1983). *Critical Theory: A Theoretical Foundation for Non-Formal Education and Action Research*. Unpublished Ph.D. dissertation. Amherst (Mass): Univ. of Massachusetts, Centre for Informal Education.
Directorate of Food Crops Economics and Processing Developemnt (1988). *Price and Quality of Rice in Indonesia*. Jakarta: Dir. of Food Crops Econ. and Proc. Dev. in coop. with SFCDP-USAID.
DirJen Pertanian Tanaman Pangan (1987). *Petunjuk Pelaksanaan Implementasi Pengendalian Hama Terpadu (PHT) di Tingkat Lapang, Musim Tanam 1987/88*. Jakarta: DirJen Pertanian Tanaman Pangan/Perlindungan Tanaman Pangan.
Felstehausen, Herman (1982). *Readings in Case Methodology*. Dept. of Landscape Architecture & Land Tenure Center, School of Natiral Resources, I Ag. Hall. Madison (Wisconsin): University of Wisconsin. 78 pp.
Feuerstein, Marie-Thérèse (1986). *Partners in Evaluation: Evaluating Development and Community Programmes with Participants*. London: Macmillan. 196 pp.
Fink, A. and J. Kosecoff (1985). *How to Conduct Surveys: A Step-by-Step Guide*. Beverly Hills: Sage. 119 pp.
Flint, Marie Louise and Robert Van den Bosch (1981). *Introduction to Integrated Pest Management*. New York: Plenum Press. 240 pp.
Foley, Sean (1987). *The Ecological Transition in Bali*. Ph.D. Disertation, Australian National University. 254 pp.
Food and Agriculture Organisation of the United Nations (1968). *Report of the First Session of the FAO Panel of Experts on Integrated Pest Control*. FAO Meeting Report No. PL/1967/M/7. Rome: FAO. 19 pp.
Food and Agriculture Organisation of the United Nations (1985), *Production Yearbook, Volume 39*. Rome: Food and Agriculture Organization.
Food and Agriculture Organisation of the United Nations (1988). *Integrated Pest Management in Rice in Indonesia. Status after Three Crop Seasons, Perspectives for Farmers' Training*. Jakarta: Food and Agriculture Organisation. 13 pp.
Fox, James J. (1983). *Equal Rights and Equal Duties*. Seminar Paper on 'Man, Land and Labour on Java'. Research School of Pacific Studies, Australian National University.
Fox, J.J. (1991). *Managing the ecology of rice production in Indonesia*. In: Hardjono, J. (op. cit.).
Freire, Paolo (1972). *Pedagogy of the Oppressed*. Harmondsworth (UK): Penguin Books. 153 pp.
Gallagher, Kevin DeWayne (1988). *Effects of Host Resistance on the Micro-evolution of the Rice Brown Planthopper, Nilaparvata Lugens* (Stal). Unpublished Ph.D. dissertation. Berkeley: Univ. of California, Graduate Division.
Gallagher, Kevin (1990). *The 'Model'*. Yogyakarta: IPM Secretariat. Unpublished paper.
Gibbons, David S., Rodolphe de Koninck and Ibrahim Hasan (1980). *Agricultural Modernization, Poverty and Inequality. The Distributional Impact of the Green Revolution in Regions of Malaysia and Indonesia*. Westmead, Farnborough: Teakfield Ltd. 255 pp.
Glaser, Edward M., Harold H. Abelson, Kathalee N. Garrison (1983). *Putting Knowledge to Use. Facilitating the Diffusion of Knowledge and the Implementation of Planned Change*. San Fransisco: Jossey-Bass.

Goebbels, P. (1988). Peasant Farmer Agricultural Self-development. The World Neighbours Experience in West Africa. *ILEIA Newsletter* 4(3): 11-14.
Goodell, Grace (1984). Challenges to International Pest Management Research and Extension in the Third World: Do We Really Want IPM to Work? *Bulletin of the ESA* (30)3:18-26.
Griliches, Zvi (1957). Hybrid Corn: An Exploration in the Economics of Technological Change. *Econometrica,* 25: 501-522.
Groot, A.E. en C.A.M. Boon (1992). *Monitoring en Evaluatie: Een Werkplan voor de Ontwikkeling van Monitoring en Evaluatie in Gendergevoelige Procesmatige Ontwikkelingsprogramma's.* Vakgroep Voorlichtingskunde in samenwerking met SNV. Wageningen: Vakgroep Voorlichtingskunde, Landbouwuniversiteit Wageningen. 85 pp.
Guan-Soon, Lim and Ong Seng-Hock (1987). *Environmental Problems of Pesticide Usage in Malaysian Rice Fields – Perception and Future Considerations.* In: Tait, Joyce and Banpot Napompeth (op. cit.), p. 10-21.
Hansen, G. (1971). Episodes in Rural Modernisation: Problems in the Bimas Program. *Indonesia* 11: 63-81.
Hansen, Gary E., Ed. (1981). *Agricultural and Rural Development in Indonesia.* Boulder, Colorado: Westview Press. 312 pp.
Hardjono, J., Ed. (1991). *Indonesia: Resources, Ecology and Environment.* Singapore: Oxford University Press.
Hart, Gillian, Andrew Turton and Benjamin White (Eds.) with Brian Fegan and Lim Teck Ghee (1989). *Agrarian Transformations: Local Processes and the State in Southeast Asia.* Berkeley: University of California Press.
Hassan, S.A. (1984). *Testing the Side Effects of Pesticides on Beneficial Arthropods.* In: Int. Rice Research Institute (op. cit.), 117-126.
Havelock, R.G. (1973). *Planning for Innovation through Dissemination and Utilisation of Knowledge.* Ann Arbor: University of Michigan, CRUSK.
Heenan, D.P. and P.E. Bacon. (1987). *Effect of Nitrogen Fertilizer Timing on Crop Growth and Nitrogen Use Efficiency by Different Rice Varieties in Southeastern Australia.* In: International Rice Research Institute (op. cit.), p. 97-105.
Herdt, R.W., L.L. Castilla and S.K. Jayasuriya (1984). *The Economics of Insect Control on Rice in the Philippines.* In: International Rice Research Institute (op. cit.), p. 41-56.
Hogstedt, C., U.G. Ahlborg, G. Ekström. T. Kjellström, O. Pettersson, M. Akerblom (1987). *Public Health Impact of Agrochemicals.* Report prepared for the World Health Organization.
Humphreys, E., W.A. Muirhead, F.M. Melhuish, R.J.G. White and P.M. Chalk (1987). *Fertilizer Nitrogen Recovery in Mechanized Dry Seeded Rice.* In: International Rice Research Institute (op. cit.), p. 107-118.
Hüsken, Frans (1988). *Een Dorp op Java. Sociale Differentiatie in een Boerengemeenschap, 1850-1980.* Overveen: ACASEA. 301 pp.
Hüsken, Frans (1989). *Cycle of Commercialization and Accumulation in a Central Javanese Village.* In: Hart, Gillian, Andrew Turton and Benjamin White (op. cit.), p. 303-331.
Hüsken, Frans and Benjamin White (1989). *Java: Social Differentiation, Food Production, and Agrarian Control.* In: Hart, Gillian, Andrew Turton and Benjamin White (op. cit.), p. 235-263.
Hüsken, Frans and Benjamin White (1992). *State Patronage in a Divided Society Government and Pesantry in Java.* In: Dahm, Bernhard (op. cit.), p. 175-200.
Hussein, Mohd Yusof (1987). *Perception and Management of Pests and Pesticides by Malaysian Traditional Small Farmers – A Case Study.* In: Tait, Joyce and Banpot Napompeth. (op. cit), p. 142-149.
Ibusno, A.C. and M.D.G.G. Petil (1988). *Low-input Rice Production: Toward a Sustainable Rice Farming Technology.* In: Proc. of Int. Training Course on Sustainable Agriculture (Ecofarming) and On-Farm Experimentation. Los Baños, Philippines: SEARCA. p. 143-149.
Ifzal, Ali (1987). *Rice in Indonesia: Price Policy and Comparative Advantage.* Asian Development Bank Staff Paper Series.
ILEIA (1989). *Proceedings ILEAI Workshop PTD.* Leusden: ETC/ILEIA.
International Rice Research Institute (1984). *Judicious and Efficient Use of Insecticides on Rice.* Proceedings of the FAO/IRRI Workshop. Los Baños: IRRI.180 pp.
International Rice Research Institute (1985). *Women in Rice Farming.* Proceedings of a conference

on Women in Rice Farming Systems. Hants (UK): Gower Publishing Company Ltd.
International Rice Research Institute (1987). *Efficiency of Nitrogen Fertilizers for Rice.* Proceedings of the Meeting of the International Network on Soil Fertility and Fertilizer Evaluation for Rice, Griffith, New South Wales, 10-16 April 1985. Los Baños: IRRI. 260 pp.
IOCU (1987). *Violating the Pesticide Code. A Survey of Indonesia, the Philippines and Thailand.* Paper by Int. Organisation of Consumers Unions, Reg. Office for Asia and the Pacific, and Institute for Consumer Policy Research, Consumers Union. Penang: IOCU. 22 pp.
Ito, Y., K. Miyashita and K. Sekiguchi (1962). Studies on the Predators of the Rice Crop Insect Pests, Using the Insecticidal Check Method. *Jap. J. Ecology* 12(1): 1-11.
Janssen, B.H. and M.L. van Beusichem (1991). *Nutriënten in bodem-plant relaties.* Syllabus Bodemkunde en Plantevoeding. Wageningen: Landbouwuniversiteit. 353 pp.
Jatileksono, Tumari (1987). *Equity achievement in the Indonesian Rice Economy.* Yogyakarta: Gadjah Mada Univeristy Press. 168 pp.
Jick, T.D. (1979). Mixing Qualitative and Quantitative Methods: Triangulation in Action. *Admin. Science Q.* 24 (Dec.): 602-611.
Jiggins, J.L.S. (1983). Poverty-oriented Rural Development Participation and Management. A Review Article. *Development Policy Review*, Vol. 1: 219-252.
Jorgensen, Danny L. (1989). *Participant Observations. A Methodology for Human Studies.* Applied Social Research Methods Series Vol. 15. Beverly Hills CA: Sage. 133 pp.
Kaimowitz. David (1990). Moving Forces: External Pressure and the Dynamics of Technology Systems. *Knowledge in Society: The International Journal of Knowledge Transfer*, 3(3): 36-43.
Kenmore P. E. (1980). *Ecology and Outbreaks of a Tropical Insect Pest of the Green Revolution, the Brown Planthopper, Nilarparvata lugens* (Stal). Ph.D. Thesis, University of California, Berkeley, USA. 226 pp.
Kenmore, P.E. (1987). *Crop Loss Assessment in a Practical Integrated Pest Control Program for Tropical Asian Rice.* In: Teng, P.S. (op. cit.), p. 225-241.
Kenmore, P.E., J.A. Litsinger, J.P. Bandong, A.C. Santiago and M.M. Salac (1987). *Philippine Rice Farmers and Insecticides: Thirty Years of Growing Dependency and New Options for Change.* In: Tait, Joyce and Banpot Napompeth (op. cit.), p. 98-108.
Kennedy, M.M. (1979). Generalising from Single Case Studies. *Evaluation Q.* 3 (Nov.): 661-678.
Kindervatter, Suzanne. (1979). *Nonformal Education as an Empowering Process, with Case Studies from Indonesia and Thailand.* Amherst: Univ. of Massachusetts, Center for Int. Education. 294 pp.
Kiritani, K. (1972). *Integrated Control of Rice Pests.* Rev. of Plant Protection. Japan.
Kodiran dan Bambang Hudayana (1990). *Peranan Wanita dalam Sawah Surjan.* Yogyakarta: Pusat Penelitian Kependudukan, UGM.
Koningsberger, J.C. (1915). *Java Zoölogisch en Biologisch.* Buitenzorg: Drukkerij Dep. v. L.N. en H.
Kumar, Krishna (1987). *Conducting Group Interviews in Developing Countries.* A.I.D Program Design and Evaluation Methodology Report No. 8. Washington: Agency for International Development (A.I.D). 44 pp.
Krueger, Richard A. (1988). *Focus Groups. A Practical Guide for Applied Research.* Beverly Hills CA: Sage. 197 pp.
LaBelle, Thomas J. (1976). *Nonformal Education and Social Change in Latin America.* California: UCLA Latin American Center Publications.
Laboratorium Lapang PHT (1992). *Nogos: Racun Serangga yang Dijadikan Obat Gatal di Kecamatan Anjatan, Indramayu, Jawa Barat.* Jakarta: IPM Secretariat, unpublished report. 8 pp.
Lawson, T.J. (1982). *Information Flow and Crop Protection Decision Making.* In: Austin, R.B. (op. cit.), p. 21-32.
Lipton, Michael and Richard Longhurst (1989). *New Seeds and Poor People.* London: Unwin Hyman. 473 pp.
Litsinger, J.A. (1984). *Assessment of Need-based Insecticide Application for Rice.* Paper presented at MA-IRRI Technology Transfer Workshop, 19 March 1984.
Litsinger, James A. (1989). Second Generation Insect Pest Problems on High Yielding Rices. *Tropical Pest Management* 35(3): 235-242.
Loevinsohn, Michael E. (1987). Insecticide Use and Increased Mortality in Rural Central Luzon, Philippines. *The Lancet*, June 13: 1359-1362.

Luck, R.F., R. Van den Bosch and R. Garcia (1977). Chemical Insect Control, a Troubled Pest Management Strategy. *BioScience* 27: 606-611.

Manning, Chris (1988). Rural Employment Creation in Java: Lessons from the Green Revolution and Oil Boom. *Population and Development Review*, 14(1): 47-80.

Mascarenhas, James, et al. (1991). *Participatory Rural Appraisal.* Proceedings of the February 1991 Bangalore PRA Trainers Workshop. RRA Notes Number 13. London: Sustainable Agriculture Programme, Int. Inst. for Environment and Development. 143 pp.; 3 Endsleight Street, London WC1H 0DD, United Kingdom.

Matteson, P.C. (1986). *First Rice IPC Demonstration Results from Sri Lanka: Guidance for the National Implementation Program.* Paper presented at the 2nd International Conference on Plant Protection in the Tropics, 17-20 March 1986, Genting Highlands, Malaysia.

Matteson, Patricia C., Kevin D. Gallagher and Peter E. Kenmore (1992). *Extension of Integrated Pest Management for Planthoppers in Asian Irrigated Rice.* In: Denno, Robert F. and T. John Perfect (op. cit.), 57 pp.

Mears, Leon A. (1981). *The New Rice Economy of Indonesia.* Yogyakarta: Gajah Mada University Press.

Medina, Charito P. (1987). *Pest Control Practices and Pesticide Perception of Vegetable Farmers in Loo Valley, Benguet, Philippines.* In: Tait, Joyce and Banpot Napompeth (op. cit.), p. 150-157.

Minks, A.K. and P. Gruys, Eds. (1980). *Integrated Control of Insect Pests in the Netherlands.* Wageningen: PUDOC. 304 pp.

Morgan, David L. (1988). *Focus Groups as Qualitative Research.* Qualitative Research Methods Series Vol. 16. Beverly Hills CA: Sage. 85 pp.

Moris, Jon R. (1983). *Reforming Agricultural Extension and Research Services in Africa.* Discussion Paper 11, Agricultural Administration Network. London: Ag. Adm. Unit, ODI. 30 pp.

Mumford, J.D. (1987). *Analysis of Decision Making in Pest Management.* In: Teng, P.S. (op. cit.), p. 201-208.

Mumford, J.D. and G.A. Norton (1984). Economics of Decision Making in Pest Management. *Ann. Rev. Entomol.* 29: 157-174.

Mumford, J.D. and G.A. Norton (1987). *Economics of Integrated Pest Control.* In: Teng, P.S. (op. cit.), p. 191-200.

Murphy, Josette and Tim J. Marchant (1988). *Monitoring and Evaluation in Extension Agencies.* World Bank Technical Paper No. 79, Monitoring and Evaluation Series. Washington: World Bank. 81 pp.

National Research Council (1989). *Alternative Agriculture.* Committee on the Role of Alternative Farming Methods in Modern Production Agriculture, Board on Agriculture. Washington, D.C.: National Academy Press. 449 pp.

Nimma, Dan, Ed. (1979). *Communication Yearbook 3.* New Brunswick: International Communication Association.

NODAI Research Institute (1982). *Socioeconomic Study of Rice Farming and Marketing in Indonesia: A Case Study of Gadingsari, Central Java.* Tokyo: Univ. of Agriculture, DGHE-JSPS Program.

Norton, G.A. (1982). *Crop Protection Decision Making.* In: Austin, R.B. (op. cit.), p. 3-11.

Norton, G.A. and J.D. Mumford (1983). Decision Making in Pest Control. *Adv. Appl. Biol.* 8:87-119.

O'Brien, D.T., M. Sudjadi, J. Sri Adiningsih and Irawan (1987). *Economic Evaluation of Deep Placed Urea for Rice in Farmers' Fields: a Pilot Area Approach, Ngawi, East Java, Indonesia.* In: Int. Rice Research Institute (op. cit.), p. 141-159.

Oka, I.N. (1987). *Future Needs for Pesticide Management in Southeast Asia.* Paper presented at workshop on Southeast Asian Pesticide Management, Pattaya, Thailand, Febr. 1987. 12 pp.

Oka, Ida Nyoman (1990a). *Masalah Hama dalam Pembangunan Pertanian dan Sistem Pendidikan untuk Memperkuat Penerapan PHT.* Paper presented at 'Forum Komunikasi dan Seminar Ilmiah Mahasiswa Perlindungan Tanaman Se Indonesia IV, Universitas Hasanuddin, Ujung Pandang, 19-24 September 1990. 10 pp.

Oka, Ida Nyoman (1990b). *Visits to Farmers' Field Schools on Rice IPM: Impressions and Analysis.* Internal Report Steering Committee National IPM Programme. Jakarta: Nat. Programme for Development and Training of Integrated Pest Management in Rice-based Cropping Systems, Jl. Ki Mangunsarkoro 5, Jakarta 10310.

Opheffer (G.L. Gongrijp) (about 1914). *Brieven van Opheffer aan de Redactie van het Bataviaasch Handelsblad, 1911-1914.* Maastricht: N.V. Leiter-Nypels.
Patton, Michael Quinn (1986). *Utilisation-focused Evaluation.* Second Edition. Newbury Park (CA): Sage. 367 pp.
Pickett, A.D. and N.A. Patterson (1953). The Influence of Spray Programs on the Fauna of Apple Orchards in Nova Scotia IV. A Review. *Canadian Entomologist* 85(12): 472-478.
Pickett, A.D., W.L. Putman and E.J. Le Roux (1958). *Progress in Harmonizing Biological and Chemical Control of Orchard Pests in Eastern Canada.* Tenth Int. Congress of Entomology, Proceedings 3: 169-174.
Pimentel, D. (1991). *CRC Handbook of Pest Management in Agriculture, Volume 1.* Boca Raton: CRC Press. 765 pp.
Pincus, Jonathan (1991). *Farmer Field School Survey: Impact of IPM Training on Farmers' Pest Control Behavior.* Jakarta: IPM National Programme, Jl. Ki Mangunsarkoro 5, Jakarta 10310.
Pingali, P.L., P.F. Moya and L.E. Velasco (1990). Prospects for Rice Yield Improvement in the Post-Green Revolution Philippines. *Phil. Review of Economics and Business* 18(1): 85-106.
Pontius, John (1990). *Consultant's Report.* 'Training and Development of Integrated Pest Management in Rice-based Cropping Systems.' Jakarta: Food and Agriculture Organization of the United Nations.
Posavac, Emil J. and Raymond G. Carey (1989). *Program Evaluation. Methods and Case Studies.* Third ed. Englewood Cliffs (New Jersey): Prentice-Hall. 351 pp.
Program Nasional PHT (1990). *IPM Indonesia.* Information brochure. Jakarta: Nat. Programme for Development and Training of Integrated Pest Management in Rice-based Cropping Systems, Jl. Ki Mangunsarkoro 5, Jakarta 10310.
Pusat Penelitian Sosial Ekonomi Pertanian (1991). *Studi Baseline Aspek Sosial Ekonomi Pengendalian Hama Terpadu.* Bahan Seminar. Kerjasama penelitian Badan Perencanaan Pembangunan Nasional dengan Pusat Penelitian Sosial Ekonomi Pertanian. Bogor: PPSEP. 36 pp.
Reijntjes, C., B. Haverkort and A. Waters-Bayer (1992). *Farming for the Future. An Introduction to Low-External Input and Sustainable Agriculture.* London: Macmillan.
Rogers, Everett M. (1983). *Diffusion of Innovations.* 2nd ed. New York: Free Press. 453 pp.
Rogers, Everett M. and Ronny Adhikarya (1979). *Diffusion of Innovations: An Up-to-date Review and Commentary.* In: Nimma, Dan (op. cit.), p. 67-81.
Rogers, Everett M. and A. Eugene Havens (1962). Rejoinder to Griliches' 'Another False Dichotomy'. *Rural Sociology* 27: 330-332.
Röling, Niels (1986). Extension Science: Increasingly Preoccupied with Knowledge Systems. *Sociologia Ruralis,* 25: 269-290.
Röling, Niels (1988). *Extension Science: Information Systems in Agricultural Development.* Cambridge: Cambridge University Press. 233 pp.
Röling, Niels (1992a). *The Effects of Applied Agriculture Research and Extension: Issues for Knowledge Management.* Submitted to Journal of Extension Systems. 25 pp.
Röling, Niels (1992b). The Emergence of Knowledge Systems Thinking: A Changing Perception of Relationships among Innovation, Knowledge Process and Configuration. *Knowledge and Policy: The International Journal of Knowledge Transfer and Utilization* (5)1: 42-64.
Röling, Niels (1992c). *Facilitating Sustainable Agriculture: Turning Policy Models Upsode Down.* Invited paper for: 'Beyond Farmer First: Rural People's Knowledge Agricultural Research and Extension Practice'. Workshop at the Institute of Development Studies (IDS), University of Sussex, Brighton, UK, October 27-29, 1992, in collaboration with the International Institute for Environment and Development (IIED), London. 27 pp.
Röling, N., J. Ascroft and F. Wa. Chege (1976). Diffusion of Innovations, and the Issue of Equity in Rural Development. *Communication Research,* 3: 155-171.
Röling, Niels and Elske van de Fliert (1991). *Beyond the Green Revolution: Capturing Efficiencies from Integrated Agriculture. Indonesia's IPM Programme Provides a Generic Case.* Informal report. National Programme for Development and Training of Integrated Pest Management in Rice-based Cropping Systems. Jakarta: IPM Secretariat, Jl. Ki Mangunsarkoro 5, Jakarta 10310, Indonesia. 72 pp.
Röling, Niels and Elske van de Fliert (to be published 1993). *The Transformation of Extension for Sustainable Agriculture: The Case of Integrated Pest Management in Rice in Indonesia.* Submitted

to Agriculture and Human Values' special issue on 'Participation and Empowerment in Sustainable Rural Development'. 19 pp.

Röling, N. and H. de Zeeuw (1983). *Improving the Quality of Rural Poverty Alleviation.* Report of the Working Party 'The Small Farmer and Development Cooperation'. Wageningen: Int. Agricultural Centre.

Romer, Paul. M. (1990). Endogenous Technological Change. *J. of Political Economy* 98(5), Pt. 2: S71-S102.

Rosegrant, Mark W. and Prabhu L. Pingali (1991). *Sustaining Rice Productivity Growth in Asia: A Policy Perspective.* IRRI Social Science Division Papers No. 91-01. Los Baños: International Rice Research Institute. 39 pp.

Rossing, W.A.H., J. Schans and J.C. Zadoks (1985). Het Project Epipre, Projectanalyse in de Gewasbescherming. *Landbouwkundig Tijdschrift* 23: 455-473.

Rubia, E.G., B.M. Shepard, E.B. Yambao, K.T. Ingram, G. Arida and F. Penning de Vries (1989). Stem Borer Damage and Grain Yield of Flooded Rice. *J. Pl. Prot. Tropics* 6(3): 205-211.

Salac, M. and P. Cornelio (1990). *Consultants Report of the Indonesian National IPM Program Pilot Extension Season: Farmers' Field School Yield Samples.* Jakarta: National Program for the Development and Implementation of IPM in Rice-based Cropping Systems.

Santoso, Sri Budi (1992). *Proses Difusi dan Rasionalitas Petani dalam Adopsi dan Implementasi Inovasi (Kasus Penggunaan Pupuk Buatan oleh Petani di Desa Jeketro, Kecamatan Gubug, Kabupaten Grobogan, Jawa Tengah).* Skripsi Sarjana, Fakultas Sastra. Yogyakarta: UGM. 129 pp.

Sayogyo (1987). Rural Technologies and Institutions in Indonesia, *Akademika* 30 (Jan. '87): 109-117.

Scott, James C. (1989). *Moral Ekonomi Petani, Pergolakan dan Subsistensi di Asia Tenggara.* Jakarta: LP3ES.

Scriven, M. (1980). *The Logic of Evaluation.* Inverness (CA): Edgepress.

Setti, Lou (1990). *Consultant's Report.* Programme for Training and Development of Integrated Pest Management in Rice-based Cropping Systems. Jakarta: Food and Agriculture Organization of the United Nations.

Shadish, William R., Thomas D. Cook and Laura C. Leviton (1991). *Foundations of Program Evaluation. Theories and Practice.* Newbury Park: Sage. 529 pp.

Shaner, W.W., P.F. Philipp and W.R. Schmehl (1982). *Farming Systems research and development. Guidelines for Developing Countries.* Boulder, Colorado: Westview Press. 414 pp.

Siwi, Sri Suharni, M. Machmud, and Measti Mardiharini, Eds. (1990). *Indonesian Women in Rice Farming System.* Proceedings of the First National Workshop ons Women in Rice Farming System in Indonesia. Agency for Agricultural Research and Development, Central Research Institute for Food Crops and International Rice Research Institute.

Smith, J., J.A. Litsinger, J.P. Bandong, M.D. Lumaban and C.G. Dela Cruz (1989). Economic Thresholds for Insecticide Appplication to Rice: Profitability and Risk Analysis to Filipino Farmers. *J. Pl. Prot. Tropics* 6(1): 67-87.

Smith, Louis (1974). *An Aesthetic Education Workshop for Administrators: Some Implications for a Theory of Case Studies.* Paper presented at AERA. Chicago.

Soedjono, Ibnoe (1986). Koperasi Sebagai Alat Kebijaksanaan Pemerintah dan Perkembangannya. *Prisma* (15)7: 3-11.

Soejitno, J. (1991). *Insecticide Resistance in Indonesia.* Summaries of workshop on Integrated Pest Management and Insecticide Resistance Management (IPM/IRM) in Legume Crops in Asia, Chiang Mai, Thailand. Patancheru, Andhra Pradesh: ICRISAT.

Spradley, James P. (1980). *Participant Observation.* New York: Holt, Rinehart and Winston. 195 pp.

Sriskandarajah, N., R.J. Bawden and R.G. Packham (1989). *Systems Agriculture – A Paradigm for Sustainability.* Paper presented at Ninth Annual Farming Systems Research/Extension Symposium, University of Arkansas, Fayetteville, Oct. 9-11, 1989. 18 pp.

Stake, Robert E. (1987). The Case Study Method in Social Inquiry. *Educational Researcher* 7(2): 5-8.

Stern, Vernon M., Ray F. Smith, Robert van den Bosch and Kenneth S. Hagen (1959). The Integrated Control Concept. *Hilgardia* 29(2): 81-101.

Sudaryanto, Tahlim, Waluyo, A. Suparman Hadimuslihat (1988). *Dampak Program Pembangunan Pertanian Terhadap Tenaga Kerja, Peningkatan Pendapatan dan Kesejahteraan Masyarakat di Pedesaan Jawa Tengah – Struktur Produksi, Kesempatan Kerja dan Pendapatan.* Jakarta: Pusat Penelitian Agro Ekonomi, adan Penelitian dan Pengembangan Pertanian, Departemen Pertanian.

Sumayao, Blanda R. (1986). Training the Village Non-traditional Extension Audiences on Integrated Pest Management. In: *Filipino Women in Rice Farming Systems*. U.P. at Los Baños, IRRI and the Philippines, Institute for Development Studies, p. 211-216.

Suwandi, Ima (1986). Peran Pemerintah dalam Pengembangan Koperasi Unit Desa. *Prisma* (XV)7: 72-85.

Tait, Joyce and Banpot Napompeth (1987). *Management of Pests and Pesticides. Farmers' Perceptions and Practices*. Boulder: Westview Press.

Teng, P.S., Ed. (1987). *Crop Loss Assessment and Pest Management*. St. Paul, Minnesota: The American Phytopathological Society.

Thurston, H. David (1992). *Sustainable Practices for Plant Disease Management in Traditional Farming Systems*. Boulder: Westview Press. 279 pp.

Tjondronegoro, Sediono M.P. (1984). *Social Organization and Planned Development in Rural Java*. Institute of Southeast Asian Studies. Singapore Oxford University Press, 326 pp.

Van de Fliert, Elske and Patricia C. Matteson (1989). Integrated Pest Control Channels for Extension in Sri Lanka. *Journal of Extension Systems* 5: 33-47.

Van de Fliert, Elske and Patricia C. Matteson (1990). Rice Integrated Pest Control Training Needs Identified through a Farmer Survey in Sri Lanka. *J. Pl. Prot. Tropics* 7(1): 15-26.

Van de Fliert, Elske, Karel van Elsen and F. Nangsir Soenanto (1993). Integrated Rat Management: A Community Activity. Results of a pilot programme in Indonesia. *FAO Plant Prot. Bull.* 41(3) (in press).

Van de Fliert, Elske, John Pontius and Niels Röling (in prep.). *Training on Integrated Pest Management: Four Models Compared.*

Van den Bosch, Robert (1980). *The Pesticide Conspiracy*. Garden City, New York: Anchor Press/ Doubleday. 211 pp.

Waibel, H. (1986). *The Economics of Integrated Pest Control in Irrigated Rice*. Berlin: Springer-Verlag.

Waibel, Hermann (1989). *Pesticide Subsidies and the Diffusion of IPC in Rice in SE-Asia. The Case of Thailand*. Paper presented at the 14th meeting of the FAO Panel of Experts on Integrated Pest Control, Oct. 10-13. Rome: FAO. 24 pp.

Ward, William B. (1985). *Science and Rice in Indonesia*. A.I.D. Science and technology in Development Series. Boston: Oelgeschlager, Gunn & Hain. 135 pp.

Warren, D.M., L.J Slikkeveer and D. Brokensha, Eds. (1991). *Indigenous Knowledge Systems: The Cultural Dimension of Development*. London: Kegan Paul International.

Weiss, C.H. (1972). *Evaluation Research: Methods for Assessing Programme Effectiveness*. Englewood Cliffs (New Jersey): Prentice-Hall. 160 pp.

Weiss, C.H. (1983). *The Stakeholder Approach to Evaluation: Origins and Promise*. In: Bryk (op. cit.), p. 3-14.

Wells, B.D. and T.H. Johnston (1970). Differential Response of Rice Varieties to Timing of Mid Season Nitrogen Applications. *Agron. J.* 62: 606-612.

White, Benjamin and Gunawan Wiradi (1989). *Agrarian and Nonagrarian Bases of Inequality in Nine Javanese Villages*. In: Hart, Gillian, Andrew Turton and Benjamin White (op. cit.), p. 266-302.

Whitten, M.J., L.R. Brownhall, K.G. Eveleens, W. Heneveld, M.A.R. Khan, S. Li, Z. Mihyu, K. Sogawa and S.M. Yassin (1990). *Mid-term Review of FAO Inter-country Programme for the Development and Application of Integrated Pest Control in Rice in South and South-east Asia, Phase II*. Mission report, Nov. 1990. FAO Inter-country Programme, P.O. Box 1864, Manila, Philippines. 141 pp.

Widodo, Sri (1989). *Production Efficiency of Rice Farmers in Java, Indonesia*. Yogyakarta: Gadjah Mada University Press. 226 pp.

Wijeratne, Mahinda (1989). *Farmer, Extension and Research in Sri Lanka: An Empirical Study of the Agricultural Knowledge System with Special Reference to Rice Production in Matara District*. Wageningen: PUDOC. 254 pp.

Winarto, Yunita T. (1992). *Farmers' Agroecological Knowledge Construction: The Case of Integrated Pest Management among Rice Farmers in the North Coast of West Java*. Paper presented at IIEDS/ IDS Beyond Farmer First: Rural People's Knowledge Agricultural Research and Extension Practice Workshop. IDS, Univ. of Sussex, 27-29 Oct, 1992.

Winasis, Nastiti Tri (1992). *Adopsi Inovasi Program Pengendalian Hama Terpadu. Studi Kasus Sekolah Lapangan PHT di Desa Kunjeng, Kecamatan Gubug, Kabupaten Grobogan, Jawa Tengah*. Yogyakarta: Skripsi Fakultas Sastra (Antropologi), Universitas Gadjah Mada. 140 pp.

World Commission on Environment and Development (1987). *Our Common Future*. Oxford: Oxford University Press.

World Education (1992). Bagaimana Membuat Percobaan di Lahan Petani? *Si Semut Warta Program Penanggulangan Masalah Pestisida,* 4:16-18. WE, Jl. Tebet Dalam IV F/75, Jakarta 12810.

Yin, Robert K. (1981a). The Case Study as a Serious Strategy. *Knowledge: Creation, Diffusion, Utilization* 3(1): 97-114.

Yin, Robert K. (1981b). The Case Study Crisis: Some Answers. *Admin. Science Q.* 26 (March): 58-66.

Yin, Robert K. (1984). *Case Study Reserach. Design and Methods*. Applied Social Research Methods Series, Vol. 5. Beverly Hills: Sage. 159 pp.

Zadoks, J.C. (1985). On the Conceptual Basis of Crop Loss Assessment: The Threshold Theory. *Ann. Rev. Phytopathol.* (1985)23: 455-73.

Zadoks, J.C. (1989a). Does IPM give adequate attention to disease control, in particular at the farmer level? *FAO Plant Prot. Bull.* 37(4): 144-150.

Zadoks, J.C. (1989b). EPIPRE: a Computer-based Decision Support System for Pest and Disease Control in Wheat: Its Development and Implementation in Europe. *Plant Disease Epidemiology* 2: 3-39.

Zadoks, J.C. (1991). A Hundred and More Years of Plant Protection in The Netherlands. *Neth. J. Pl. Path.* 97:3-24.

Zadoks, J.C. and R.D. Schein (1979). *Epidemiology and Plant Disease Management*. New York: Oxford University Press. 427 pp.

Zelazny, B., L. Charappa and P. Kenmore (1985). Integrated Pest Control in Developing Countries. *FAO Plant Prot. Bull.* (33)4: 147-158.

About the author

Elske Johanna van de Fliert was born on 14 December 1960 in Ede, The Netherlands. In September 1979, she started her Biology studies at the State University Utrecht where she graduated in April 1987. Major subjects were Extension Science and Plant Ecology, and minor subjects were Entomology and Didactics.

Throughout her studies, three different interests prevailed: (1) working in developing countries, (2) sustainable agriculture, particularly (Integrated) Pest Management, and (3) education and extension. These three interests combined well during her M.Sc. research which she conducted in 1985-86 within the FAO Inter-country IPM Programme in Sri Lanka. Out of this experience resulted the assignment as an associate expert to the Indonesian branch of the IPM programme, where she worked from December 1988 to June 1992, leading to the doctoral research reported in this book.

Appendix I

Organisational structure of the Ministry of Agriculture, Indonesia

(as far as relevant within the context of the study)

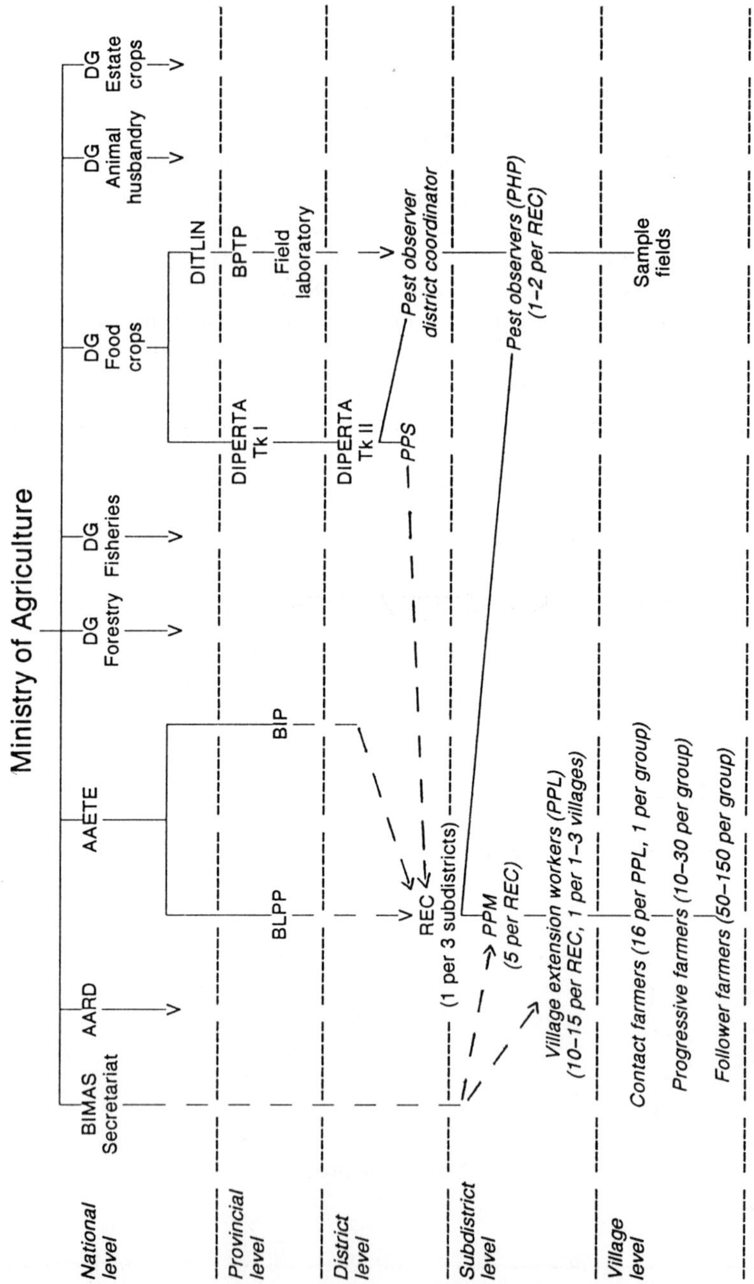

For abbreviations see list of terms. Institutions are given in roman font, persons in italic font.

Appendix II

Maps

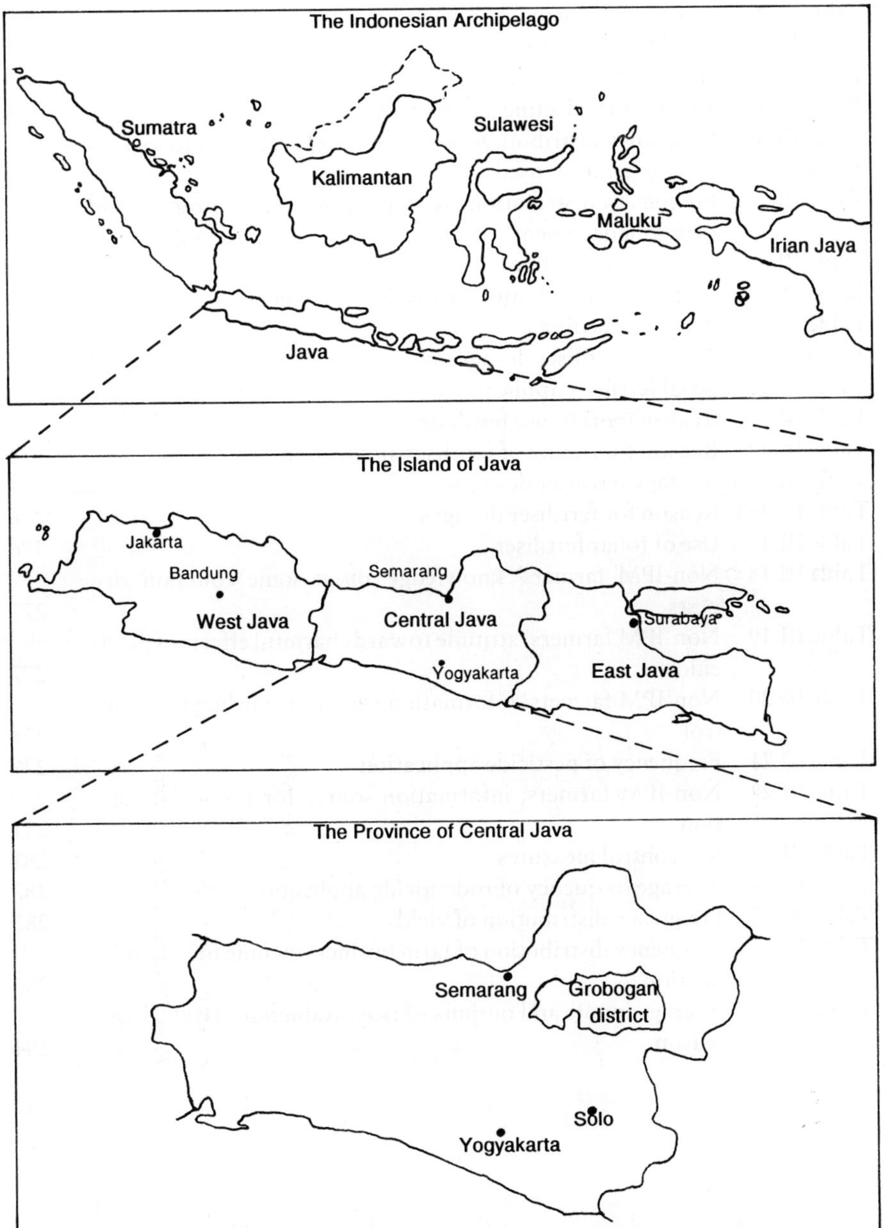

Appendix III

Tables belonging to Chapter 5

Table III.1: Rice pest occurrence and level of damage as perceived by non-IPM farmers, as % farmers reporting.

		Seneng-sari	Sugih-sari	Jaya-sari	Plosok-sari	Kudu-agung	Sumber-agung	Mulyo-agung	Lempung-agung	Total
		1989/90 wet season								
rats:	light	48%	44%	55%	29%	31%	40%	22%	55%	41%
	moderate	14%	19%	14%	28%	16%	24%	13%	11%	17%
	heavy	33%	10%	22%	13%	9%	31%	7%	5%	16%
stemborer:	light	12%	17%	24%	22%	19%	19%	6%	25%	18%
	moderate	3%	3%	29%	19%	19%	10%	4%	7%	12%
	heavy	0%	0%	16%	22%	10%	8%	1%	1%	8%
seed bug:	light	21%	17%	38%	13%	29%	29%	16%	25%	24%
	moderate	3%	2%	14%	5%	6%	8%	1%	1%	5%
	heavy	2%	0%	26%	2%	19%	5%	4%	4%	8%
gall midge		3%	0%	1%	0%	3%	3%	2%	0%	2%
caterpillars		17%	16%	16%	29%	19%	15%	20%	36%	22%
brown planthopper		0%	2%	1%	1%	4%	3%	1%	0%	2%
leaf diseases		2%	0%	2%	32%	0%	5%	6%	6%	7%
other [1)]		8%	14%	18%	9%	41%	35%	31%	14%	21%
N		*66*	*63*	*85*	*85*	*68*	*62*	*85*	*85*	*599*
		1990 dry season								
rats:	light	18%		52%		22%		31%		32%
	moderate	45%		27%		33%		31%		33%
	heavy	37%		21%		45%		29%		32%
stemborer:	light	8%		14%		9%		8%		10%
	moderate	2%		5%		3%		1%		3%
	heavy	2%		1%		1%		1%		1%
seed bug:	light	11%		34%		30%		24%		25%
	moderate	5%		13%		10%		5%		8%
	heavy	2%		11%		1%		1%		4%
gall midge		0%		1%		1%		0%		1%
caterpillars		10%		11%		22%		18%		15%
brown planthopper		0%		0%		4%		1%		1%
leaf diseases		39%		42%		46%		25%		37%
other [1]		6%		19%		18%		15%		15%
N		*62*		*85*		*67*		*85*		*299*

[1)] 'Other' includes mole cricket, grasshopper, other hoppers, crabs, unidentified pests.
Totals exceed 100% due to multiple response.

Table III.1 ctd.

	Seneng-sari	Sugih-sari	Jaya-sari	Plosok-sari	Kudu-agung	Sumber-agung	Mulyo-agung	Lempung-agung	Total
1990/91 wet season									
no pest problems	0%	2%	0%	1%	0%	5%	22%	6%	5%
rats: light	58%	42%	43%	30%	16%	26%	18%	13%	30%
moderate	7%	2%	5%	1%	2%	2%	1%	4%	3%
heavy	5%	2%	1%	7%	3%	2%	0%	0%	2%
deadhearts: light	2%	2%	3%	1%	5%	0%	9%	1%	3%
moderate	0%	0%	3%	0%	0%	0%	5%	0%	1%
heavy	0%	0%	0%	0%	0%	2%	4%	1%	1%
whiteheads: light	17%	25%	21%	7%	11%	16%	1%	13%	13%
moderate	2%	7%	5%	1%	7%	5%	1%	3%	4%
heavy	10%	9%	4%	5%	0%	5%	0%	8%	5%
seed bug: light	17%	29%	16%	11%	3%	14%	17%	8%	14%
moderate	22%	35%	39%	19%	7%	10%	9%	14%	19%
heavy	58%	25%	31%	57%	62%	31%	7%	35%	37%
gall midge	2%	5%	0%	0%	0%	0%	0%	3%	1%
caterpillars	2%	5%	0%	0%	0%	0%	0%	1%	1%
brown planthopper	8%	11%	0%	1%	0%	0%	0%	1%	2%
leaf diseases	24%	36%	31%	12%	13%	12%	10%	35%	22%
other[1)]	3%	7%	5%	0%	5%	5%	4%	5%	4%
N	*59*	*55*	*80*	*74*	*61*	*58*	*82*	*80*	*549*
1991 dry season									
no pest problems	15%		21%		20%		20%		19%
rats: light	40%		57%		52%		57%		52%
moderate	20%		8%		8%		8%		11%
heavy	5%		6%		12%		9%		8%
deadhearts: light	5%		7%		5%		8%		6%
moderate	4%		0%		0%		1%		1%
heavy	0%		1%		0%		4%		2%
whiteheads: light	9%		10%		0%		0%		5%
moderate	0%		0%		0%		1%		0%
heavy	0%		0%		0%		0%		0%
seed bug: light	5%		19%		13%		5%		11%
moderate	2%		1%		0%		1%		1%
heavy	0%		0%		2%		0%		0%
gall midge	2%		3%		3%		0%		2%
caterpillars	0%		6%		2%		5%		3%
brown planthopper	0%		1%		3%		1%		2%
leaf diseases	11%		7%		18%		21%		14%
other pests	5%		4%		10%		11%		8%
water shortage	20%		4%		0%		1%		6%
flood	0%		0%		0%		1%		0%
N	*55*		*72*		*60*		*76*		*263*

1) 'Other' includes mole cricket, grasshopper, other hoppers, crabs, unidentified pests.
Totals exceed 100% due to multiple response.

Table III.2: Quality of water supply as perceived by non-IPM farmers, as % farmers reporting.

		Seneng-sari	Sugih-sari	Jaya-sari	Plosok-sari	Kudu-agung	Sumber-agung	Mulyo-agung	Lempung-agung	Total
Irrigation:	sufficient	82%	94%	98%	100%	93%	50%	100%	87%	89%
	insufficient	18%	5%	4%	0%	4%	6%	0%	1%	4%
Rainfed:	sufficient	5%	0%	0%	0%	0%	3%	0%	8%	2%
	insufficient	0%	2%	0%	0%	4%	44%	0%	9%	7%
N		*66*	*63*	*85*	*85*	*68*	*62*	*85*	*85*	*599*

Totals exceed 100% due to multiple response by farmers operating under more than one water supply system.

Table III.3: Rice variety planted by non-IPM farmers, as % farmers reporting.

	Seneng-sari	Sugih-sari	Jaya-sari	Plosok-sari	Kudu-agung	Sumber-agung	Mulyo-agung	Lempung-agung	Total
	1989/90 wet season								
IR 64	89%	89%	98%	32%	76%	48%	19%	71%	64%
Semeru	14%	10%	5%	72%	43%	68%	78%	49%	43%
Krueng Aceh	3%	0%	0%	0%	1%	5%	18%	12%	5%
Cisedane	0%	5%	0%	0%	0%	0%	0%	0%	1%
other[1)]	5%	2%	2%	5%	1%	2%	1%	0%	2%
N	*66*	*63*	*85*	*85*	*68*	*62*	*85*	*85*	*599*
	1990 dry season								
IR 64	94%		89%		79%		12%		66%
Semeru	10%		13%		22%		53%		26%
Krueng Aceh	3%		1%		6%		58%		19%
Cisedane	0%		1%		0%		0%		0%
other[1)]	3%		4%		0%		1%		2%
N	*62*		*85*		*67*		*85*		*299*
	1990/91 wet season								
IR 64	98%	95%	96%	58%	92%	71%	68%	85%	82%
Semeru	0%	0%	4%	66%	16%	47%	45%	24%	26%
Krueng Aceh	2%	0%	0%	0%	3%	0%	13%	1%	3%
Cisedane	3%	5%	4%	1%	0%	2%	0%	3%	2%
Wai Siputih	0%	4%	9%	0%	0%	2%	0%	0%	2%
other[1)]	3%	0%	1%	4%	0%	3%	2%	0%	2%
N	*59*	*55*	*80*	*74*	*61*	*58*	*82*	*80*	*549*
	1991 dry season								
IR 64	96%		99%		92%		84%		92%
Semeru	0%		0%		13%		9%		6%
Krueng Aceh	13%		0%		3%		12%		7%
Cisedane	0%		1%		0%		0%		0%
Wai Siputih	0%		4%		2%		0%		2%
other[1)]	4%		3%		5%		5%		4%
N	*55*		*72*		*60*		*76*		*263*

[1)] 'Other' includes IR 36, Cisanggarung, glutinous (traditional) rice.
Totals exceed 100% due to multiple response.

Table III.4: Reason for selecting a rice variety, as % non-IPM farmers reporting.

	1989/90 wet season	1990 dry season	1990/91 wet season	1991 dry season	especially for variety
high yield	58%	31%	34%	34%	Semeru, IR 64
quality of rice	44%	35%	40%	44%	IR 64
pest tolerance	42%	35%	28%	28%	IR 64, Semeru
uniform planting	24%	8%	15%	20%	–
good marketing	11%	12%	11%	12%	–
suitability for area/season	7%	6%	5%	4%	–
growth duration	4%	5%	0%	2%	–
other[1)]	9%	16%	19%	21%	–
N	*599*	*299*	*549*	*263*	

[1)] 'Other' includes low fertiliser demand, availability, experimentation, good storage.
Total exceed 100% due to multiple response.

Table III.5: Frequency distribution of seedbed preparation practices by non-IPM farmers in the 1991 dry season, as % farmers reporting.

	Senengsari	Jayasari	Kuduagung	Mulyoagung	Total
Size of seedbed (m^2/ha field)					
<300	49%	79%	68%	58%	64%
300-500 (recommendation)	24%	18%	18%	21%	20%
>500	27%	3%	13%	21%	16%
Quantity of seedbed (kg/m^2)					
<0.07	5%	3%	3%	8%	5%
0.07-0.15 (recommendation)	29%	18%	15%	22%	21%
0.15-0.5	45%	47%	53%	49%	49%
>0.5	20%	32%	28%	21%	25%
Quantity of seed/field (kg/ha)					
<35	2%	25%	0%	5%	9%
35-45 (recommendation)	5%	31%	7%	8%	13%
46-80	64%	40%	63%	61%	56%
>80	29%	4%	30%	26%	22%
N	*55*	*72*	*60*	*76*	*263*

Table III.6: Seed generation used by non-IPM farmers in wet (1990/92) and dry (1991) season, as % farmers reporting.

	Seneng-sari	Sugih-sari	Jaya-sari	Plosok-sari	Kudu-agung	Sumber-agung	Mulyo-agung	Lempung-agung	Total
	1990/91 wet season								
certified seed	12%	7%	13%	12%	2%	10%	1%	8%	8%
1st progeny	5%	20%	24%	22%	28%	28%	38%	21%	24%
2nd progeny	25%	33%	46%	35%	49%	52%	65%	43%	44%
3rd progeny	24%	29%	19%	22%	16%	24%	13%	28%	21%
later progenies	41%	15%	13%	39%	16%	10%	12%	13%	19%
N	*59*	*55*	*80*	*74*	*61*	*58*	*82*	*80*	*549*
	1991 dry season								
certified seed	11%		0%		3%		4%		4%
1st progeny	13%		64%		13%		38%		34%
2nd progeny	11%		25%		33%		26%		24%
3rd progeny	78%		13%		22%		29%		33%
later progenies	0%		13%		43%		12%		17%
N	*55*		*72*		*60*		*76*		*263*

Totals exceed 100% due to multiple response.

Table III.7: Frequency distribution of time spent on monitoring and water management by non-IPM farmers as classified by farm size.

	<0.1 ha	0.1-0,24 ha	0.25-0.49 ha	0.50-0.99 ha	≥1.00 ha	Total
in hours/week/farmer:						
≤7	86%	76%	69%	60%	55%	67%
8-14	10%	17%	22%	28%	23%	22%
15-21	0%	4%	6%	3%	9%	5%
21-28	5%	1%	2%	6%	9%	4%
28-35	0%	0%	0%	1%	1%	1%
>35	0%	1%	0%	2%	3%	1%
in hours/week/ha:						
≤7	19%	17%	29%	37%	64%	32%
8-14	5%	21%	28%	35%	25%	27%
15-21	0%	6%	12%	13%	8%	10%
21-28	5%	11%	10%	4%	4%	8%
28-35	24%	10%	6%	4%	0%	6%
>35	48%	35%	15%	8%	0%	17%
N	*21*	*140*	*202*	*156*	*77*	*596*

Table III.8: Method of weeding by non-IPM farmers, as % farmers reporting.

	Seneng-sari	Sugih-sari	Jaya-sari	Plosok-sari	Kudu-agung	Sumber-agung	Mulyo-agung	Lempung-agung	Total
Handweeding	100%	98%	98%	100%	93%	63%	97%	89%	93%
Mechanical weeder	0%	5%	2%	0%	7%	37%	4%	11%	7%
N	*66*	*63*	*85*	*85*	*68*	*62*	*85*	*85*	*599*

Totals might exceed 100% due to multiple response.

Table III.9: Frequency distribution of weeding frequency by non-IPM farmers in (1989/90) wet and (1990) dry season, as % farmers reporting.

	wet season								
	Seneng-sari	Sugih-sari	Jaya-sari	Plosok-sari	Kudu-agung	Sumber-agung	Mulyo-agung	Lempung-agung	Total
no weeding	0%	0%	0%	0%	0%	0%	0%	0%	0%
1 time/season	32%	22%	7%	55%	28%	31%	20%	28%	28%
2 times	64%	75%	85%	41%	71%	50%	75%	53%	64%
3 times	3%	3%	7%	4%	1%	18%	2%	13%	6%
>3 times	2%	0%	1%	0%	0%	2%	2%	6%	2%
N	*66*	*63*	*85*	*85*	*68*	*62*	*85*	*85*	*599*
	dry season								
no weeding	0%		1%		1%		0%		1%
1 time/season	11%		6%		25%		29%		18%
2 times	77%		76%		64%		61%		70%
3 times	8%		15%		7%		8%		10%
>3 times	3%		1%		1%		1%		2%
N	*62*		*85*		*67*		*85*		*299*

Table III.10: Time of weeding by non-IPM farmers, 1989/90 wet season.

a. Average time of weeding (in DAT).

	Seneng-sari	Sugih-sari	Jaya-sari	Plosok-sari	Kudu-agung	Sumber-agung	Mulyo-agung	Lempung-agung	Total
1st weeding:									
average	23	22	19	22	23	22	18	19	21
st. deviation	7	8	5	6	11	9	5	6	7
N	*66*	*63*	*85*	*85*	*68*	*62*	*85*	*85*	*599*
2nd weeding:									
average	40	36	36	38	36	40	30	32	35
st. deviation	13	9	7	10	11	10	6	6	10
N	*45*	*49*	*79*	*37*	*48*	*46*	*66*	*59*	*429*

b. Frequency distribution, as % farmers reporting.

	Seneng-sari	Sugih-sari	Jaya-sari	Plosok-sari	Kudu-agung	Sumber-agung	Mulyo-agung	Lempung-agung	Total
1st weeding:									
≤15 DAT	24%	32%	42%	26%	31%	32%	51%	47%	36%
16-30 DAT	59%	57%	53%	67%	60%	53%	47%	48%	55%
>30 DAT	17%	11%	5%	7%	9%	15%	2%	5%	8%
N	*66*	*63*	*85*	*85*	*68*	*62*	*85*	*85*	*599*
2nd weeding:									
≤30 DAT	24%	35%	25%	32%	42%	20%	71%	47%	38%
31-45 DAT	51%	59%	70%	46%	46%	57%	27%	49%	51%
>45 DAT	24%	6%	5%	22%	13%	24%	2%	3%	11%
N	*45*	*49*	*79*	*37*	*48*	*46*	*66*	*59*	*429*

Table III.11: Average fertiliser dosage on seedbed applied by non-IPM farmers (users only), 1991 dry season.

	in kg/ha rice field		in kg/ha seedbed	
	urea	TSP	urea	TSP
average	19	15	1,167	1,054
standard deviation	13	11	1,400	1,356
N	*258*	*211*	*258*	*211*

Table III.12: Basal fertiliser application by non-IPM farmers, as % farmers reporting.

a. Practice of basal fertilisation

	Seneng-sari	Jaya-sari	Kudu-agung	Mulyo-agung	Total
Never practised	42%	61%	42%	42%	47%
Ever tried out	49%	26%	52%	43%	42%
Often practise	9%	13%	7%	14%	11%
N	*55*	*72*	*60*	*76*	*263*

b. Reason for (not) applying basal fertilisation

	Seneng-sari	Jaya-sari	Kudu-agung	Mulyo-agung	Total
Reason for applying:					
fast growth, better yield	7%	6%	7%	16%	9%
to obtain better soil structure	0%	0%	0%	4%	1%
advice from fellow farmers	0%	0%	0%	4%	1%
other[1)]	4%	10%	0%	5%	5%
Reason for not applying:					
uneconomical re. expenses	15%	19%	18%	22%	19%
uneconomical re. labour	15%	14%	20%	5%	13%
not common practice	25%	21%	13%	12%	17%
no additional result	13%	11%	17%	21%	16%
useless[2)]	15%	6%	32%	5%	13%
shortage of money	11%	31%	5%	0%	12%
shortage of time	0%	6%	0%	7%	3%
other[3)]	24%	6%	10%	12%	12%
N	*55*	*72*	*60*	*76*	*263*

1) 'Other' includes advice of extension worker or fellow farmers, and experience.

2) 'Useless' because fertiliser is washed away by rain, trampled into the soil, or sticks on the feet of planters.

3) 'Other' includes irritation on feet of planter, not suitable with local conditions, to discourage fast growth of seedlings to discourage fast growth of seedlings, to discourage fast weed growth.

Totals exceed 100% due to multiple response.

Table III.13: Reason (not) to use fertiliser, as % non-IPM farmers reporting (N=263).

Reasons to use urea:	for smooth leaf growth	65%
	for green colour of leaves	37%
	for good tillering	33%
	other	5%
Reasons to use TSP:	for heavy grains	66%
	for long-lasting green colour	35%
	to prevent lodging	14%
	for better soil structure	8%
	for stronger roots	5%
	for good tillering	3%
	other	5%
Reasons to use KCl:	for heavy grains	11%
	to prevent lodging	5%
	for higher pest resistance	1%
	other	4%
Reasons not to use KCl:	not convinced of result	31%
	is not necessary, regarding soil fertility, or use TSP	29%
	shortage of money	24%
	is not a common practice	5%
	is not available in the market	3%
	other	2%
Reasons to use ZA:	for smooth leave growth	1%
	other	1%
Reasons not to use ZA:	not convinced of result	51%
	is not necessary regarding soil fertility or use of urea	27%
	shortage of money	22%
	is not a common practice	8%
	is not available in the market	2%
	other	2%

Totals exceed 100% due to multiple response

Table III.14: Reason by non-IPM farmers for timing of fertiliser application, as % farmers reporting.

	Seneng-sari	Jaya-sari	Kudu-agung	Mulyo-agung	Total
1st application:					
plant/roots just start to live	16%	29%	60%	36%	35%
to encourage tillering	31%	17%	12%	29%	22%
to encourage fast growth/green colour	25%	24%	2%	24%	19%
experience/common practice	4%	6%	3%	4%	4%
after weeding	5%	4%	0%	3%	3%
other[1]	5%	19%	17%	14%	14%
do not know/no response	31%	19%	7%	8%	16%
2nd application:					
to encourage panicle initiation	16%	28%	25%	34%	27%
right moment for replenishment	15%	21%	23%	41%	26%
after weeding	22%	22%	15%	11%	17%
to encourage tillering	7%	7%	2%	7%	6%
experience/common practice	4%	3%	18%	4%	7%
water shortage	24%	0%	0%	3%	6%
other[1]	7%	32%	15%	9%	16%
do not know/no response	18%	8%	13%	7%	11%
3rd application:					
to encourage panicle initiation	13%	26%	8%	8%	14%
for uniform growth	7%	13%	0%	3%	6%
right moment for replenishment	9%	13%	3%	3%	7%
after weeding	9%	3%	3%	0%	3%
other[1]	11%	6%	2%	4%	5%
do not know/no response	62%	54%	87%	84%	72%
N	*55*	*72*	*60*	*76*	*263*

[1] 'Other' includes to facilitate weeding, advice from fellow farmers, shortage of money.
Totals exceed 100% due to multiple response.

Table III.15: Average fertiliser dosages (in kg/ha) by non-IPM farmers classified by village, over four season.

		1989/90 wet season								
		Seneng-sari	Sugih-sari	Jaya-sari	Plosok-sari	Kudu-agung	Sumber-agung	Mulyo-agung	Lempung-agung	Total
urea	average	312	249	232	229	262	264	286	244	258
	st. dev.	156	123	119	90	108	148	102	118	123
	N	*66*	*63*	*85*	*83*	*67*	*62*	*85*	*85*	*596*
TSP	average	271	197	199	200	215	230	250	209	221
	st. dev.	141	94	91	96	96	116	98	101	107
	N	*66*	*63*	*85*	*83*	*67*	*62*	*85*	*85*	*596*
KCl	average	122	135	111	90	97	85	106	80	109
(users	st. dev.	59	73	58	53	47	40	73	57	60
only)	*N*	*20*	*20*	*21*	*6*	*21*	*16*	*6*	*3*	*113*
ZA	average	114	129	117	77	76	81	52	55	95
(users	st. dev.	71	59	79	64	48	34	12	52	62
only)	*N*	*9*	*9*	*6*	*6*	*11*	*11*	*2*	*2*	*56*

Table III.15 ctd.

		Seneng-sari	Sugih-sari	Jaya-sari	Plosok-sari	Kudu-agung	Sumber-agung	Mulyo-agung	Lempung-agung	Total
		1990 dry season								
urea	average	319		239		257		278		271
	st. dev.	136		116		93		93		114
	N	*62*		*85*		*67*		*85*		*299*
TSP	average	246		195		216		254		227
	st. dev.	110		85		88		107		101
	N	*62*		*85*		*67*		*85*		*299*
KCl (users only)	average	123		109		87		123		111
	st. dev.	64		57		47		58		59
	N	*21*		*27*		*11*		*7*		*66*
ZA (users only)	average	61		94		55		109		79
	st. dev.	25		81		19		72		59
	N	*5*		*4*		*4*		*4*		*17*
		1990/91 wet season								
urea	average	328	321	255	248	282	272	306	337	293
	st. dev.	124	140	189	92	107	102	139	173	144
	N	*59*	*55*	*80*	*74*	*61*	*58*	*82*	*80*	*549*
TSP	average	264	261	230	198	231	226	246	288	243
	st. dev.	130	129	192	97	101	107	146	156	141
	N	*59*	*55*	*80*	*74*	*61*	*58*	*82*	*80*	*549*
KCl (users only)	average	127	133	77	107	83	100	106	115	109
	st. dev.	56	98	58	73	45	61	55	134	76
	N	*20*	*18*	*9*	*9*	*15*	*15*	*11*	*7*	*104*
ZA (users only)	average	–	93	44	70	84	113	61	163	100
	st. dev.	–	52	0	26	40	57	45	83	62
	N	*0*	*6*	*1*	*3*	*9*	*7*	*2*	*5*	*33*
		1991 dry season								
urea	average	345		270		302		309		304
	st. dev.	180		106		135		139		142
	N	*55*		*72*		*60*		*76*		*263*
TSP	average	286		222		241		250		248
	st. dev.	130		110		117		153		131
	N	*55*		*72*		*60*		*76*		*263*
KCl (users only)	average	99		107		68		115		96
	st. dev.	51		76		31		67		59
	N	*15*		*7*		*12*		*12*		*46*
ZA (users only)	average	–		23		89		–		67
	st. dev.	–		0		56		–		56
	N	*0*		*1*		*2*		*0*		*3*

Table III.16: Reason for fertiliser dosages classified by dosage (kg/ha) applied, as % non-IPM farmers reporting.

	urea				TSP			
	< 200	200-299	≥ 300	Total	< 100	100-199	≥ 200	Total
experience/trial	34%	39%	34%	36%	19%	32%	35%	32%
considering conditions	9%	11%	16%	13%	15%	10%	6%	8%
shortage of money	16%	6%	5%	8%	11%	3%	4%	5%
common practice	0%	7%	2%	3%	4%	4%	3%	4%
advice of ext. worker	3%	0%	1%	1%	0%	0%	0%	0%
advice of fellow farmers	0%	0%	1%	0%	0%	1%	0%	0%
for heavy grains	–	–	–	–	4%	10%	15%	12%
as complement only	–	–	–	–	11%	5%	0%	3%
other	8%	7%	11%	9%	4%	0%	1%	1%
do not know	31%	36%	36%	35%	48%	35%	37%	37%
N	*64*	*107*	*91*	*262*	*27*	*91*	*145*	*263*

Totals exceed 100% due to multiple response.

Table III.17: Use of foliar fertiliser by non-IPM farmers over four seasons, as % farmers reporting.

	Seneng-sari	Sugih-sari	Jaya-sari	Plosok-sari	Kudu-agung	Sumber-agung	Mulyo-agung	Lempung-agung	Total
1989/90	0%	0%	6%	0%	1%	8%	0%	1%	2%
N	*66*	*63*	*85*	*85*	*68*	*62*	*85*	*85*	*599*
1990	3%		15%		9%		1%		7%
N	*62*		*85*		*67*		*85*		*299*
1990/91	0%	4%	10%	1%	13%	28%	2%	3%	6%
N	*59*	*55*	*80*	*74*	*61*	*58*	*82*	*80*	*549*
1991	0%		10%	0%	7%		3%		5%
N	*55*		*72*		*60*		*76*		*263*

Table III.18: Non-IPM farmers' knowledge about some common rice pests, 1989/90 wet season classified by village and 1990/91 wet season total average, as % farmers reporting.

	1989/90									1990/91
	Seneng-sari	Sugih-sari	Jaya-sari	Plosok-sari	Kudu-agung	Sumber-agung	Mulyo-agung	Lempung-agung	Total	Total
Brown planthopper:										
know symptoms	73%	67%	76%	53%	91%	68%	81%	87%	75%	74%
know appearance	56%	67%	81%	54%	87%	66%	75%	87%	72%	68%
do not know	26%	32%	18%	45%	7%	26%	18%	8%	22%	21%
Stemborer deadheart:										
know symptoms	85%	76%	95%	82%	97%	92%	84%	92%	88%	81%
know appearance	35%	46%	54%	33%	19%	42%	20%	22%	34%	35%
do not know	14%	22%	5%	18%	3%	6%	16%	7%	11%	15%
Stemborer whitehead:										
know symptoms	94%	84%	99%	99%	100%	90%	95%	94%	95%	95%
know appearance	56%	65%	69%	61%	57%	48%	53%	58%	59%	63%
do not know	6%	16%	1%	1%	0%	8%	5%	6%	5%	3%
Rice seed bug:										
know symptoms	98%	95%	99%	100%	100%	100%	99%	100%	99%	97%
know appearance	98%	87%	100%	100%	96%	98%	95%	100%	97%	96%
do not know	0%	5%	0%	0%	0%	0%	1%	0%	1%	1%
Rice leaffolder:										
know symptoms	12%	8%	2%	12%	3%	2%	0%	0%	5%	10%
know appearance	8%	8%	2%	8%	0%	0%	0%	0%	3%	11%
do not know	88%	92%	98%	88%	97%	98%	100%	100%	95%	85%
Bacterial leaf blight:										
know symptoms	12%	0%	1%	0%	0%	2%	1%	0%	2%	1%
know appearance	5%	0%	0%	0%	0%	0%	0%	0%	1%	1%
do not know	88%	100%	99%	100%	100%	98%	99%	100%	98%	99%
N	*66*	*63*	*85*	*85*	*68*	*62*	*85*	*85*	*599*	*549*

Table III.19: Non-IPM farmers' attitude towards harmful effects of pesticides, as % farmers reporting.

	Seneng-sari	Sugih-sari	Jaya-sari	Plosok-sari	Kudu-agung	Sumber-agung	Mulyo-agung	Lempung-agung	Total
Effects on animals?									
yes	42%	25%	54%	20%	34%	63%	24%	38%	37%
no	58%	75%	46%	80%	66%	37%	76%	62%	63%
Effects on humans?									
yes	56%	25%	61%	21%	46%	53%	35%	46%	43%
no	44%	75%	39%	79%	54%	47%	65%	54%	57%
N	*66*	*63*	*85*	*85*	*68*	*62*	*85*	*85*	*599*

Table III.20: Non-IPM farmers' information source regarding pest control, as % farmers reporting.

	Seneng-sari	Sugih-sari	Jaya-sari	Plosok-sari	Kudu-agung	Sumber-agung	Mulyo-agung	Lempung-agung	Total
fellow farmers	56%	44%	60%	74%	51%	66%	62%	55%	59%
extension workers	11%	38%	29%	9%	25%	21%	19%	20%	21%
own experience	8%	8%	5%	15%	6%	6%	6%	5%	7%
other [1)]	26%	14%	18%	6%	16%	21%	12%	22%	17%
no response	17%	0%	1%	0%	1%	6%	1%	7%	4%
N	*66*	*63*	*85*	*85*	*68*	*62*	*85*	*85*	*599*

[1)] 'Other' includes mass media (16%), KUD employee (2%), pesticide salesman (2%), other villagers (2%).

Totals exceed 100% due to multiple response.

Table III.21: Frequency of pesticide application by non-IPM farmers over four seasons.

a. Average frequency in number of applications per season.

		Seneng-sari	Sugih-sari	Jaya-sari	Plosok-sari	Kudu-agung	Sumber-agung	Mulyo-agung	Lempung-agung	Total
		1989/90 wet season								
sprays:	average	0.8	1.2	2.9	2.0	2.2	1.7	1.1	1.6	1.7
	st. dev.	1.6	1.1	1.7	1.7	1.3	1.4	1.1	1.1	1.5
granules:	average	0.1	0.1	0.6	0.3	0.4	0.7	0.6	1.0	0.5
	st. dev.	0.3	0.4	0.7	0.6	0.8	1.0	0.9	1.0	0.8
total:	average	0.9	1.4	3.4	2.3	2.6	2.3	1.7	2.7	2.2
	st. dev.	1.7	1.1	1.9	1.9	1.4	1.6	1.4	1.8	1.8
N		*66*	*63*	*85*	*85*	*68*	*62*	*85*	*85*	*599*
		1990 dry season								
sprays:	average	0.9	1.3	2.1	1.9	2.0	1.2	1.6	1.8	1.6
	st. dev.	1.0	0.9	1.3	1.7	1.0	1.3	1.3	1.6	1.4
granules:	average	0.5	0.6	0.7	0.5	1.3	0.7	1.3	0.9	0.8
	st. dev.	0.7	0.8	0.8	0.7	1.0	1.0	1.0	1.0	0.9
total:	average	1.3	1.9	2.9	2.4	3.3	1.9	2.9	2.7	2.5
	st. dev.	1.2	1.2	1.6	2.2	1.5	1.9	1.8	2.3	1.9
N		*62*	*55*	*85*	*74*	*67*	*58*	*85*	*80*	*566*
		1990/91 wet season								
sprays:	average	0.6	1.6	1.8	2.4	2.0	1.5	0.6	1.5	1.5
	st. dev.	0.8	1.4	1.6	1.8	1.3	1.2	0.9	1.1	1.4
granules:	average	0.3	0.7	0.6	0.6	1.2	1.0	0.4	1.0	0.7
	st. dev.	0.6	0.8	0.9	0.8	1.3	1.0	0.6	1.1	1.0
total:	average	0.9	2.3	2.4	3.0	3.2	2.5	1.0	2.5	2.2
	st. dev.	1.1	1.5	2.0	2.3	2.2	1.7	1.2	1.8	2.0
N		*59*	*55*	*80*	*74*	*61*	*58*	*82*	*80*	*549*
		1991 dry season								
sprays:	average	0.3		1.2		1.2		0.4		0.8
	st. dev.	0.5		1.3		1.2		0.8		1.1
granules:	average	0.2		0.6		1.2		1.4		0.9
	st. dev.	0.5		0.9		1.0		0.9		1.0
total:	average	0.5		1.8		2.4		1.8		1.7
	st. dev.	0.7		1.6		1.9		1.4		1.6
N		*55*		*72*		*60*		*76*		*263*

Table III.21 ctd.

b. Frequency distribution of number of applications, as % farmers reporting.

	Seneng-sari	Sugih-sari	Jaya-sari	Plosok-sari	Kudu-agung	Sumber-agung	Mulyo-agung	Lempung-agung	Total
1989/90 wet season									
no applications	55%	25%	1%	20%	7%	15%	26%	15%	20%
1 application	21%	30%	7%	11%	10%	15%	20%	9%	15%
2 applications	15%	30%	28%	20%	28%	31%	26%	27%	26%
3 applications	8%	10%	27%	32%	40%	19%	18%	15%	21%
4 applications	0%	3%	16%	12%	6%	13%	7%	16%	10%
>4 applications	2%	2%	20%	6%	9%	8%	4%	16%	9%
N	*66*	*63*	*85*	*85*	*68*	*62*	*85*	*85*	*599*
1990 dry season									
no applications	35%	9%	6%	27%	3%	41%	9%	20%	18%
1 application	23%	31%	12%	9%	7%	5%	13%	11%	13%
2 applications	21%	29%	27%	16%	24%	12%	20%	25%	22%
3 applications	18%	22%	24%	20%	21%	19%	18%	16%	20%
4 applications	3%	7%	18%	15%	22%	12%	25%	9%	14%
>4 applications	0%	2%	14%	12%	22%	10%	15%	19%	13%
N	*62*	*55*	*85*	*74*	*67*	*58*	*85*	*80*	*566*
1990/91 wet season									
no applications	41%	7%	21%	15%	8%	17%	48%	19%	23%
1 application	37%	33%	13%	12%	15%	12%	26%	14%	19%
2 applications	15%	18%	24%	18%	18%	19%	16%	23%	19%
3 applications	2%	22%	14%	18%	15%	21%	4%	16%	13%
4 applications	3%	13%	11%	15%	26%	21%	6%	11%	13%
>4 applications	2%	7%	18%	23%	18%	10%	1%	18%	12%
N	*59*	*55*	*80*	*74*	*61*	*58*	*82*	*80*	*549*
1991 dry season									
no applications	64%		28%		22%		20%		32%
1 application	25%		21%		7%		21%		19%
2 applications	9%		21%		33%		38%		26%
3 applications	2%		17%		12%		8%		10%
4 applications	0%		6%		15%		9%		8%
>4 applications	0%		8%		12%		4%		6%
N	*55*		*72*		*60*		*76*		*263*

Table III.22: Non-IPM farmers' information source for pesticide selection in 1990/91 wet season, as % farmers reporting.

	Seneng-sari	Sugih-sari	Jaya-sari	Plosok-sari	Kudu-agung	Sumber-agung	Mulyo-agung	Lempung-agung	Total
own experience	24%	55%	46%	54%	34%	34%	17%	30%	36%
fellow farmers	8%	9%	10%	20%	3%	10%	7%	20%	11%
salesmen	10%	11%	9%	8%	16%	3%	2%	5%	8%
extension workers	3%	2%	6%	1%	10%	12%	7%	9%	6%
in KUD package	2%	2%	0%	0%	16%	5%	0%	0%	3%
other[1)]	0%	0%	3%	1%	7%	7%	6%	11%	5%
no response	53%	22%	26%	15%	13%	28%	60%	25%	31%
N	*59*	*55*	*80*	*74*	*61*	*58*	*82*	*80*	*549*

[1)] 'Other' includes villages officials, other villagers

Table III.23: Rat control measures by non-IPM farmers over four seasons, as % farmers reporting.

	1989/90 wet season								
	Seneng-sari	Sugih-sari	Jaya-sari	Plosok-sari	Kudu-agung	Sumber-agung	Mulyo-agung	Lempung-agung	Total
Preventive measures:									
rat drives	35%	13%	42%	13%	21%	21%	7%	18%	21%
poison baiting	29%	3%	42%	0%	6%	18%	2%	15%	15%
fumigation	17%	0%	18%	0%	0%	3%	0%	1%	5%
other[1)]	5%	2%	0%	1%	0%	0%	0%	0%	1%
Curative measures:									
rat drives	65%	29%	79%	49%	22%	47%	42%	49%	49%
poison baiting	67%	40%	52%	48%	28%	60%	38%	38%	46%
fumigation	32%	2%	38%	0%	0%	5%	1%	1%	10%
other[1)]	2%	0%	0%	0%	0%	0%	0%	0%	0%
N	*66*	*63*	*85*	*85*	*68*	*62*	*85*	*85*	*599*
	1990 dry season								
Preventive measures:									
rat drives	27%		33%		6%		1%		17%
poison baiting	0%		9%		3%		1%		4%
fumigation	11%		9%		0%		0%		5%
other[1)]	0%		1%		0%		0%		0%
Curative measures:									
rat drives	82%		79%		61%		87%		78%
poison baiting	92%		91%		94%		69%		86%
fumigation	35%		33%		3%		4%		18%
other[1)]	2%		1%		0%		0%		1%
N	*62*		*85*		*67*		*85*		*299*

Table III.23 ctd.

	Seneng-sari	Sugih-sari	Jaya-sari	Plosok-sari	Kudu-agung	Sumber-agung	Mulyo-agung	Lempung-agung	Total
1990/91 wet season									
Preventive measures:									
rat drives	12%	5%	6%	7%	0%	2%	1%	4%	5%
poison baiting	5%	4%	14%	5%	2%	5%	1%	1%	5%
fumigation	0%	0%	0%	0%	0%	0%	0%	1%	0%
Curative measures:									
rat drives	39%	36%	3%	12%	5%	10%	2%	11%	13%
poison baiting	47%	33%	36%	18%	20%	17%	5%	16%	23%
fumigation	7%	0%	8%	0%	0%	0%	1%	1%	2%
other[1)]	0%	0%	0%	0%	0%	0%	0%	4%	1%
N	*59*	*55*	*80*	*74*	*61*	*58*	*82*	*80*	*549*
1991 dry season									
Preventive measures:									
rat drives	27%		28%		15%		18%		22%
poison baiting	15%		21%		5%		7%		12%
fumigation	0%		6%		0%		0%		2%
Curative measures:									
rat drives	29%		33%		58%		38%		40%
poison baiting	29%		50%		47%		37%		41%
fumigation	5%		10%		2%		4%		5%
N	*55*		*72*		*60*		*76*		*263*

1) 'Other' includes plastic fencing of seedbed.

Table III.24: Average frequency of rodenticide application in number of applications per season by non-IPM farmers over four seasons.

	Seneng-sari	Sugih-sari	Jaya-sari	Plosok-sari	Kudu-agung	Sumber-agung	Mulyo-agung	Lempung-agung	Total
1989/90 wet season:									
average	4.1	0.6	2.8	0.6	0.6	1.3	0.5	0.5	1.4
st. dev.	9.5	0.8	4.3	0.8	1.0	2.1	0.6	0.6	3.9
N	*66*	*63*	*85*	*85*	*68*	*62*	*85*	*85*	*599*
1990 dry season:									
average	8.5		5.7		8.2		6.1		6.4
st. dev.	8		5.1		6.4		5.7		6.5
N	*62*		*85*		*67*		*85*		*299*
1990/91 wet season:									
average	3.6	2.1	2.3	1.3	0.6	1.7	0.3	0.3	1.5
st. dev.	6.1	4.9	4.1	3.8	1.6	4.4	1.2	0.8	3.8
N	*59*	*55*	*80*	*74*	*61*	*58*	*82*	*80*	*549*
1991 dry season:									
average	2.0		2.3		2.5		2.6		2.4
st. dev.	3.5		3.5		5.0		5.1		4.4
N	*55*		*72*		*60*		*76*		*263*

Table III.25: Frequency distribution of yields by IPM farmers over four seasons, as % farmers reporting.

	Seneng-sari	Sugih-sari	Jaya-sari	Plosok-sari	Kudu-agung	Sumber-agung	Mulyo-agung	Lempung-agung	Total
1989/90 wet season									
< 1.00 ton/ha	0%	0%	0%	0%	0%	0%	0%	0%	0%
1.00-2.99 tons/ha	8%	13%	23%	34%	14%	13%	2%	8%	15%
3.00-4.99 tons/ha	20%	31%	58%	38%	48%	35%	21%	35%	36%
5.00-6.99 tons/ha	59%	30%	14%	22%	33%	32%	45%	37%	33%
≥7.00 tons/ha	13%	26%	5%	7%	5%	20%	31%	20%	16%
N	*61*	*61*	*81*	*77*	*66*	*60*	*84*	*84*	*574*
1990 dry season									
< 1.00 ton/ha	0%		2%		1%		0%		1%
1.00-2.99 tons/ha	10%		11%		19%		6%		11%
3.00-4.99 tons/ha	47%		30%		40%		34%		37%
5.00-6.99 tons/ha	35%		41%		37%		36%		38%
≥7.00 tons/ha	8%		16%		1%		24%		13%
N	*62*		*83*		*67*		*85*		*297*
1990/91 wet season									
< 1.00 ton/ha	3%	0%	0%	19%	20%	0%	1%	4%	6%
1.00-2.99 tons/ha	28%	29%	20%	42%	41%	22%	5%	22%	25%
3.00-4.99 tons/ha	55%	54%	37%	26%	31%	38%	23%	44%	37%
5.00-6.99 tons/ha	14%	15%	35%	11%	7%	31%	45%	23%	24%
≥7.00 tons/ha	0%	2%	8%	3%	2%	9%	26%	8%	8%
N	*58*	*52*	*79*	*74*	*61*	*58*	*80*	*79*	*541*
1991 dry season									
< 1.00 ton/ha	0%		0%		0%		0%		0%
1.00-2.99 tons/ha	2%		4%		4%		1%		3%
3.00-4.99 tons/ha	23%		27%		30%		18%		24%
5.00-6.99 tons/ha	60%		42%		54%		51%		51%
≥7.00 tons/ha	15%		27%		12%		30%		22%
N	*52*		*67*		*57*		*71*		*247*

Table III.26: Frequency distribution of farm business income for rice production by non-IPM farmers in 1989/90 wet season, as % farmers reporting.

	owners	tenants	share-croppers	mixed status	total average
≤0 Rp/ha	4%	14%	11%	7%	7%
1-200,000 Rp/ha	6%	24%	38%	9%	15%
201-400,000 Rp/ha	19%	26%	33%	29%	25%
401-600,000 Rp/ha	27%	16%	16%	30%	24%
601-800,000 Rp/ha	21%	8%	1%	18%	15%
≥800,000 Rp/ha	25%	12%	1%	8%	15%
N	*253*	*111*	*90*	*128*	*582*

Table III.27: Average inputs and outputs of rice production by non-IPM farmers as classified by tenure status, 1990/91 wet season.

		owners	tenants	share-croppers	mixed status	total average
General						
number of farmers		246	105	47	150	548
average farm size (ha)		0.50	0.45	0.40	0.90	0.59
Output						
yield (tons/ha)		3.93	4.12	4.08	4.18	4.05
gross return (Rp 1,000/ha)[1]		751	781	421	863	758
Costs (Rp 1,000/ha)						
land:	opportunity cost [2]	378	0	0	*	*
	tax	17	0	0	9	10
	rent	0	395	0	167	121
	subtotal	395	395	0	*	*
labour:	family labour[3]	124	112	146	79	111
	hired labour[4]	181	168	165	171	174
	subtotal	305	280	311	250	285
purchased inputs:	seed[5]	28	29	32	26	28
	fertiliser	131	132	128	128	130
	pest control[6]	23	25	18	21	22
	irrigation	3	4	2	3	3
	remaining expenditures[7]	0	1	0	1	1
	subtotal	185	191	180	179	184
Paid-out costs (Rp 1,000/ha)		383	754	345	526	489
Farm business income of rice production						
(Rp 1,000/ha) of which return to:		368	27	76	337	269
	land	378	0	0	*	*
	family labour	124	112	146	79	111
	capital and management	–134	–85	–70	*	*

[1] Where possible, returns are computed using real prices. For stored product and family consumption an opportunity cost of Rp 230/kg dry, unhusked paddy is used. Share for harvest labour is subtracted when wage is paid in kind. Gross return for sharecroppers includes only farmer's share (usually 50%), therefore no cost for land is stated.

[2] Owners' opportunity cost for land is equated with the average rent for rice land with land tax subtracted.

[3] Family labour at market wage rate.

[4] Includes harvest labour only when wages were paid in money, not in kind.

[5] Price of purchased seed or own seed at market price.

[6] Includes chemicals, rent for sprayer, and rat control implements.

[7] Includes contributions for collective activities, and costs for transportation of inputs.

Appendix IV

Ten weeks IPM farmer field school in Senengsari

Week I: Opening of the IPM farmer field school
The opening ceremony of the IPM farmer field school in the subdistrict Sumbersari is organised for four villages together at the Community Centre in Sugihsari. In addition to the 100 trainees and the trainers (the pest observer and two extension workers), the meeting is attended by the four village heads, the subdistrict head ('*camat*'), Rural Extension Centre (REC) staff, the head of the District Agriculture Service, the supervisor from the FTF, all in all quite an official company.

After an official welcome by the REC head, everybody is sent to the rice fields to do the Ballot Box pre-test, a direct acquaintance with the field-oriented approach of the IPM training. Back in the Community Centre, the compulsory official speeches are pronounced, after which a new approach to agricultural extension is introduced to the farmers by the pest observer: experiential learning which they will soon experience themselves.

Week II
The field school participants gather first in the Village Centre where the pest observer explains the concept of IPM, and the definition of an ecosystem. In the field, the farmers are divided into five subgroups. They are given the exercise to enter the field, look at and note down whatever insects and damage symptoms they come across, and collect insects in a plastic bag, which is done right away. When ready with the field observations, they sit down under a row of shade trees on the small field road through the rice fields. The pest observer explains how to make the drawings of the agroecosystem: rice plants in the mid, natural enemies on the left, pests on the right, all flanked by necessary information on environmental conditions. Each subgroup received a copy of the leaflet with colour pictures of rice pests and natural enemies, which is used to compare with the insects caught in the plastic bags. A conclusion is discussed and written under the drawings. The pest observer attends to all groups, and helps the farmers to explain what they have discovered in the field. He does not give direct answers to the questions. The extension worker is around, but does not seem to know exactly what is expected from him.

The five groups present their drawings and conclusions which brings on much response from the other members. One farmer (a young village official with secondary education) gives an almost scientific explanation about photosynthesis (in Javanese). His group has put arrows in the drawings to indicate how energy flows through the system, which causes a lively discussion about what the direction of the arrows should be. Another farmer tells convincedly that the farmers' enemies (pests) have to be killed. This immediately raises the question what the natural enemies should eat if all the pests are killed. After a short discussion about pros and cons they seek the answer from the trainers. The Field Leader I, who just joined the session for supervision, answers that they actually know most of the answers to their questions themselves, because they have a lot of experience in the rice field. The answer causes amazement in the group, and no reaction. The pest observer summarises the exercise, and elaborates on the importance of self-discovery about features in the field.

Next, the special topic for this week, on rat population growth, is introduced. The farmers are given the exercise to calculate the growth of a rat population during one year, starting with one couple in September, and assuming that every litter has six babies that can reproduce within three months. When the farmers start the exercise they have a lot of questions to the pest observer and extension worker, who only give directions, no full answers. In each group there are two or at most three farmers who are seriously working on the calculation and understand what they are doing. At first, the others listen and try to follow, but quit after some time finding the calculations too complicated. The presentations of the growth tables, however, attract everybody's attention again. Although no group succeeds to make the correct calculation, by doing the exercise they all start to understand the danger of rats and the right moment to control them, i.e. before the population has expanded.

The pest observer and extension worker use only Javanese both when speaking with farmers individually, and in front of the whole group. The pest observer shows an energetic attitude, although he seemed somewhat uncertain in summarising and processing the results of the agroecosystem analysis. The extension worker tries to step in wherever he can. The atmosphere in the group is very open. There is a lot of joking and laughter. Although the group is quite heterogeneous (regarding age, education, and status in the village), this did not seem to hinder any of the participants to enter the discussions. The first field school day has been attractive enough for the participants to be motivated and to come again the next week.

Week III

Early morning, the group leaves right away for the observation field. The pest observer explains first how the farmers should best sample the field: walk diagonally through the field, observe 30 hills per group, one hill at every ten steps, look above the plant canopy first, then at the plant base, then the leaves, and count the number of pest insects and natural enemies. All farmers step into the field, do the observations, and next work out the agroecosystem analysis the way they had learned the week before. The discussions are focused on the question whether the numbers of pests and natural enemies are balanced enough not to take a control measure. One group suggest that carbofuran granules should be applied ('According to the recommendation 17 kg/ha, but the more the better!'). The whole group, however, decides that it is not necessary yet to spray or use granules, but to continue monitoring the development of the pest populations.

As a break, the farmers are invited to tear a piece of paper in such a way that a ring is created, causing hilarity and fun. One farmer succeeds quite fast, and tells that he discovered this by accident when he was a child. The pest observer asks the farmers for the use of the exercise and lists the answers: (1) to invent something new, (2) to recall something, and (3) to be patient.

The last activity of the day is the special topic on prevention of damage to the crop, explaining more about the first principle of IPM: grow a healthy crop. With input from the farmers, the pest observer lists the various factors in the rice crop ecosystem (soil, water, crop, weeds, wind, sunlight), and asks the farmers to fill out a table describing the management, result of the management, and appropriate time of management for each factor. There are lively discussions in all groups. It takes a while, though, before they put something on paper. One farmer exclaims: 'Wah, it is not so difficult to grow rice, but to write it down is something else. Theory is different from practice!' After the presentations and discussion, the pest observer gives a summary and conclusion, mentioning that not all factors are included in the table yet. Pests, natural enemies, and humans are missing. This causes a discussion about the effect of pesticides on natural enemies. A couple of people express their concern about a reduction of pesticides, although they realise that the natural enemies might die by pesticides.

Week IV

This week, the pest observer is not present because he has to attend the FTF workshop. The extension worker distributes observation forms, and explains briefly how to use the form which is a copy of the observations sheet used by the pest observers in their routine observations for the department of Agriculture. Field observations, agroecosystem analysis, and presentations are done as usual.

A special topic called 'Carbofuran, monocrotophos and natural enemies' is done with natural enemies collected in the field by the farmers. The natural enemies (insects and small frogs) are put in plastic cups, the extension worker adds an insecticide solution (monocrotophos) and granules (carbofuran) in the cups, and the farmers count after how many minutes the natural enemies die. The groups present the results of the experiment about the effect of carbofuran and monocrotophos on natural enemies. Most natural enemies died, especially the frogs, and fastest with the monocrotophos treatment. The first conclusion by the farmers is that monocrotophos is a better insecticide, because it gives a better and faster kill. The extension worker gives a very clear explanation about the difference between systemic and contact insecticides. He elaborates on the bad effects of insecticides to the environment, especially natural enemies, and on the need for observation before a decision on a pest control measure is made. The farmers, then, realise that monocrotophos is not good to encourage natural enemies.

The issue of rat control is touched again in order to motivate the farmers to organise themselves to start collective rat management.

A group dynamics exercise is organised in which the farmers have to sit in two rows, and pass on a sentence to the one sitting behind while whispering in his ear. The result is compared with the original sentence, which, of course, changed a lot, most strikingly from Indonesian to Javanese. The meaning of the game is summarised with input from the farmers. The extension worker gives a good, clear conclusion about what has to be kept in mind when information is disseminated.

The extension worker performs much better compared to the weeks before, now that he had a clear role as trainer/facilitator. He has had a decent preparation with the pest observer, and does very well on his own. His explanations are clear, he is able to create a good climate for learning: a relaxed but inspiring atmosphere.

Week V

The pest observer gives an introduction on how to fill out the observation sheet. The farmers have to copy the table format from the form into their notebooks. They are given a formula to calculate the percentage of damage or insect occurrence per rice hill. The observation field is now divided into two plots: IPM and local package. The latter plot has been given a carbofuran application which is obvious from the many dead earthworms at the soil surface. Each group has to monitor both plots. Agroecosystem analyses now contain percentages of damage or occurrence next to the drawings of insects. The concentration of the participants during the presentations becomes less, as if they are in a hurry to finish.

Another game is played with the farmers: they are tied one-to-one with a raffia string on the wrists and have to try to get loose without cutting the string.The use of this game is listed together with the farmers: be precise, do not be stressed, be patient, be collaborative, try to solve a problem, be an experimenter.

The special topic for this week is on prevention of rat damage to the crop for which the same table format as in week III is used to be filled out on rat damage. The groups present the results of the exercise. There is not much discussion, because they are in a hurry to finish the meeting and go home.

The role of the pest observer as a facilitator is less obvious this week than the weeks before. He is more led than leading. He does not show the committed and self-assured attitude of the first weeks. The group, actually, needs to be steered in managing the time schedule. The field sampling and agroecosystem analysis take too much time. In fact, the groups finish earlier, but nobody jumps in to continue the programme. The farmers are waiting for the pest observer and extension worker to lead the group, and the pest observer is waiting for the farmers to give any obvious sign that they were ready, probably afraid of being too pushing in his role as facilitator.

Week VI

Field monitoring, agroecosystem analysis and presentations are done as usual. Special topic is about life cycles and food webs. The pest observer draws a food chain on a newsprint with input from the farmers: 1. rice plants –> 2. life cycle of brown planthopper –> 3. dragonflies –> 4. spiders –> 5. frogs –> 6. snakes –> 7. birds of prey –> 8. humans, after which the life cycle of the brown planthopper is explained. A second example of the food chain forrats is discussed, and a third one with rice seed bug as the central pest.

Next, the key elements of IPM are introduced: (1) grow a healthy crop, (2) balanced fertilisation, (3) monitoring, and (4) efficient use of control measures. The pest observer gives an example of how the rice seed bug can be controlled without pesticides by trapping the bugs with burned rice field crabs. He further elaborates on the best moment to spray, if necessary, and the best moment for soil preparation regarding the life cycle of pests.

Some fields in the area are infested by something causing the leaves turn red. The pest observer explains that these symptoms are probably caused by a bacterial disease (bacterial leaf streak), as a result of too much nitrogen fertiliser, or too much water. He advises not to spray, but to reduce the water in the fields and take care of good, balanced fertilisation of the crop.

Week VII

The farmers get together at the Village Centre, where a discussion arises about the water problem in the rice fields that the village is coping with. They raise the question why the village government does not report this shortage to the subdistrict irrigation committee.

The farmers, and also some of the village officials who are participating in the field school, blame the village leader for not being firm in his policy. After an hour discussion, they go to the fields, and do their regular observations, agroecosystem analysis, and presentations. There is quite a lot of rat damage in the demonstration plot. The drawings of the agroecosystem are made by only a few farmers per group. It looks as if the others lost interest, perhaps because of the water problems. The quality of the drawings is no longer improving as in the previous weeks. There is not much discussion.

A game cheers up the trainees. The group is divided into two small groups. One subgroup has to make questions, the other answers. Standing in two rows in front of each other, one person reads aloud his question, and the person opposite him reads his answer, which, of course, rarely fits to the question. This causes a lot of laughter. The evaluation of the game, however, is quite weak, but the game had served its function.

After a brief explanation about the concept of economic threshold level, the special topic for this session, the pest observer dictates a fairly abstract definition which is written down by all the farmers: 'Economic threshold level is the value for the balance between plant damage that is caused by pest organisms, and the control cost'. He gives examples of the damage threshold levels for rice leaffolder (25%), rice seed bug (5%, or 2-3 insects/m^2), and stemborer (5% whiteheads). He works out how yield loss and control cost must be calculated. The pest observer lectures, but tries to get a lot of input from the farmers by questioning them about prices of commodity and inputs, and involves them in the calculations. His examples and explanations are very clear. However, only a part of the group is actively involved, in the sense that they respond to the questions, probably due to the abstract level of this topic. A group exercise in which farmers calculate yield loss and pest control cost should give clarity. Only a minority of the trainees is actively involved. In spite of the clear explanation by the pest observer, the majority apparently does not grasp the topic.

Week VIII

In the field observations and agroecosystem analysis, the farmers have found that pest and natural enemy populations are in balance. Most pest damage comes from rat attack. A few groups mention that rats should be controlled in this week at the latest, considering what they have learned about rat population growth in the first week of the field school. The extension worker, in charge of the programme for this week since the pest observer is in the FTF workshop, notices that some fields in the village are infested by bacterial leaf streak, explains how it occurs, what should be done to prevent and control this disease, and recommends that the farmers should reduce the water level in the fields, and not to spray the expensive bactericides.

Roots, plant vessels and systemic pesticides is the title for the special topic for this week. The experiment to demonstrate the absorption ability of plant roots is done with broadleaved weeds which were collected from the rice field by some farmers, and a straw as control. The roots of the plants and the straw are put in a solution of water with a red colouring. A few farmers help the extension worker executing the experiment, observing the result, and one of them gives a presentation. Most of the participants, however, are not (actively) involved. The extension worker gives a summary on the experiment and its meaning, related to the working of systemic pesticides.

Week IX

The farmers arrive at the field locations and start the observations right away. Six new participants are present, most young fellows who are representing their fathers or neighbours who could not attend. They seem to know exactly what they have to do. They have the notebooks of the ones they are representing with them. Two of them told me they were given a briefing about the field school before they came. Agroecosystem analysis is done as usual. The first presentation focuses on the rat damage observed and rat control. The presenter suggests that collective rat drives should be organised immediately. This raises a hot discussion about cooperation and collectivity in the village. Repeatedly it is said that 'the people' are not willing to participate. It comes to the topic that there is not enough enthusiasm because of feelings of jealousy among the small farmers towards the big farmers. Finally, they express that there is no response 'from above' (the village leader in particular) to the farmers' problems. The pest observer opens a discussion about the role which the IPM field school group could, and should, have in organising collective rat control activities in the village, repeating the arguments mentioned above but nothing firm comes up. Attendants are in low spirits.

The Field Leader II, visiting the group in his role of supervisor, gives a talk on how the environment can be kept healthy by a need-based pesticide use. Then, they go on to the special topic for this week: to make an insect collection. Insect pests and natural enemies collected during the field observation are identified and put in small bottles. At the start, there are eight farmers who pay attention to the exercise, but soon only three farmers are left behind to finish it. The rest is sitting aside and chatting. When the collection is ready, the pest observer discusses the bottled insects one by one, asking the farmers how they call the insect, and what it eats. He gives interesting details about the various insects. There is a good response from the group. At the end of the session, several farmers are recruited to help prepare the field day that will take place in the tenth week.

Some farmers tell the village study assistant that for the last couple of weeks they would like a change in the field school program, knowing now what the agroecosystem is like, and how to sample the field. This explains, in addition to the problems of rat attack and water shortage that the farmers are facing, why the atmosphere has been less lively during the last weeks, especially for the agroecosystem presentations. Response and discussion have diminished. The farmers, however, can not make clear what additional lessons or materials they want. Contrary to this is the fact that many participants do not show an increased interest in the special topics, rather decreased attention is visible. It looks as if they are already tired by the time the special topics are done. Some participants say that they regret that the field school is almost over. They would not mind to continue for some more time.

Week X: Field day

A field day is organised to present the results of the IPM farmers field school to the surrounding community. The four villages in the subdistrict Sumbersari having received IPM training hold their field day together in the house of the village head of Rejosari, one of the other IPM villages of this pest observer. The pest observer expects about 100 people for the field day, including most of the IPM participants from the four villages, as well as the community of Rejosari, several government officials, and REC staff. The attendance is disappointing, from the side of the IPM farmers and of other villagers, especially because none of the village heads involved are present. A field day per village would probably have resulted in more commitment from the IPM trainees, and more interest from the other villagers and village officials.

After a brief welcome to the participants by the pest observer everybody is invited to go to the field, which is opposite the field day location. One of the IPM trainees from Rejosari explains the purpose and use of the observation plots during the IPM field school. The way how to monitor for pests and natural enemies is demonstrated by another farmer. An exhibition is put up of agroecosystem analyses and posters made in the field schools, insect collection, and insect gardens. Each part is presented by a farmer from Rejosari. The programme runs very smoothly, and the presentations are clear and well structured. The farmers look convinced of what they are telling.

The pest observer gives a short talk to motivate the farmers to disseminate the IPM technology to other farmers in the village. In a discussion led by the REC head, visiting farmers are invited to ask questions, which are answered by the IPM farmers. The main topic discussed is the use of pesticides and the effect on the environment. The discussion comes several times back to the question whether pesticides really can be reduced or even be abolished. Even among several IPM farmers there does not seem to be a solid conviction yet.

The IPM certificates are handed out to the trainees by the REC head. He expresses that the IPM field school will not officially be closed, because IPM has to continue, and be extended to all farmers.

Appendix V

Tables belonging to Chapter 7

Table V.1: Reason for selecting a rice variety over four seasons, as % IPM farmers reporting.

	1989/90 before	1990 training	1990/91 after	1991 after
high yield	52%	44%	48%	57%
quality of rice	37%	33%	33%	43%
pest tolerance	43%	54%	53%	34%
uniform planting	7%	4%	2%	7%
good marketing	13%	15%	8%	21%
suitability for area/season	7%	10%	6%	14%
growth duration	6%	8%	0%	2%
N	*94*	*48*	*93*	*42*

Totals exceed 100% due to multiple response

Note: Reasons mentioned seem to be influenced by the conditions experienced in the previous seasons. In the 1991 season when high yields were obtained and low pest pressure occurred, a shift in reasons for varietal selection is visible. This shift is supposed to result partly of a changed perception of farmers under changing conditions, rather than purely reflecting changing opinions.

Table V.2: Fertiliser dosages.

a. Average dosages (in kg/ha), N = number of farmers reporting.

		1989/90 wet season					1990 dry season		
		Seneng-sari	Sugih-sari	Kudu-agung	Sumber-agung	Total	Seneng-sari	Kudu-agung	Total
urea	average	297	223	248	255	255	296	227	260
	st. dev.	131	70	126	110	114	139	66	113
	N	*23*	*25*	*21*	*25*	*94*	*23*	*25*	*48*
TSP	average	302	175	180	223	220	244	180	210
	st. dev.	131	53	101	96	111	107	62	92
	N	*23*	*25*	*21*	*25*	*94*	*23*	*25*	*48*
KCl (users only)	average	123	113	94	106	108	132	78	108
	st. dev.	62	42	44	65	52	105	35	86
	N	*10*	*18*	*17*	*9*	*54*	*15*	*12*	*27*
ZA (users only)	average	128	103	99	85	104	117	61	86
	st. dev.	80	48	45	46	57	60	30	54
	N	*9*	*14*	*14*	*6*	*43*	*5*	*6*	*11*
		1990/91 wet season					1991 dry season		
urea	average	330	291	292	246	290	351	272	312
	st. dev.	99	114	150	66	115	115	122	125
	N	*23*	*24*	*23*	*23*	*93*	*21*	*21*	*42*
TSP	average	287	194	228	231	235	308	178	243
	st. dev.	114	85	110	59	100	106	70	111
	N	*23*	*24*	*23*	*23*	*93*	*21*	*21*	*42*
KCl (users only)	average	175	114	66	104	115	110	76	93
	st. dev.	64	53	37	59	64	56	38	51
	N	*9*	*18*	*8*	*11*	*46*	*7*	*7*	*14*
ZA (users only)	average	–	110	73	54	86	–	57	57
	st. dev.	–	67	27	33	58	–	24	24
	N	*0*	*14*	*6*	*8*	*28*	*0*	*3*	*3*

Table V.2 ctd.

b. Frequency distribution, as % farmers reporting.

		1989/90 wet season					1990/91 wet season				
		Seneng-sari	Sugih-sari	Kudu-agung	Sumber-agung	Total	Seneng-sari	Sugih-sari	Kudu-agung	Sumber-agung	Total
urea	< 200 kg/ha	26%	48%	33%	36%	36%	4%	13%	39%	35%	23%
	200-299	26%	32%	48%	36%	35%	30%	54%	13%	43%	35%
	300-399	30%	20%	10%	16%	19%	48%	13%	22%	22%	26%
	≥ 400	17%	0%	10%	12%	10%	17%	21%	26%	0%	16%
TSP	< 200 kg/ha	22%	64%	76%	48%	52%	22%	58%	52%	39%	43%
	200-299	30%	36%	10%	36%	29%	35%	21%	17%	52%	31%
	300-399	30%	0%	5%	12%	12%	35%	21%	22%	9%	22%
	≥ 400	17%	0%	10%	4%	7%	9%	0%	9%	0%	4%
KCl	0 kg/ha	57%	28%	19%	64%	43%	61%	25%	65%	52%	51%
	1-49	4%	4%	14%	8%	7%	0%	8%	17%	9%	9%
	50-99	13%	28%	33%	16%	22%	4%	29%	9%	17%	15%
	100-149	17%	32%	29%	4%	20%	13%	17%	9%	13%	13%
	> 150	9%	8%	5%	8%	7%	22%	21%	0%	9%	13%
ZA	0 kg/ha	61%	44%	33%	76%	54%	100%	42%	74%	65%	70%
	1-49	0%	8%	10%	8%	6%	0%	13%	9%	13%	9%
	50-99	17%	20%	29%	8%	18%	0%	17%	13%	17%	12%
	100-149	13%	16%	24%	4%	14%	0%	13%	4%	4%	5%
	> 150	9%	12%	5%	4%	7%	0%	17%	0%	0%	4%
N		*23*	*25*	*21*	*25*	*94*	*23*	*24*	*23*	*23*	*93*

Table V.3: IPM farmers' knowledge about some common rice pests, before (1989/90 wet season) and after (1990/91 wet season) IPM training, as % farmers reporting.

	Senengsari		Sugihsari		Kuduagung		Sumberagung		Total	
	before	after	before	after	before	after	before	after	before	after
Brown planthopper:										
know symptoms	87%	96%	76%	83%	90%	87%	88%	83%	85%	87%
know appearance	74%	91%	76%	83%	86%	87%	84%	91%	80%	88%
do not know	9%	4%	24%	8%	10%	13%	12%	9%	14%	9%
Stemborer deadheart:										
know symptoms	91%	96%	84%	96%	90%	91%	100%	83%	91%	91%
know appearance	43%	70%	40%	63%	24%	52%	64%	48%	44%	58%
do not know	9%	0%	16%	4%	10%	4%	0%	13%	9%	5%
Stemborer whitehead:										
know symptoms	91%	96%	92%	100%	100%	96%	96%	83%	95%	94%
know appearance	74%	87%	64%	88%	71%	96%	84%	91%	73%	90%
do not know	9%	0%	8%	0%	0%	0%	4%	9%	5%	2%
Rice seed bug:										
know symptoms	100%	100%	96%	100%	100%	91%	100%	100%	99%	98%
know appearance	100%	100%	92%	100%	90%	100%	100%	100%	96%	100%
do not know	0%	0%	0%	0%	0%	0%	0%	0%	0%	0%
Rice leaffolder:										
know symptoms	30%	52%	12%	46%	24%	43%	0%	30%	16%	43%
know appearance	26%	39%	8%	29%	5%	35%	0%	22%	10%	31%
do not know	70%	43%	88%	50%	76%	48%	100%	65%	84%	52%
Bacterial leaf blight:										
know symptoms	17%	0%	0%	25%	5%	17%	0%	4%	5%	12%
know appearance	0%	4%	0%	8%	0%	4%	4%	0%	1%	4%
do not know	83%	96%	100%	75%	95%	83%	96%	96%	94%	87%
N	*23*	*23*	*25*	*24*	*21*	*23*	*25*	*23*	*94*	*93*

Note 1: Bacterial leaf blight was an unknown disease in the baseline season, because it hardly occurred in the area. The small knowledge increase after training is due to the fact that this disease hardly occurred during the training season either. Only in Sugihsari and Kuduagung low infection levels were observed, resulting in higher knowledge levels among field school graduates in these villages.

Note 2: Knowledge on stemborers by Sumberagung IPM farmers rather seems to decrease after training. Although the training process was not observed in this village, it is suspected that misunderstandings among some field school graduates occurred due to language problems. The trainers used the national language (Indonesian), whereas many of the Sumberagung farmers exclusively speak Javanese.

Table V.4: IPM farmers' knowledge about natural enemies, before (1989/90) and after (1990/91) IPM training, as % farmers reporting.

	Senengsari		Sugihsari		Kuduagung		Sumberagung		Total	
	before	after	before	after	before	after	before	after	before	after
Mention: snakes	78%	74%	40%	63%	29%	39%	56%	17%	51%	48%
birds	35%	26%	20%	0%	29%	39%	20%	17%	26%	20%
frogs	30%	65%	36%	83%	38%	61%	20%	52%	31%	66%
spiders	22%	70%	20%	67%	19%	74%	24%	96%	21%	76%
beetles	0%	13%	0%	54%	0%	9%	0%	26%	0%	26%
dragonflies	9%	26%	12%	42%	14%	57%	20%	61%	14%	46%
other[1)]	26%	48%	8%	21%	33%	26%	28%	22%	23%	29%
Do not know	4%	0%	28%	0%	24%	0%	20%	0%	19%	0%
N	*23*	*23*	*25*	*24*	*21*	*23*	*25*	*23*	*94*	*93*

Totals exceed 100% due to multiple response.
[1)] 'Other before' includes cat, dog, duck, salamander, chicken, mongoose, fish, crab and cricket; 'Other after' additionally includes waterstrider.

Table V.5: Frequency of pesticide application.

a. Average number of applications per season, N = number of farmers reporting.

		1989/90 wet season					1990 dry season				
		Seneng-sari	Sugih-sari	Kudu-agung	Sumber-agung	Total	Seneng-sari	Sugih-sari	Kudu-agung	Sumber-agung	Total
sprays:	average	0.7	1.8	2.2	1.8	1.6	0.2	1.1	1.2	1.1	0.9
	st. dev.	1.5	1.0	1.0	1.2	1.3	0.5	1.1	1.1	1.5	1.2
granules:	average	0.4	0.4	0.3	0.7	0.5	0.2	1.1	1.1	1.1	0.9
	st. dev.	0.6	0.7	0.5	0.8	0.7	0.4	1.0	0.9	1.5	1.1
total:	average	1.1	2.1	2.5	2.5	2.1	0.4	2.2	2.3	2.3	1.8
	st. dev.	1.4	1.0	1.2	1.6	1.4	0.6	1.6	1.8	2.6	2.0
N		*23*	*25*	*21*	*25*	*94*	*23*	*24*	*25*	*23*	*95*
		1990/91 wet season					1991 dry season				
sprays:	average	0.1	0.8	1.5	2.6	1.2	0.0		0.5		0.3
	st. dev.	0.3	0.9	0.9	1.7	1.4	0.2		0.7		0.5
granules:	average	0.1	1.3	1.3	2.0	1.2	0.2		1.0		0.6
	st. dev.	0.3	1.0	1.0	1.3	1.2	0.5		1.0		0.9
total:	average	0.2	2.0	2.8	4.6	2.4	0.3		1.6		0.9
	st. dev.	0.5	1.4	1.7	2.3	2.3	0.5		1.2		1.1
N		*23*	*24*	*23*	*23*	*93*	*21*		*21*		*42*

Table V.5 ctd.

b. Frequency distribution, as % farmers reporting.

	1989/90 wet season					1990 dry season				
	Seneng-sari	Sugih-sari	Kudu-agung	Sumber-agung	Total	Seneng-sari	Sugih-sari	Kudu-agung	Sumber-agung	Total
no applications	48%	8%	4%	20%	19%	65%	13%	16%	43%	34%
1 application	22%	12%	12%	12%	14%	30%	29%	28%	4%	23%
2 applications	17%	52%	36%	8%	29%	4%	21%	12%	9%	12%
3 applications	4%	12%	32%	28%	19%	0%	17%	16%	13%	12%
4 applications	0%	16%	8%	24%	12%	0%	13%	16%	17%	12%
> 4 applications	9%	0%	8%	8%	6%	0%	8%	12%	13%	8%
N	*23*	*25*	*25*	*25*	*98*	*23*	*24*	*25*	*23*	*95*
	1990/91 wet season					1990 dry season				
no applications	83%	8%	9%	4%	26%	76%		24%		50%
1 application	13%	42%	17%	0%	18%	19%		29%		24%
2 applications	4%	21%	17%	13%	14%	50%		19%		12%
3 applications	0%	8%	22%	9%	10%	0%		24%		12%
4 applications	0%	13%	22%	35%	17%	0%		5%		2%
> 4 applications	0%	8%	13%	39%	15%	0%		0%		0%
N	*23*	*24*	*23*	*23*	*93*	*21*		*21*		*42*

Table V.6: Type of chemical pest control measure applied, as % farmers reporting.

	Sugihsari				Sumberagung			
	IPM		Non-IPM		IPM		Non-IPM	
	1989/90	1990/91	1989/90	1990/91	1989/90	1990/91	1989/90	1990/91
Preventive measures:								
pesticide sprays	56%	38%	38%	58%	28%	65%	26%	40%
pesticide granules	4%	67%	8%	51%	20%	78%	24%	52%
Curative measures:								
pesticide sprays	28%	21%	48%	33%	48%	78%	66%	41%
pesticide granules	4%	4%	0%	4%	16%	13%	23%	12%
No control measures	8%	8%	25%	7%	20%	4%	15%	17%
N	*25*	*24*	*63*	*55*	*25*	*23*	*62*	*58*

	Total over four IPM villages							
	IPM				Non-IPM			
	1989/90	1990	1990/91	1991	1989/90	1990	1990/91	1991
Preventive measures:								
pesticide sprays	33%	8%	43%	17%	26%	30%	53%	1%
pesticide granules	17%	40%	55%	38%	12%	43%	44%	44%
Curative measures:					0%	0%	0%	0%
pesticide sprays	38%	33%	31%	7%	55%	52%	28%	20%
pesticide granules	13%	13%	6%	5%	7%	19%	6%	3%
No control measures	19%	40%	26%	50%	25%	19%	18%	47%
N	*98*	*47*	*93*	*42*	*259*	*129*	*418*	*115*

Totals exceed 100% due to multiple response.

Table V.7: IPM dissemination channels and information sources in 1990/91 wet season, as % of non-IPM farmers aware of IPM.

	IPM villages				Non-IPM villages				Total
	Seneng-sari	Sugih-sari	Kudu-agung	Sumber-agung	Jaya-sari	Plosok-sari	Mulyo-agung	Lempung-agung	
IPM field school graduates	39%	33%	17%	27%	0%	0%	0%	0%	17%
extension worker	4%	24%	25%	19%	61%	30%	48%	38%	31%
village assistant/investigator	22%	5%	13%	4%	18%	10%	26%	0%	13%
village officials	22%	29%	0%	19%	14%	20%	17%	25%	18%
radio/TV/newspaper	0%	10%	8%	8%	4%	30%	17%	25%	10%
neighbours/other farmers	0%	5%	38%	19%	0%	10%	9%	0%	11%
other[1]	13%	0%	0%	4%	4%	10%	0%	13%	4%
N	*23*	*21*	*24*	*26*	*28*	*10*	*23*	*8*	*163*

Totals exceed 100% due to multiple response.
[1] 'Other' includes other villagers, invitation letters for interviews.

Table V.8: Perception about IPM in 1990/91 wet season, as % of non-IPM farmers aware of IPM.

	IPM villages				Non-IPM villages				Total
	Seneng-sari	Sugih-sari	Kudu-agung	Sumber-agung	Jaya-sari	Plosok-sari	Mulyo-agung	Lempung-agung	
pest control - correct[1]	39%	38%	29%	15%	32%	10%	17%	13%	26%
pest control - incorrect[2]	4%	14%	8%	8%	43%	20%	13%	25%	17%
monitoring/identification[3]	26%	29%	13%	15%	11%	0%	17%	50%	18%
agronomy[4]	4%	10%	13%	8%	14%	10%	4%	0%	9%
other[5]	0%	5%	0%	8%	0%	30%	0%	0%	4%
not clear/do not know	35%	29%	46%	38%	25%	40%	52%	63%	39%
N	*23*	*21*	*24*	*26*	*28*	*10*	*23*	*8*	*163*

Totals exceed 100% due to multiple response.
[1] Includes collective rat control, aspects of need-based and correct pesticide use, integrated measures.
[2] Includes collective control measures (spraying), preventive measures, government pesticide aid.
[3] Includes field observations, identification and function of pests/natural enemies, (collective) decision making.
[4] Includes good cultivation, synchronised and timely planting, balanced fertilisation, resistant varieties.
[5] Includes farmer meeting/training, collective problem solving, drawing of rice plants.

List of terms

Indonesian terms in the text are preferably translated to English. Explanations of these terms are given here, some from Indonesian to English and from English to Indonesian.

AAETE	Agency for Agricultural Education, Training and Extension
AARD	Agency for Agricultural Research and Development
AIC	Agricultural Information Centre = Balai Informasi Pertanian (BIP)
ani-ani	traditional rice knife
arisan	revolving lottery fund
BIMAS	Bimbingan Masal = Mass Guidance (Food Crops Intensification Programmes)
BIP	Balai Informasi Pertanian = Agricultural Information Centre
BLPP	Balai Latihan Pegawai Pertanian = Agricultural In-Service Training Centre
BND	BIMAS Nasional yang Disempurnakan = National Improved BIMAS
BPP	Balai Penyuluhan Pertanian = Rural Extension Centre (REC)
bengkok	village government salary lands
becak	tricycle
borongan	limited group of labourers hired to complete a certain operation
BPTP	Balai Perlindungan Tanaman Pangan = Food Crop Protection Centre
BULOG	Badan Urusan Logistik = National Logistics Agency
bupati	head of district (kabupaten)
camat	head of subdistrict (kecamatan)
contact farmer	kontak tani = head of the farmer group
DEPTAN	Departemen Tanaman Pangan = Department of Food Crops (at national level)
desa	village
DG	Directorate General
DIPERTA Tk I	Dinas Pertanian Tanaman Pangan Tingkat I = Agriculture Service (at province level)
DIPERTA Tk II	Dinas Pertanian Tanaman Pangan Tingkat II = Agriculture Service (at district level)
district	kabupaten

DITLIN	Direktorat Perlindungan Tanaman Pangan = Directorate of Food Crops Protection
dry season	musim ketigo = second rice-growing season
dusun = dukuh	hamlet
FAO	Food and Agriculture Organisation of the United Nations
farmer group	kelompok tani
FFS	IPM farmer field school = SLPHT
FGI	focus group interview
Field Leader I/II	Pemandu Lapang I/II = PL I/II = principal trainers of the IPM programme
FTF	Field Training Facility = IPM regional training centre
INPRES	Instruksi Presiden = Presidential Decree
INSUS	Intensifikasi Khusus = Special Intensification
intermediate season	musim labu = secondary food crops season
IPM	Integrated Pest Management
IRRI	International Rice Research Institute, Los Baños, The Philippines
JATENG	Jawa Tengah = Central Java
kabupaten	district
KANWIL	Kantor Wilayah = Provincial Office
kecamatan	subdistrict
kelompok tani	farmer group as designated by the extension system
kontak tani	contact farmer or head of the farmer group
KUD	Koperasi Unit Desa = village unit cooperation
mantri tani	agriculture officer at subdistrict office
musim rendhengan	wet (rice) season
musim ketigo	dry (rice) season
musim labu	intermediate (secondary food crops) season
NAEP	National Agricultural Extension Programme
NFE	Nonformal Education
NGO	Nongovernmental Organisation
obat	medicine
palawija	secondary food crops
penebas	rice trader who buys the standing crop
pest observer	PHP
PHP	Pengamat Hama dan Penyakit = pest observer
PHT	Pengendalian Hama Terpadu = Integrated Pest Management
PL I/II	Pemandu Lapang I/II = Field Leader I/II, principal trainers of the National IPM programme
POSKO	Pos Komando = commando post at subdistrict level responsible to take action against brown planthopper outbreak

PPL	Penyuluh Pertanian Lapangan = village extension worker
PPM	Penyuluh Pertanian Madya = mid-level extension worker
REC	Rural Extension Centre = BPP
REPELITA	Rencana Pembangunan Lima Tahun = Five-year Development Plan
RT	Rukun tetangga = small neighbourhood
Rp	rupiah
rupiah	Indonesian money, Rp 1,980 = US$ 1.- (mid 1991)
RW	Rukun warga = large neighbourhood
SD	sekolah dasar = primary school
secondary food crops	palawija
selapan	period of five weeks
SLPHT	IPM farmer field school = FFS
SMP	sekolah menengah pertama = junior high school
SMA	sekolah menengah atas = high school
subdistrict	kecamatan
tebasan	selling of the standing rice crop
tricycle	becak
T&V system	Training-and-Visit Extension System
UN	United Nations
USAID	United States Agency for International Development
village	desa
village extension worker	PPL
wet season	musim rendhengan = first rice-growing season

Overview study villages

Village profile characteristics	Study intensity	IPM training	Village name	Subdistrict map
within reach, conventional, extravagant, indifferent policies reasonable extension worker, no KUD influence	intensive	IPM	Senengsari	Sumbersari 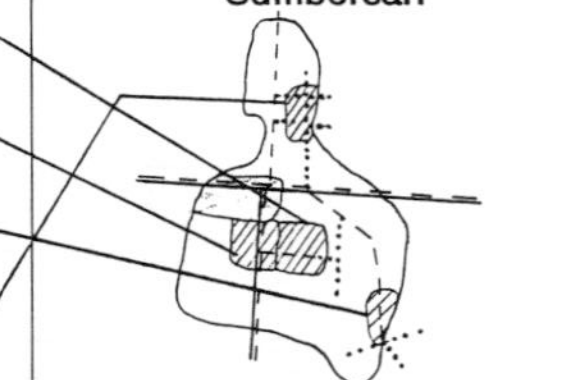
on main road, urban, progressive policies reasonable extension worker, strong KUD influence	control	IPM	Sugihsari	
far but well attainable, progressive policies committed extension worker, no KUD influence	intensive	non-IPM	Jayasari	
isolated, traditional, indifferent policies indifferent extension worker, no KUD influence	control	non-IPM	Plosoksari	
within reach, village-like, orderly and fairly progressive policies corrupt extension worker, strong KUD influence	intensive	IPM	Kuduagung	Sumberagung 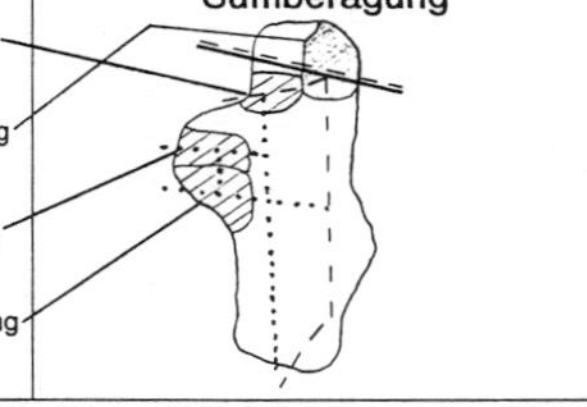
on main road, partly urban partly village-like, fairly progressive policies corrupt extension worker, strong KUD influence	control	IPM	Sumberagung	
remote, traditional, cohesive community, consistent policies committed extension worker, no KUD influence	intensive	non-IPM	Mulyoagung	
remote, isolated, indifferent policies committed extension worker, no KUD influence	control	non-IPM	Lempungagung	

Legenda: village area; subdistrict town; main road; secondary road; field track

Overview study seasons

Season profile characteristics	Study activities	Study phase	Wet/ dry	Study season
moderate climatical conditions, rat attack, moderate yields	survey	baseline	wet	1989/90
moderate climatical conditions, rat attack, moderate yields	survey, observations, group interviews	IPM training	dry	1990
bad climatical conditions, late water supply, stemborer outbreak, low yields	survey, observations, group interviews, record keeping	1st post-training	wet	1990/91
good climatical conditions, no pest damage of importance, good yields	survey, observations, group interviews	2nd post-training	dry	1991
good climatical conditions, no pest damage of importance, very good yields	observations, group interviews	3rd post-training	wet	1991/92